0. INTRODUCTION

Let (X,σ) be a subshift and let φ be an automorphism of (X,σ). Analysis of the dynamical system (X,φ) is one of the major subjects of symbolic dynamics. If (X,σ) is a full shift, then φ is a 1-dimensional cellular automaton map. Hence analysis of the dynamical behavior of a 1-dimensional cellular automaton is included in the subject. It is well known (see [H]) that if φ is an expansive automorphism of (X,σ), then (X,φ) is topologically conjugate to a subshift and if φ is a positively expansive endomorphism of (X,σ), then (X,φ) is topologically conjugate to a one-sided subshift. It is natural to ask whether φ is expansive or not, or positively expansive or not. When φ is an expansive automorphism or a positively expansive endomorphism, the best solution for the problem of analyzing (X,φ) is to *subshift-identify* (X,φ), that is, to identify (up to topological conjugacy) the subshift or the one-sided subshift to which (X,φ) is topologically conjugate. Whether φ is expansive or not, the topological entropy $h(\varphi)$ is an important piece of information.

Works analyzing the dynamical behavior of endomorphisms and automorphisms of subshifts in those points have been rare. There are some examples of nontrivial computation of the topological entropy of endomorphisms of topological Markov shifts as found in [C], [L], [BK1], [M], and [S]. But there are no nontrivial subshift-identifications of the dynamical systems (X,φ) defined by expansive automorphisms φ of topological Markov shifts (X,σ) in the literature (except for the case where φ is the composition of a simple automorphism and a power of σ [N3]). Boyle and Krieger proved in [BK1], however, that for every automorphism φ of every topological Markov shift (X,σ) and for all integers n greater than a coding bound for φ and φ^{-1}, $(X,\varphi\sigma^n)$ is topologically conjugate to a topological Markov shift, and specified the dimension group triple of this topological Markov shift. In particular, they proved that if the inverse of the zeta function of (X,σ) is an irreducible polynomial and φ is shiftless, then $(X,\varphi\sigma^n)$ is shift equivalent to (X,σ^n) for all n greater than a coding bound for φ and φ^{-1}. They also showed that if (X,σ) is a topological Markov shift such that the inverse of its zeta function is an irreducible polynomial and if φ is an automorphism of (X,σ) of finite order, then $(X,\varphi\sigma^n)$ is shift equivalent to (X,σ^n) for all integers $n\neq 0$. Furthermore, they gave some basic results on the dynamical behavior of automorphisms of subshifts (see Lemmas 2.6, 2.11, and 2.14 of [BK1]).

In this memoir, we introduce the notion of a "textile system". Textile systems are natural and useful machines to analyze the dynamical behavior of endomorphisms and

Received by the editors March 25, 1992.

automorphisms of topological Markov shifts. A textile system T is defined by an ordered pair of graph-homomorphisms p and q of a graph Γ into a graph G (with some additional constraint), and written

$$T = (p, q : \Gamma \to G).$$

It weaves "textiles". The textiles weaved by T define the "textile shifts" $(U_T, \sigma_T^{(i,j)})$, $(i, j) \in \mathbf{Z}^2$, the "woof shift" (X_T, σ_T), the "warp shift" (X_{T^*}, σ_{T^*}) and the "bias shifts" $(\check{X}_T, \check{\sigma}_T^{(i,j)})$, $i, j \in \mathbf{Z}$. The textile system T has its "dual" T^*, which is the most important notion in the theory of textile systems (the definition will be given in Section 2). T^* weaves the textiles that are the transposes of the textiles that T weaves. Hence the woof shift of T^* is the warp shift of T and the warp shift of T^* is the woof shift of T. Let $\xi : (X_\Gamma, \sigma_\Gamma) \to (X_G, \sigma_G)$ and $\eta : (X_\Gamma, \sigma_\Gamma) \to (X_G, \sigma_G)$ be the factor maps given by p and q, respectively, where $(X_\Gamma, \sigma_\Gamma)$ and (X_G, σ_G) are the topological Markov shifts defined by Γ and G, respectively. Throughout this memoir, a "factor map" is assumed to be not necessarily onto. Let X_n and Z_n be defined for all integers $n \geq 0$ as follows:

$$X_0 = X_G, \quad Z_0 = X_\Gamma,$$

$$X_n = \xi(Z_{n-1}) \cap \eta(Z_{n-1}) \quad \text{and} \quad Z_n = \xi^{-1}(X_n) \cap \eta^{-1}(X_n), \qquad n \in \mathbf{N}.$$

Then we have the woof shift (X_T, σ_T) by

$$X_T = \bigcap_{n=0}^{\infty} X_n \quad \text{and} \quad \sigma_T = \sigma_G | X_T$$

and another subshift (Z_T, ς_T) by

$$Z_T = \bigcap_{n=0}^{\infty} Z_n \quad \text{and} \quad \varsigma_T = \sigma_\Gamma | Z_T.$$

We say that T is "nondegenerate" if $X_T = X_0$ (i.e., both ξ and η are onto) and "finitely saturated" if $X_T = X_n$ for some nonnegative integer n. We have two onto factor maps $\xi_T : (Z_T, \varsigma_T) \to (X_T, \sigma_T)$ and $\eta_T : (Z_T, \varsigma_T) \to (X_T, \sigma_T)$ by restricting ξ and η, and ξ_T and η_T naturally define the onto factor maps $\tilde{\xi}_T : (\tilde{Z}_T, \tilde{\varsigma}_T) \to (\tilde{X}_T, \tilde{\sigma}_T)$ and $\tilde{\eta}_T : (\tilde{Z}_T, \tilde{\varsigma}_T) \to (\tilde{X}_T, \tilde{\sigma}_T)$ between the one-sided subshifts. If T is "one-sided 1-1", i.e., ξ_T is 1-1, then an onto endomorphism of the woof shift (X_T, σ_T) is defined by

$$\varphi_T = \eta_T \xi_T^{-1}.$$

If $\tilde{\xi}_T$ is 1-1, then an onto endomorphism $\tilde{\varphi}_T$ of the one-sided woof shift $(\tilde{X}_T, \tilde{\sigma}_T)$ is defined by

$$\tilde{\varphi}_T = \tilde{\eta}_T \tilde{\xi}_T^{-1}.$$

If T is "1-1", i.e., both ξ_T and η_T are 1-1, then φ_T is an automorphism of (X_T, σ_T) and if both $\tilde{\xi}_T$ and $\tilde{\eta}_T$ are 1-1, then $\tilde{\varphi}_T$ is an automorphism of $(\tilde{X}_T, \tilde{\sigma}_T)$.

Recent Titles in This Series

546 **Masakazu Nasu,** Textile systems for endomorphisms and automorphisms of the shift, 1995
545 **John L. Lewis and Margaret A. M. Murray,** The method of layer potentials for the heat equation in time-varying domains, 1995
544 **Hans-Otto Walther,** The 2-dimensional attractor of $x'(t) = -\mu x(t) + f(x(t-1))$, 1995
543 **J. P. C. Greenlees and J. P. May,** Generalized Tate cohomology, 1995
542 **Alouf Jirari,** Second-order Sturm-Liouville difference equations and orthogonal polynomials, 1995
541 **Peter Cholak,** Automorphisms of the lattice of recursively enumerable sets, 1995
540 **Vladimir Ya. Lin and Yehuda Pinchover,** Manifolds with group actions and elliptic operators, 1994
539 **Lynne M. Butler,** Subgroup lattices and symmetric functions, 1994
538 **P. D. T. A. Elliott,** On the correlation of multiplicative and the sum of additive arithmetic functions, 1994
537 **I. V. Evstigneev and P. E. Greenwood,** Markov fields over countable partially ordered sets: Extrema and splitting, 1994
536 **George A. Hagedorn,** Molecular propagation through electron energy level crossings, 1994
535 **A. L. Levin and D. S. Lubinsky,** Christoffel functions and orthogonal polynomials for exponential weights on [-1,1], 1994
534 **Svante Janson,** Orthogonal decompositions and functional limit theorems for random graph statistics, 1994
533 **Rainer Buckdahn,** Anticipative Girsanov transformations and Skorohod stochastic differential equations, 1994
532 **Hans Plesner Jakobsen,** The full set of unitarizable highest weight modules of basic classical Lie superalgebras, 1994
531 **Alessandro Figà-Talamanca and Tim Steger,** Harmonic analysis for anisotropic random walks on homogeneous trees, 1994
530 **Y. S. Han and E. T. Sawyer,** Littlewood-Paley theory on spaces of homogeneous type and the classical function spaces, 1994
529 **Eric M. Friedlander and Barry Mazur,** Filtrations on the homology of algebraic varieties, 1994
528 **J. F. Jardine,** Higher spinor classes, 1994
527 **Giora Dula and Reinhard Schultz,** Diagram cohomology and isovariant homotopy theory, 1994
526 **Shiro Goto and Koji Nishida,** The Cohen-Macaulay and Gorenstein Rees algebras associated to filtrations, 1994
525 **Enrique Artal-Bartolo,** Forme de Jordan de la monodromie des singularités superisolées de surfaces, 1994
524 **Justin R. Smith,** Iterating the cobar construction, 1994
523 **Mark I. Freidlin and Alexander D. Wentzell,** Random perturbations of Hamiltonian systems, 1994
522 **Joel D. Pincus and Shaojie Zhou,** Principal currents for a pair of unitary operators, 1994
521 **K. R. Goodearl and E. S. Letzter,** Prime ideals in skew and q-skew polynomial rings, 1994
520 **Tom Ilmanen,** Elliptic regularization and partial regularity for motion by mean curvature, 1994
519 **William M. McGovern,** Completely prime maximal ideals and quantization, 1994

(*Continued in the back of this publication*)

MEMOIRS
of the
American Mathematical Society

Number 546

Textile Systems for Endomorphisms and Automorphisms of the Shift

Masakazu Nasu

March 1995 • Volume 114 • Number 546 (second of 4 numbers) • ISSN 0065-9266

American Mathematical Society
Providence, Rhode Island

1991 *Mathematics Subject Classification.*
Primary 54H20; Secondary 58F08.

Library of Congress Cataloging-in-Publication Data

Nasu, Masakazu, 1941–
Textile systems for endomorphisms and automorphisms of the shift / Masakazu Nasu.
p. cm. – (Memoirs of the American Mathematical Society, ISSN 0065-9266; no. 546)
"March 1995, volume 114."
Includes bibliographical references and index.
ISBN 0-8218-2606-9
1. Topological dynamics. 2. Differentiable dynamical systems. I. Title. II. Series.
QA3.A57 no. 546
[QA611.5]
510 s–dc20
[514]

94-43210
CIP

Memoirs of the American Mathematical Society

This journal is devoted entirely to research in pure and applied mathematics.

Subscription information. The 1995 subscription begins with Number 541 and consists of six mailings, each containing one or more numbers. Subscription prices for 1995 are $369 list, $295 institutional member. A late charge of 10% of the subscription price will be imposed on orders received from nonmembers after January 1 of the subscription year. Subscribers outside the United States and India must pay a postage surcharge of $25; subscribers in India must pay a postage surcharge of $43. Expedited delivery to destinations in North America $30; elsewhere $92. Each number may be ordered separately; *please specify number* when ordering an individual number. For prices and titles of recently released numbers, see the New Publications sections of the *Notices of the American Mathematical Society.*

Back number information. For back issues see the *AMS Catalog of Publications.*

Subscriptions and orders should be addressed to the American Mathematical Society, P. O. Box 5904, Boston, MA 02206-5904. *All orders must be accompanied by payment.* Other correspondence should be addressed to Box 6248, Providence, RI 02940-6248.

Memoirs of the American Mathematical Society is published bimonthly (each volume consisting usually of more than one number) by the American Mathematical Society at 201 Charles Street, Providence, RI 02904-2213. Second-class postage paid at Providence, Rhode Island. Postmaster: Send address changes to Memoirs, American Mathematical Society, P. O. Box 6248, Providence, RI 02940-6248.

Printed in the United States of America.
This volume was printed directly from author-prepared copy.
∞ The paper used in this book is acid-free and falls within the guidelines established to ensure permanence and durability.
♲ Printed on recycled paper.

10 9 8 7 6 5 4 3 2 1   00 99 98 97 96 95

TABLE OF CONTENTS

ABSTRACT

We introduce the notion of a textile system. Using this, we study the dynamical properties of endomorphisms and automorphisms of topological Markov shifts including one-sided ones. The dynamical properties of automorphisms of sofic systems are also studied.

1991 Mathematics Subject Classification: Primary 54H20; Secondary 58F08

Keywords: Symbolic dynamics, Textile system, Topological Markov shift, Sofic system, Dynamical behavior of automorphisms, Dynamical behavior of endomorphisms, Strong shift equivalence, Shift equivalence.

DEDICATED TO THE MEMORY OF MY MOTHER, KUNIKO NASU

A textile system T has the following properties. If T is 1-1, then φ_T is expansive if and only if T^* is 1-1; if both T and T^* are 1-1, then (X_T, φ_T) is topologically conjugate to (X_{T^*}, σ_{T^*}) and hence (X_{T^*}, φ_{T^*}) is topologically conjugate to (X_T, σ_T) (Theorem 2.5). If T is one-sided 1-1, then φ_T is positively expansive if and only if both $\tilde{\xi}_{T^*}$ and $\tilde{\eta}_{T^*}$ are 1-1; if $\tilde{\xi}_T$ and $\tilde{\eta}_T$ are 1-1, then $\tilde{\varphi}_T$ is expansive if and only if T^* is one-sided 1-1; if T is one-sided 1-1 and both $\tilde{\xi}_{T^*}$ and $\tilde{\eta}_{T^*}$ are 1-1, then (X_T, φ_T) is topologically conjugate to $(\tilde{X}_{T^*}, \tilde{\sigma}_{T^*})$, and $(\tilde{X}_{T^*}, \tilde{\varphi}_{T^*})$ is topologically conjugate to (X_T, σ_T) (Theorem 2.11). If $\tilde{\xi}_T$ is 1-1, then $\tilde{\varphi}_T$ is positively expansive if and only if $\tilde{\xi}_{T^*}$ is 1-1; if both $\tilde{\xi}_T$ and $\tilde{\xi}_{T^*}$ are 1-1, then $(\tilde{X}_T, \tilde{\varphi}_T)$ is topologically conjugate to $(\tilde{X}_{T^*}, \tilde{\sigma}_{T^*})$ and hence $(\tilde{X}_{T^*}, \tilde{\varphi}_{T^*})$ is topologically conjugate to $(\tilde{X}_T, \tilde{\sigma}_T)$ (Theorem 2.12). Every endomorphism φ of a topological Markov shift restricted to its eventual image is equal to φ_T for some one-sided 1-1 and nondegenerate textile system T and we can easily construct this T, if φ is given as a block map. Thus the question of whether a given automorphism or a given onto endomorphism of a topological Markov shift is expansive or not, or positively expansive or not, and the subshift-identification problem for expansive automorphisms and positively expansive onto endomorphisms of topological Markov shifts including one-sided ones, are reduced to the problem of analyzing the dual textile system T^* of a given 1-1 or one-sided 1-1, nondegenerate textile system T. This problem is generally difficult. But when the warp shift (X_{T^*}, σ_{T^*}) of T is a subshift of finite type (i.e., a subshift which is topologically conjugate to a topological Markov shift), we can identify (X_{T^*}, σ_{T^*}) and check the injectivity of $\xi_{T^*}, \eta_{T^*}, \tilde{\xi}_{T^*}$ and $\tilde{\eta}_{T^*}$. As a consequence, we know that if φ is an automorphism of a topological Markov shift (X, σ) and if (X, φ) is topologically conjugate to a topological Markov shift, we can subshift-identify (X, φ) (Theorem 2.10). An example of very nontrivial subshift-identification will be given in Section 10.

Similarly, for automorphisms of one-sided topological Markov shifts, onto endomorphisms of topological Markov shifts, and onto endomorphisms of one-sided topological Markov shifts, we can subshift-identify the dynamical systems defined by them when these are conjugate to topological Markov shifts or one-sided topological Markov shifts. If T is a 1-1 and nondegenerate textile system, then its warp shift is a subshift of finite type if and only if T^* is 1-1 and finitely saturated (Theorem 2.7). But it is not known whether or not there are commuting dynamical systems (X, σ) and (X, τ) such that one is a topological Markov shift and the other is conjugate to a subshift which is not conjugate to any topological Markov shift (Question 2a). Similar questions are not answered even for automorphisms of one-sided topological Markov shifts, onto endomorphisms of topological Markov shifts, and onto endomorphisms of one-sided topological Markov shifts (Questions 3.a and 3.c).

For a one-sided 1-1 textile system T, the topological entropy $h(\sigma_{T^*})$ of the warp shift of T is equal to $h(\varphi_T)$ under some condition (Corollary 2.14). Hence the computation of the topological entropy of the warp shifts of one-sided 1-1 textile systems is useful to compute the topological entropy of endomorphisms of topological Markov shifts including one-sided ones. We can compute the topological entropy of the warp shifts of nondegenerate textile systems when they are subshifts of finite type (Proposition 2.15).

If T is a 1-1 textile system, then the textile shift $(U_T, \boldsymbol{\sigma}_T^{(k,n)})$ is topologically conjugate to $(X_T, \varphi_T^k\sigma_T^n)$ for all $k, n \in \mathbf{Z}$ (Proposition 2.4). The textile shift $(U_T, \boldsymbol{\sigma}_T^{(k,n)})$ is expansive if and only if the natural factor map $\check{\theta}_T^{(k,n)}$ of $(U_T, \boldsymbol{\sigma}_T^{(k,n)})$ onto the (k,n)-bias shift $(\check{X}_T^{(k,n)}, \check{\sigma}_T^{(k,n)})$, is a conjugacy (Proposition 2.3). These easy observations are fundamental to study the dynamical behavior of the automorphisms $\varphi^k\sigma^n$, $k, n \in \mathbf{Z}$, for a given automorphisms φ of a topological Markov shift (X, σ).

In Section 3, we study the textile systems with some resolving properties and endomorphisms and automorphisms of topological Markov shifts defined by those textile systems. Every onto endomorphism $\tilde{\varphi}$ of a one-sided topological Markov shift is equal to $\tilde{\varphi}_T$ for some one-sided 1-1 and nondegenerate textile system $T = (p, q : \Gamma \to G)$ with p left resolving. Every automorphism $\tilde{\varphi}$ of a one-sided topological Markov shift is equal to $\tilde{\varphi}_T$ for some 1-1 textile system $T = (p, q : \Gamma \to G)$ with both p and q left resolving. A textile system $T = (p, q : \Gamma \to G)$ with both p and q left resolving is said to be "LL" and an endomorphism defined by a one-sided 1-1 LL textile system is called an "LL endomorphism". A textile system $T = (p, q : \Gamma \to G)$ with p left resolving and q right resolving, is said to be "LR" and an "LR endomorphism" is also defined by a one-sided 1-1 LR textile system. LR textile systems and LL ones are nondegenerate. The resolving properties are inherited to dual textile systems in an interesting way (Proposition 3.1). In particular, the dual of an LR textile system is LR, so that an LR textile system is completely analyzable. If T is a one-sided 1-1 and nondegenerate textile system with p left resolving, then $h(\varphi_T)(= h(\tilde{\varphi}_T)) = h(\sigma_{T^*})$ (see Theorem 2.13 and Proposition 3.4). In particular, for an LR endomorphism φ, $h(\varphi)(= h(\tilde{\varphi}))$ is easily computed (see Theorem 6.31). As stated, the dynamical system (X, φ) defined by an onto endomorphism φ of a topological Markov shift (X, σ) can be subshift-identified when it is conjugate to a one-sided topological Markov shift, but such an endomorphism φ is very special: it is "essentially given" (i.e., $\varphi = \psi^{-1}\varphi_T\psi$ for some topological conjugacy $\psi : (X, \sigma) \to (X_T, \sigma_T)$) by a one-sided 1-1 and nondegenerate textile system $T = (p, q : \Gamma \to G)$ with q biresolving and T^* 1-1 (Theorem 3.9). Moreover, if (X, σ) is an irreducible and aperiodic nontrivial topological Markov shift and if φ is an onto endomorphism of (X, σ) such that (X, φ) is topologically conjugate to a one-sided topological Markov shift, then φ is N-to-one and (X, φ) is topologically conjugate to the one-sided full N-shift for some $N \geq 3$ (Theorem 3.12). If $(\tilde{X}, \tilde{\sigma})$ is a nontrivial one-sided topological Markov shift having an expansive automorphism $\tilde{\varphi}$ such that $\tilde{\varphi}^n$ is topologically transitive for all $n \in \mathbf{N}$, then $(\tilde{X}, \tilde{\sigma})$ is topologically conjugate to the one-sided full N-shift for some $N \geq 3$ (Theorem 3.12). If $\tilde{\varphi}$ is an automorphism of a one-sided topological Markov shift $(\tilde{X},\tilde{\sigma})$ such that $(\tilde{X},\tilde{\varphi})$ is topologically conjugate to a topological Markov shift, then we can subshift-identify $(\tilde{X},\tilde{\varphi})$ (Theorem 3.9). But we know no example of an automorphism $\tilde{\varphi}$ of a one-sided full shift $(\tilde{X},\tilde{\sigma})$ such that $(\tilde{X},\tilde{\varphi})$ is conjugate to an irreducible and aperiodic topological Markov shift which is not topologically conjugate to any full shift (Question 3.b). We can subshift-identify the dynamical system defined by an onto endomorphism of a one-sided topological Markov shift when it is conjugate to a one-sided topological Markov shift, but such an endomorphism is "essentially LR" (i.e., essentially given by an LR textile system) (Theorem 3.13). Hence two commuting dynamical systems (X,σ) and (X,τ) which are conjugate to one-sided topological Markov shifts, are essentially equal to two one-sided topological Markov shifts whose defining matrices commute (see Proposition 6.1).

In Section 4, we introduce the notions of a "sofic textile system" and its dual, which are natural generalizations of those of a textile system and its dual. For a "1-1" sofic textile system $\mathcal{T}$, the automorphism $\varphi_{\mathcal{T}}$ of $(X_{\mathcal{T}},\sigma_{\mathcal{T}})$ is expansive if and only if there is $n \in \mathbf{N}$ such that $(\mathcal{T}^{[n]})^*$ is 1-1, where $\mathcal{T}^{[n]}$ is the "higher block system" of order n of $\mathcal{T}$; if $(\mathcal{T}^{[n]})^*$ is 1-1, then $(X_{\mathcal{T}},\varphi_{\mathcal{T}})$ is topologically conjugate to $(X_{(\mathcal{T}^{[n]})^*},\sigma_{(\mathcal{T}^{[n]})^*})$ (Theorem 4.1). For every automorphism φ of a sofic system, there is a 1-1 and nondegenerate sofic textile system $\mathcal{T}$ such that $\varphi_{\mathcal{T}} = \varphi$ (Proposition 4.4).

In Section 5, we review the structure of topological conjugacies between sofic systems given in [N2] and give some refinement. Any topological conjugacy ϕ between subshifts is factorized into a composition of the form

$$\phi = \kappa_n\zeta_n\kappa_{n-1}\zeta_{n-1}\cdots\kappa_1\zeta_1\kappa_0$$

with $n \geq 1$, where each of $\kappa_0,\cdots,\kappa_n$ is a symbolic conjugacy and each of $\zeta_1,\cdots,\zeta_n$ is a forward or backward bipartite conjugacy. We say that ϕ is "forward" if all ζ_i's are forward. (Hence if the number of backward ζ_i's is l, then $\phi\sigma_1^m$ is a forward conjugacy for all $m \geq l$.) A κ-ζ factorization $\phi = \kappa_n\zeta_n\kappa_{n-1}\cdots\kappa_1\zeta_1\kappa_0$ of a forward conjugacy ϕ between sofic systems $(X_{\mathcal{M}},\sigma_{\mathcal{M}})$ and $(X_{\mathcal{N}},\sigma_{\mathcal{N}})$ whose defining canonical representation matrices are $\mathcal{M}$ and $\mathcal{N}$, respectively, with $n \geq 1$, and a "canonical specified strong shift equivalence"

$$\mathcal{M} \overset{k_0}{\simeq} \mathcal{P}_1\mathcal{Q}_1,\ \mathcal{Q}_1\mathcal{P}_1 \overset{k_1}{\simeq} \mathcal{P}_2\mathcal{Q}_2,\ \cdots,\ \mathcal{Q}_{n-1}\mathcal{P}_{n-1} \overset{k_{n-1}}{\simeq} \mathcal{P}_n\mathcal{Q}_n,\ \mathcal{Q}_n\mathcal{P}_n \overset{k_n}{\simeq} \mathcal{N}$$

for representation matrices, uniquely correspond to each other.

In Section 6, we study LR textile systems and sofic LR textile systems in detail. I said above that textile systems are natural and useful machines to analyze the dynamical behavior of endomorphisms and automorphisms of topological Markov shifts. This would especially be made understandable by the natural relation of the forward automorphisms of topological Markov shifts to the 1-1 LR textile systems: the structure of forward automorphisms of topological Markov shifts given by the specified strong shift equivalences naturally induces the structure of 1-1 LR textile systems. That relation is extended to the natural one of the forward automorphisms of sofic systems to the 1-1 LR sofic textile systems. As mentioned above also, Boyle and Krieger proved in [BK1] that for every automorphism φ of every topological Markov shift (X, σ) and for every n greater than a coding bound for φ and φ^{-1}, $(X, \varphi\sigma^n)$ is topologically conjugate to a topological Markov shift and specified the dimension group triple of this topological Markov shift. The problem of identifying (up to topological conjugacy) the topological Markov shift, which was a motivation of this research, is solved more generally for the sofic case.

Each LR sofic textile system $\mathcal{T}$ having $\mathcal{M}$ and $\mathcal{N}$ as the defining representation matrices of $(X_{\mathcal{T}}, \sigma_{\mathcal{T}})$ and $(X_{\mathcal{T}^*}, \sigma_{\mathcal{T}^*})$, respectively, is uniquely associated with a "specified equivalence" of the form

$$\mathcal{M}\mathcal{N} \overset{k}{\simeq} \mathcal{N}\mathcal{M}$$

and conversely each specified equivalence of this form defines an LR sofic textile system associated with it. If a 1-1 LR sofic textile system $\mathcal{T}$ is associated with a specified equivalence $\mathcal{M}\mathcal{N} \overset{k}{\simeq} \mathcal{N}\mathcal{M}$, then for all $l \geq 0$ and $m \geq 1$, $(X_{\mathcal{T}}, \varphi_{\mathcal{T}}^l \sigma_{\mathcal{T}}^m)$ is topologically conjugate to the sofic system whose defining representation matrix is $\mathcal{N}^l\mathcal{M}^m$; if $\varphi_{\mathcal{T}}$ is expansive, then $(X_{\mathcal{T}}, \varphi_{\mathcal{T}})$ is topologically conjugate to a sofic system (Corollary 6.20). In particular, if $\mathcal{T}$ is a textile system T, then $\mathcal{M}$ and $\mathcal{N}$ above are given as the symbolic representations $\tilde{M}$ and $\tilde{N}$ of the defining (nonnegative integral) matrices M and N of (X_T, σ_T) and (X_{T^*}, σ_{T^*}), respectively, and if T^* is 1-1, then (X_T, φ_T) is topologically conjugate to the topological Markov shift whose defining matrix is N, and otherwise, φ_T is not expansive (Corollary 6.5).

These results are directly applied to forward automorphisms of sofic systems. If φ is a forward automorphism of a sofic system and has the κ -ζ factorization given by a specified strong shift equivalence

$$\mathcal{M} \overset{k_0}{\simeq} \mathcal{P}_1\mathcal{Q}_1,\ \mathcal{Q}_1\mathcal{P}_1 \overset{k_2}{\simeq} \mathcal{P}_2\mathcal{Q}_2,\ \cdots,\ \mathcal{Q}_{n-1}\mathcal{P}_{n-1} \overset{k_{n-1}}{\simeq} \mathcal{P}_n\mathcal{Q}_n,\ \mathcal{Q}_n\mathcal{P}_n \overset{k_n}{\simeq} \mathcal{M}$$

with $n \geq 1$, then this specified strong shift equivalence naturally induces the specified equivalence of the form

$$\mathcal{M}\mathcal{P} \overset{k}{\simeq} \mathcal{P}\mathcal{M}$$

with

$$\mathcal{P} = \mathcal{P}_1 \cdots \mathcal{P}_n,$$

the sofic textile system $\mathcal{T}$ associated with this specified equivalence is 1-1, and $\varphi_{\mathcal{T}} = \varphi$ (Corollary 6.24), so that $(X_{\mathcal{M}}, \varphi^l\sigma_{\mathcal{M}}^m)$ is topologically conjugate to the sofic system whose defining representation matrix is $\mathcal{P}^l\mathcal{M}^m$ for all $l \geq 0$ and $m \geq 1$. The complete statements of the main results of Section 6 including this are given as Theorems 6.10 and 6.23.

By the above, a forward automorphism of a sofic system is a sofic LR automorphism. We find that for an automorphism φ of a topological Markov shift, φ is LR if and only if φ is forward (Theorem 6.29). We also find that the entropy $h(\varphi)$ is equal to $-\log R(\varphi)$ for every LR endomorphism φ of every irreducible topological Markov shift, where $R(\varphi)$ is the number considered in [B2] and [BK1] (Theorem 6.31). This is an extension of a result of Boyle and Krieger on the entropy of automorphisms of topological Markov shifts given in [BK1].

In Section 7, we introduce the notion of a "resolvable textile system" and using this we characterize the textile systems in some class that are "essentially resolving", for some types of resolving properties (Theorems 7.19 and 7.20). In particular, we present a necessary and sufficient condition for a one-sided 1-1 and nondegenerate textile system T to have the property that $\varphi_T \in C_i$, for each $i = 1, 2, 3$, where C_1 is the class of endomorphisms that are essentially given by one-sided 1-1 and nondegenerate textile systems $T = (p, q : \Gamma \to G)$ with p left resolving, C_2 is the class of essentially LR endomorphisms, and C_3 is the class of endomorphisms that are essentially given by one-sided 1-1 and nondegenerate textile systems $T = (p, q : \Gamma \to G)$ with q biresolving (Theorem 7.21). We find that for an onto endomorphism φ of a topological Markov shift and for $i = 1, 2, 3$, if $\varphi^n \in C_i$ for some $n \in \mathbf{N}$, then $\varphi \in C_i$ (Theorem 7.22). We also find that for an onto endomorphism φ of a topological Markov shift (X, σ) and for $i = 1, 2, 3$, if $\varphi^{(n)} \in C_i$ for some $n \in \mathbf{N}$, then $\varphi \in C_i$ (Theorem 7.23), where $\varphi^{(n)}$ denotes the endomorphism of (X, σ^n) which is naturally induced by φ.

We find that if φ is an LR endomorphism of a topological Markov shift, then a one-sided 1-1 LR textile system with $\varphi_T = \varphi$, is unique (Corollary 7.25).

We are interested in classifying the dynamical systems $(X, \varphi^k \sigma^n)$, $k, n \in \mathbf{Z}$, for a given automorphism φ of a topological Markov shift (X, σ). Clearly expansiveness depends only on the directions n/k. The property of being conjugate to a topological Markov shift, also depends only on the directions n/k, by the result of Boyle and Krieger (Lemma 2.6 of [BK1]) that the root of a topological Markov shift is conjugate to a topological Markov shift. Moreover, by the result stated above, the property of being an essentially LR automorphism of (X, σ), also depends only on the directions n/k. Let (X, σ) be a topological Markov shift with $h(\sigma) > 0$ and let φ be an automorphism of (X, σ). For $\tau = \varphi^{k_0} \sigma^{n_0}$ with $(n_0, k_0) \in \mathbf{Z}^2$ such that (X, τ) is conjugate to a topological Markov shift, we define

$$K_\varphi(\tau) = \{(n, k) \in \mathbf{Z}^2 \mid \varphi^k \sigma^n \text{ is an essentially LR automorphism of } (X, \tau)\}$$

and we define $CK_\varphi(\tau)$ to be the convex cone generated by $K_\varphi(\tau)$ in $\mathbf{R}^2$, which is called an "ELR cone" for φ. The ELR cones for φ have the following properties (Theorem 8.16): (i) each ELR cone is a convex cone which is either pointed or equal to a closed half-plane; (ii) each ELR cone has its interior; (iii) the interiors of any two distinct

ELR cones are disjoint; (iv) for any $(n,k) \in \mathbf{Z}^2$, $(X, \varphi^k\sigma^n)$ is conjugate to a topological Markov shift if and only if (n,k) is an interior point of some ELR cone; (v) if a lattice point (n,k) in a ELR cone is on the boundary of the cone, then $\varphi^k\sigma^n$ is nonexpansive; (vi) for $(n,k) \in \mathbf{Z}^2$ and for $(n',k') \in \mathbf{Z}^2$ such that $(X, \varphi^{k'}\sigma^{n'})$ is conjugate to a topological Markov shift, $\varphi^k\sigma^n$ is an essentially LR automorphism of $(X, \varphi^{k'}\sigma^{n'})$ if and only if (n,k) and (n',k') belong to the same ELR cone; (vii) Every (some) ELR cone is pointed if and only if there is no $(n,k) \in \mathbf{Z}^2 - \{(0,0)\}$ such that $\varphi^k\sigma^n$ is of finite order; (viii) if the topological Markov shift (X,σ) is irreducible, then entropy function is additive on each ELR cone.

In Section 9, we introduce the notions of "similarity" and "weak similarity" for topological conjugacies for sofic systems. Closely related materials for topological Markov shifts were studied by Wagoner in [Wag1], [Wag2], and [Wag3], but our approach is different. The notions of simple automorphisms and inert automorphisms for topological Markov shifts are naturally generalized to those for sofic systems. If two automorphisms of a sofic system (Y,σ) are forward and associated with the same shift equivalence, then $(Y, \varphi_1\sigma^l)$ and $(Y, \varphi_2\sigma^l)$ are topologically conjugate for all integers $l > 0$, and if φ_1 and φ_2 are weakly similar, then for all sufficiently large integers l, $(Y, \varphi_1\sigma^l)$ and $(Y, \varphi_2\sigma^l)$ are topologically conjugate and for all sufficiently large integers l, $(Y, \varphi_1\sigma^{-l})$ and $(Y, \varphi_2\sigma^{-l})$ are topologically conjugate (Proposition 9.18).

In particular if φ_1 and φ_2 are expansive forward automorphisms of a topological Markov shift (X,σ) and are associated with the same shift equivalence, then (X,φ_1) and (X,φ_2) are topologically conjugate. Hence the dynamical behavior of expansive forward automorphisms is far from depending on the specifications of the specified strong shift equivalences giving their κ-ζ factorizations. But the situation is drastically different for automorphisms which are not expansive-forward. An example showing this is given in Section 10.

There are many problems waiting for further research. What are the possible subshifts given as the woof shifts (X_T, σ_T) of degenerate textile systems T? Similar questions should be made for various subclasses of textile systems T. (According to Question 2.a, one of the classes should be that of 1-1 and infinitely saturated textile systems T with T^* 1-1 and nondegenerate.) How is the dynamics of textile shifts ? The study of this problem for 1-1 and nondegenerate textile systems is the same as the study of the dynamical systems $(X, \varphi^k\sigma^n)$, $k, n \in \mathbf{Z}$, for automorphisms φ of topological Markov shifts (X,σ). Further properties of ELR cones are especially intriguing (see Questions 8a and 8b). For general textile systems, however, the woof shifts (the warp shifts) and the textile shifts are subjects of symbolic dynamics as their own right. Sofic textile systems should also be further studied.

This work was begun in the late summer of 1987. Part of the results (on automorphisms of the shift) of this memoir were presented at the Workshop on Ergodic Theory and Symbolic Dynamics held at the University of Washington in the summer of 1989 and the Symposium on Ergodic Theory and Related Topics held at Kyoto Seminar House in January 1990. Part of the results on more extensive materials of this memoir were also presented at the Conference on Arithmetic and Symbolic Dynamics held at CIRM in September 1991. I thank the people who gave me the benefit of helpful discussions and conversations on some occasions including those during the development of the work. I am particularly grateful to Mike Boyle for discussions and comments which were useful, especially to add and revise Section 9, and for his recommendation on which the title of this memoir was chosen, to Brian Marcus for his suggestive conversation which was useful to obtain Theorem 9.4 and for discussions which were useful to revise Section 9, to Yoichiro Takahashi for conversations which prompted the enlargement of application of textile systems to endomorphisms of the shift, and to Susan Williams who let me know the words "woof" and "warp". I benefited from the trip as a researcher abroad of the Ministry of Education, Science, and Culture of Japan during the period December 1990 - January 1991. I am grateful to the University of Washington, University of Maryland, and IBM Almaden Research Center for their hospitality during my visit to them.

The manuscript was typeset by Mrs. Miyoko Kawamura, whom I especially wish to thank for her patience. Special thanks are also due to Hiroshi Kamabe who taught her how to use TeX, provided some macros for the manuscript and drew all figures of this memoir. Without their cooperation and TeX, the manuscript would not have prepared so soon and so beautifully.

I thank the referees for several useful suggestions to improve the presentation.

1. PRELIMINARIES

Let A be an alphabet (i.e., a finite nonempty set of symbols) Let X_A be the set of all bisequences $(a_i)_{i\in\mathbf{Z}}$ with $a_i \in A$. We define a metric d on X_A as follows. Let $x = (a_i)_{i\in\mathbf{Z}}$ and $y = (b_i)_{i\in\mathbf{Z}}$ be elements of X_A. Then $d(x, y) = 0$ if $x \neq y$, and $d(x, y) = 1/(1 + k)$ if $x \neq y$, where

$$k = \min\{|i| \mid i \in \mathbf{Z}, a_i \neq b_i\}.$$

Then with this metric,X_A is compact. The homeomorphism $\sigma_A : X_A \to X_A$ defined by

$$\sigma_A((a_i)_{i\in\mathbf{Z}}) = (a_{i+1})_{i\in\mathbf{Z}}, \qquad (a_i)_{i\in\mathbf{Z}} \in X_A.$$

is called the *shift transformation.* The dynamical system (X_A, σ_A) is called the *full shift* over A. Let X be a closed σ_A-invariant subset of X_A. Let

$$\sigma = \sigma_A|X.$$

Then we have a dynamical system (X, σ), which is called a *subshift* over A.

Let G be a graph. Throughout this memoir, a graph will mean a directed graph which may have multiple arcs and multiple loops. Let A_G and V_G denote the arc-set of G and the vertex-set of G, respectively. Let $i_G : A_G \to V_G$ and $t_G : A_G \to V_G$ be the mappings such that for $a \in A_G$, $i_G(a)$ and $t_G(a)$ are the initial and terminal vertices of a, respectively. We say that G is *nondegenerate,* when both i_G and t_G are onto. Let X_G be the set of all bisequences

$$(a_i)_{i\in\mathbf{Z}}, \qquad a_i \in A_G$$

with

$$t_G(a_i) = i_G(a_{i+1}), \qquad i \in \mathbf{Z}.$$

Let

$$\sigma_G = \sigma_{A_G}|X_G.$$

Then we have a subshift (X_G, σ_G) over A_G, which is called the *topological Markov shift* defined by G.

For a graph G, let M_G denote the *adjacency matrix* of G. For example, if G is given by

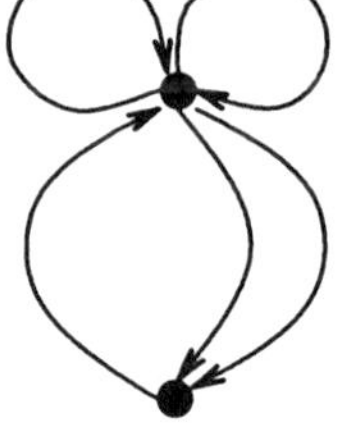

then M_G is

$$\begin{pmatrix} 2 & 2 \\ 1 & 0 \end{pmatrix}.$$

For a square nonnegative integral matrix M, (X_M, σ_M) will denote (X_G, σ_G), where G is the graph with $M_G = M$. We call G the *defining graph* of (X_G, σ_G) and M the *defining matrix* of (X_M, σ_M).

Let (X_1, τ_1) and (X_2, τ_2) be topological dynamical systems. A continuous map $\phi : X_1 \to X_2$ such that $\phi\tau_1 = \tau_2\phi$, is called a *factor map* of (X_1, τ_1) into (X_2, τ_2). We do not assume that a factor map is onto. If $\bar{\tau}_2 : \phi(X_1) \to \phi(X_1)$ is the restriction of τ_2, then the dynamical system

$$(\phi(X_1), \bar{\tau}_2)$$

is called a *factor* of (X_1, τ_1). For a factor map ϕ of (X_1, τ_1) into (X_2, τ_2), if ϕ is a homeomorphism, then ϕ is called a *topological conjugacy* and we say that (X_1, τ_1) and (X_2, τ_2) are *topologically conjugate*. A factor map of a topological dynamical system into itself is called an *endomorphism* of the topological dynamical system. A topological conjugacy of a topological dynamical system onto itself is called an *automorphism* of the topological dynamical system.

Let (X, σ) be a subshift over an alphabet A. Let $n \in \mathbf{N}$. Let $L_n(X)$ denote the set of all n-blocks or words of length n that appear on some bisequence in X, that is,

$$L_n(X) = \{a_0 a_1 \cdots a_{n-1} \mid (a_i)_{i \in \mathbf{Z}} \in X, a_i \in A\}.$$

Let (X_i, σ_i) be a subshift over an alphabet A_i for $i = 1, 2$. Then by the theorem of Curtis, Hedlund, and Lyndon [H], a mapping $\phi : X_1 \to X_2$ is a factor map of (X_1, σ_1) into (X_2, σ_2) if and only if for some nonnegative integers m and n, ϕ is a *block map of* (m, n) *type*, i.e., there is a mapping $f : L_{m+n+1}(X_1) \to A_2$ by which ϕ is given in such a way that if for $(a_i)_{i \in \mathbf{Z}} \in X_1$ with $a_i \in A_1$,

$$\phi((a_i)_{i \in \mathbf{Z}}) = (f(a_{-m+i} \cdots a_{n+i}))_{i \in \mathbf{Z}}.$$

In particular, a block map of (0,0) type is called a *1-block map*.

For graphs Γ and G, a *homomorphism* of Γ into G, written by $h : \Gamma \to G$, is a pair of mappings $h_A : A_\Gamma \to A_G$ and $h_V : V_\Gamma \to V_G$ such that

$$i_G h_A = h_V i_\Gamma \quad \text{and} \quad t_G h_A = h_V t_\Gamma.$$

We call h_A the *arc-map* of h and h_V the *vertex-map* of h. If G is nondegenerate, then h_A determines h_V uniquely. Therefore we shall often use h instead of h_A. A homomorphism $h : \Gamma \to G$ is said to be *onto* if both of h_A and h_V are onto. A graph-homomorphism $h : \Gamma \to G$ naturally gives a 1-block factor map of $(X_\Gamma, \sigma_\Gamma)$ into (X_G, σ_G).

A λ-*graph* $\mathcal{G}$ over an alphabet A is a pair of a graph G and a *labeling* (onto mapping) $\lambda : A_G \to A$ which assigns a symbol from A to each arc of G. It will be denoted by $\mathcal{G} = (G, \lambda)$. We call G the *support* of $\mathcal{G}$. For a λ-graph $\mathcal{G} = (G, \lambda)$, let

$$X_{\mathcal{G}} = \{(\lambda(a_i))_{i\in\mathbf{Z}} \mid (a_i)_{i\in\mathbf{Z}} \in X_G\}.$$

Then we have a subshift $(X_{\mathcal{G}}, \sigma_{\mathcal{G}})$, which is a factor of the topological Markov shift (X_G, σ_G) since λ gives a 1-block factor map of (X_G, σ_G) onto $(X_{\mathcal{G}}, \sigma_{\mathcal{G}})$. By definition, a *sofic system* is a factor of a topological Markov shift [We]. It is well known that every sofic system is given as $(X_{\mathcal{G}}, \sigma_{\mathcal{G}})$ for some λ-graph $\mathcal{G}$. For a λ-graph $\mathcal{G} = (G, \lambda)$, the 1-block factor map of (X_G, σ_G) onto $(X_{\mathcal{G}}, \sigma_{\mathcal{G}})$ given by λ is called the *sofic cover* defined by $\mathcal{G}$.

Let (X, σ) be a subshift. Let n be a positive integer. Define

$$X^{[n]} = \{(a_i \cdots a_{i+n-1})_{i\in\mathbf{Z}} \mid (a_i)_{i\in\mathbf{Z}} \in X\}.$$

Then we have a subshift $(X^{[n]}, \sigma^{[n]})$ over $L_n(X)$ which is called the *higher block system of order* n of (X, σ) [AM]. Clearly $(X^{[n]}, \sigma^{[n]})$ is topologically conjugate to (X, σ) for all $n \in \mathbf{N}$. Define

$$X^{(n)} = \{(a_{in} \cdots a_{(i+1)n-1})_{i\in\mathbf{Z}} \mid (a_i)_{i\in\mathbf{Z}} \in X\}.$$

Then we have a subshift $(X^{(n)}, \sigma^{(n)})$ over $L_n(X)$, which will be called the *n-th power system* of (X, σ). It is easy to see that $(X^{(n)}, \sigma^{(n)})$ is topologically conjugate to (X, σ^n). For an endomorphism φ of (X, σ), $(\varphi)^{(n)}$ will denote the endomorphism of $(X^{(n)}, \sigma^{(n)})$ which is naturally induced by φ. Note that $\sigma^{(n)} = (\sigma^n)^{(n)}$. But we shall use $\varphi^{(n)}$ to mean $(\varphi)^{(n)}$ wherever no confusion arises.

Let G be a graph. Let $n \in \mathbf{N}$. Let $L_n(G)$ denote the set of all *paths* of length n of G, that is,

$$L_n(G) = \{a_1 \cdots a_n \mid a_i \in A_G,\ t_G(a_j) = i_G(a_{j+1}) \text{ for } j = 1, \cdots, n-1\}$$

For $n \geq 2$, let $G^{[n]}$ denote the graph such that

$$A_{G^{[n]}} = L_n(G), \quad V_{G^{[n]}} = L_{n-1}(G)$$

and $i_{G^{[n]}}$ and $t_{G^{[n]}}$ are defined by

$$i_{G^{[n]}}(a_1 \cdots a_n) = a_1 \cdots a_{n-1} \quad \text{and} \quad t_{G^{[n]}}(a_1 \cdots a_n) = a_2 \cdots a_n$$

for $a_1 \cdots a_n \in A_{G^{[n]}}$ with $a_i \in A_G$. For convenience we define $L_0(G) = V_G$ and $G^{[1]} = G$. We call $G^{[n]}$ the *higher block graph of order* n of G. We define

$$L(G) = \bigcup_{n=0}^{\infty} L_n(G).$$

Let G and H be two graphs with $V_G = V_H$. We define the *product* GH of G and H as the graph such that

$$\begin{aligned} V_{GH} &= V_G, \\ A_{GH} &= \{ab \mid a \in A_G, b \in A_H, t_G(a) = i_H(b)\} \end{aligned}$$

and i_{GH} and t_{GH} are defined by

$$i_{GH}(ab) = i_G(a) \quad \text{and} \quad t_{GH}(ab) = t_G(b), \quad ab \in A_{GH}.$$

Clearly we have $M_{GH} = M_G M_H$.

If G is a graph, n is a positive integer and (X, σ) is the topological Markov shift defined by G, then $(X^{[n]}, \sigma^{[n]})$ is the topological Markov shift defined by $G^{[n]}$ and $(X^{(n)}, \sigma^{(n)})$ is the topological Markov shift defined by $G^n = \overbrace{G \cdots G}^{n}$.

For a graph G, the *transpose* of G is the graph obtained from G by reversing the direction of the arcs, which will be denoted by G^{-1}. For $n \in \mathbf{N}$, G^{-n} means $(G^{-1})^n$.

Let $\mathcal{G} = (G, \lambda)$ and $\mathcal{H} = (H, \lambda')$ be two λ-graphs with $V_G = V_H$. We define the *product* $\mathcal{GH}$ of $\mathcal{G}$ and $\mathcal{H}$ as the λ-graph (GH, λ'') such that for $ab \in A_{GH}$ with $a \in A_G$ and $b \in A_H$, $\lambda''(ab) = \lambda(a)\lambda'(b)$. The *transpose* $\mathcal{G}^{-1}$ of $\mathcal{G}$ is defined as $\mathcal{G}^{-1} = (G^{-1}, \lambda)$. The definitions of $\mathcal{G}^n$ and $\mathcal{G}^{-n}$ should be clear for $n \in \mathbf{N}$.

Let $y = (x_i)_{i \in \mathbf{Z}}$ be a bisequence of elements x_i from some set. Let $\rightarrow y$ denote the bisequence written in the form

$$\cdots \quad x_{-1}, x_0, x_1, \quad \cdots$$

and let $\downarrow y$ denote the bisequence written in the form

$$\begin{matrix} \vdots \\ x_{-1} \\ x_0 \\ x_1 \\ \vdots \end{matrix} .$$

The bisequences $\rightarrow y$ and $\downarrow y$ are said to be in the *row form* and in the *column form*, respectively. Each of them is the *transpose* of the other.

Let A be an alphabet. A two-dimensional arrangement $(a_{ij})_{i,j \in \mathbf{Z}}$ of elements a_{ij} from A is called a *configuration* over A. If $x_i = \rightarrow (a_{ij})_{j \in \mathbf{Z}}$ for $i \in \mathbf{Z}$ and $y_j = \downarrow (a_{ij})_{i \in \mathbf{Z}}$ for $j \in \mathbf{Z}$, then each of $\downarrow (x_i)_{i \in \mathbf{Z}}$ and $\rightarrow (y_j)_{j \in \mathbf{Z}}$ represents $(a_{ij})_{i,j \in \mathbf{Z}}$. We call x_i, $i \in \mathbf{Z}$, and $y_j, j \in \mathbf{Z}$, the rows and the columns of $(a_{ij})_{i,j \in \mathbf{Z}}$. We call $(a_{ji})_{i,j \in \mathbf{Z}}$ the *transpose* of $(a_{ij})_{i,j \in \mathbf{Z}}$. For a subset E of $\mathbf{Z}^2$, an arrangement $(a_{ij})_{(i,j) \in E}$ of elements a_{ij} from A will be called an E-*configuration* over A or a *subconfiguration* over A.

2. TEXTILE SYSTEMS

A *textile system* T over a graph G is an ordered pair of graph-homomorphisms $p : \Gamma \to G$ and $q : \Gamma \to G$ such that for $\alpha \in A_\Gamma$, the quadruple

$$(i_\Gamma(\alpha), t_\Gamma(\alpha), p(\alpha), q(\alpha))$$

uniquely determines α. Write

$$T = (p, q : \Gamma \to G).$$

Let ξ and η be the 1-block factor maps of $(X_\Gamma, \sigma_\Gamma)$ into (X_G, σ_G) given by p and q, respectively. A *textile weaved* by T is a configuration over A_Γ of the form

$$\downarrow (\to z_i)_{i \in \mathbf{Z}},$$

where $z_i, i \in \mathbf{Z}$, are elements of X_Γ such that

$$\eta(z_{i-1}) = \xi(z_i) \quad \text{for all } i \in \mathbf{Z}.$$

If $t = \downarrow (\to z_i)_{i \in \mathbf{Z}}$ is a textile weaved by T, then

$$(\xi(z_i))_{i \in \mathbf{Z}}$$

will be called the *text* of t or a *text generated* by T.

Let $T = (p, q : \Gamma \to G)$ be a textile system. We define a graph G^T by

$$A_{G^T} = V_\Gamma,$$

$$V_{G^T} = V_G,$$

$$i_{G^T} = p_V \quad \text{and} \quad t_{G^T} = q_V,$$

and we define a graph Γ^T by

$$A_{\Gamma^T} = A_\Gamma,$$

$$V_{\Gamma^T} = A_G,$$

$$i_{\Gamma^T} = p_A \quad \text{and} \quad t_{\Gamma^T} = q_A.$$

We can define two graph-homomorphisms $p^T : \Gamma^T \to G^T$ and $q^T : \Gamma^T \to G^T$ by

$$(p^T)_A = i_\Gamma \quad , \quad (p^T)_V = i_G$$

and

$$(q^T)_A = t_\Gamma \quad , \quad (q^T)_V = t_G.$$

Thus we have a textile system

$$T^* = (p^T, q^T : \Gamma^T \to G^T),$$

which will be called the *dual* of T. It directly follows from the definition that

$$(T^*)^* = T.$$

Let

$$t = \downarrow (\to z_i)_{i \in \mathbf{Z}} = (\alpha_{ij})_{i,j \in \mathbf{Z}}$$

be a textile weaved by T with $\alpha_{ij} \in A_\Gamma = A_{\Gamma^T}$, $i, j \in \mathbf{Z}$. Then

$$z_i = (\alpha_{ij})_{j \in \mathbf{Z}}, \quad i \in \mathbf{Z},$$

and $z_i \in X_\Gamma$. Let

$$z_j^* = (\alpha_{ij})_{i \in \mathbf{Z}}, \quad j \in \mathbf{Z}.$$

Then $z_j^* \in X_{\Gamma^T}$ because

$$\eta(z_{i-1}) = \xi(z_i) \quad \text{for all } i \in \mathbf{Z}$$

implies that

$$q(\alpha_{i-1,j}) = p(\alpha_{ij}) \quad \text{for all } i \in \mathbf{Z},$$

so that

$$t_{\Gamma^T}(\alpha_{i-1,j}) = i_{\Gamma^T}(\alpha_{ij}) \quad \text{for all } i \in \mathbf{Z}.$$

Let ξ^* and η^* be the 1-block factor maps given by p^T and q^T, respectively. Since for all $i \in \mathbf{Z}$,

$$q^T(\alpha_{i,j-1}) = t_\Gamma(\alpha_{i,j-1}) = i_\Gamma(\alpha_{ij}) = p^T(\alpha_{ij})$$

for all $j \in \mathbf{Z}$, we have

$$\eta^*(z_{j-1}^*) = \xi^*(z_j^*) \quad \text{for all } j \in \mathbf{Z}$$

so that

$$\downarrow (\to z_j^*)_{j \in \mathbf{Z}} = (\alpha_{ji})_{i,j \in \mathbf{Z}}$$

is a textile weaved by T^*. This and the duality imply that a configuration over $A_\Gamma = A_{\Gamma^T}$ is a textile weaved by T if and only if its transpose is a textile weaved by T^*. We recall that $\alpha_{ij} \in A_\Gamma = A_{\Gamma^T}$ is uniquely determined by the quadruple

$$\begin{aligned} & (i_\Gamma(\alpha_{ij}), t_\Gamma(\alpha_{ij}), p(\alpha_{ij}), q(\alpha_{ij})) \\ = \; & (p^T(\alpha_{ij}), q^T(\alpha_{ij}), p(\alpha_{ij}), q(\alpha_{ij})). \end{aligned}$$

We can identify α_{ij} with this quadruple, which is described as the square in Fig. 2.1.

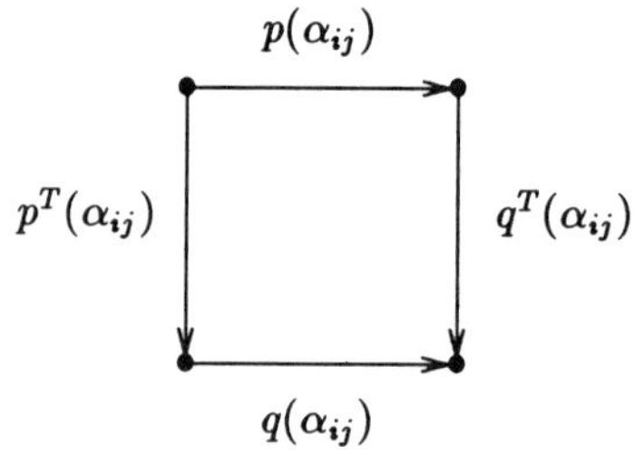

Figure 2.1.

This square represents a piece of the text of t combined with the text of t^*. The corners represent vertices of G (or G^T), the horizontal sides represent arcs of G, and the vertical sides represent arcs of G^T. We can identify α_{ij} with this square. Therefore if $t =\downarrow (\rightarrow z_i)_{i\in\mathbf{Z}}$ is a textile weaved by T with $z_i \in X_\Gamma$ and $t^* =\downarrow (\rightarrow z_j^*)_{j\in\mathbf{Z}}$ is its transpose with $z_j^* \in X_{\Gamma^T}$, then the text

$$(x_i)_{i\in\mathbf{Z}} = (\eta(z_{i-1}))_{i\in\mathbf{Z}} = (\xi(z_i))_{i\in\mathbf{Z}}$$

of t and the text

$$(x_j^*)_{j\in\mathbf{Z}} = (\eta^*(z_{j-1}^*))_{j\in\mathbf{Z}} = (\xi^*(z_j^*))_{j\in\mathbf{Z}}$$

of t^* uniquely determine t, like the woof and the warp weave a textile. See Fig. 2.2. We call x_i the *i-th thread* of t for $i \in \mathbf{Z}$ and x_j^* the *j-th dual thread* of t for $j \in \mathbf{Z}$.

Let E be a subset of $\mathbf{Z}^2$. A subconfiguration $(\alpha_{ij})_{(i,j)\in E}$ with $\alpha_{ij} \in A_\Gamma = A_{\Gamma^T}$ is said to be *well weaved* by T if the following conditions (i) and (ii) are satisfied: (i) if both (i,j) and $(i+1,j)$ are in E, then $q(\alpha_{ij}) = p(\alpha_{i+1,j})$; (ii) if both (i,j) and $(i,j+1)$ are in E then $q^T(\alpha_{ij}) = p^T(\alpha_{i,j+1})$.

Here we give a remark due to one of the referees.

A textile system $T = (p,q : \Gamma \to G)$ may be considered as a Wang tiling such that its tiles are the arcs $\alpha \in A_\Gamma$ whose left and right vertical edge colors are $i_\Gamma(\alpha)$ and $t_\Gamma(\alpha)$ and whose top and bottom horizontal edge colors are $p_A(\alpha)$ and $q_A(\alpha)$. (Wang tiles are square tiles with colored edges. In a Wang tiling, the tiles must be placed edge-to-edge (without reflection or rotation) so that the edge colors match.) Then a textile weaved by T is a Wang tiling of the plane. By this view, we know, for example, that given a textile system T, the question of whether there exists a textile weaved by T is undecidable. (See [GS], Chapter 11, and [Wan] for information on Wang tilings.)

Let $T = (p,q : \Gamma \to G)$ be a textile system. Let $n \in \mathbf{N}$. We define a textile system $T^{[n]} = (p^{[n]}, q^{[n]} : \Gamma^{[n]} \to G^{[n]})$ by

$$p^{[n]}(\alpha_1 \cdots \alpha_n) = p(\alpha_1)\cdots p(\alpha_n)$$

and

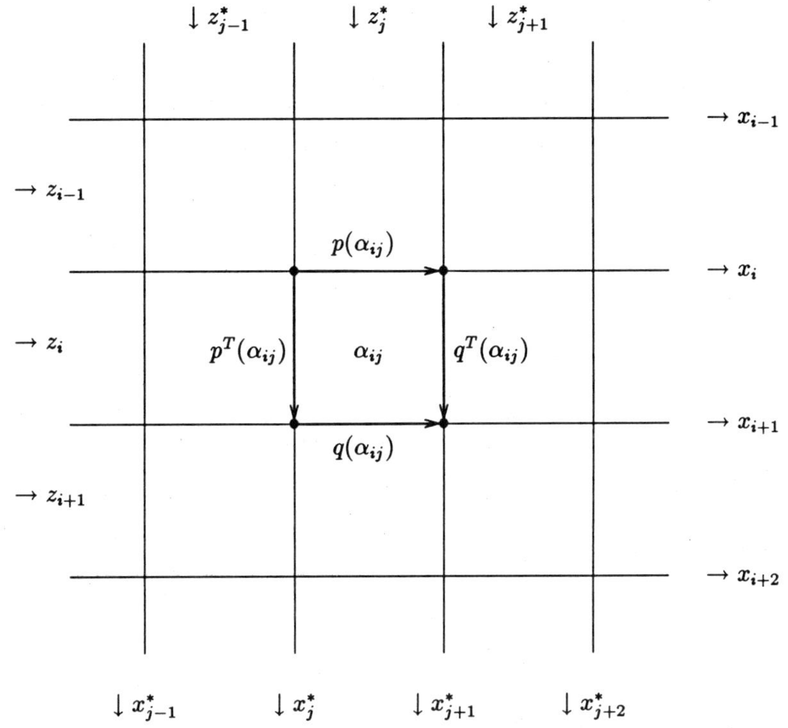

Figure 2.2.

$$q^{[n]}(\alpha_1 \cdots \alpha_n) = q(\alpha_1) \cdots q(\alpha_n),$$

where $\alpha_1 \cdots \alpha_n \in A_{\Gamma^{[n]}}$ with $\alpha_i \in A_\Gamma$, $i = 1, \cdots, n$. We shall call $T^{[n]}$ the *higher block system of order n* of T.

Let $T_1 = (p_1, q_1 : \Gamma_1 \to G_1)$ and $T_2 = (p_2, q_2 : \Gamma_2 \to G_2)$ be two textile systems. Let $T_1^* = (p_1^{T_1}, q_1^{T_1} : \Gamma_1^{T_1} \to G_1^{T_1})$ and $T_2^* = (p_2^{T_2}, q_2^{T_2} : \Gamma_2^{T_2} \to G_2^{T_2})$. Assume that $G_1^{T_1} = G_2^{T_2}$ (alternatively, $V_{\Gamma_1} = V_{\Gamma_2}, V_{G_1} = V_{G_2}, (p_1)_V = (p_2)_V$ and $(q_1)_V = (q_2)_V$). Then we can define a textile system T_1T_2 which will be called the *product* of T_1 and T_2. Consider the product $\Gamma_1\Gamma_2$ of the graphs Γ_1 and Γ_2. Every arc of $\Gamma_1\Gamma_2$ is of the form $\alpha_1\alpha_2$, where $\alpha_1 \in A_{\Gamma_1}, \alpha_2 \in A_{\Gamma_2}$ and $t_{\Gamma_1}(\alpha_1) = i_{\Gamma_2}(\alpha_2)$. Let us define an equivalence relation on $A_{\Gamma_1\Gamma_2}$ as follows: for any two arcs $\alpha_1\alpha_2$ and $\alpha_1'\alpha_2'$ in $A_{\Gamma_1\Gamma_2}$, they are equivalent if $i_{\Gamma_1}(\alpha_1) = i_{\Gamma_1}(\alpha_1'), t_{\Gamma_2}(\alpha_2) = t_{\Gamma_2}(\alpha_2'), p_1(\alpha_1) = p_1(\alpha_1'), p_2(\alpha_2) = p_2(\alpha_2'), q_1(\alpha_1) = q_1(\alpha_1')$ and $q_2(\alpha_2) = q_2(\alpha_2')$. For $\alpha_1\alpha_2 \in A_{\Gamma_1\Gamma_2}$, let $\overline{\alpha_1\alpha_2}$ denote the equivalence class containing $\alpha_1\alpha_2$. We define a graph $\overline{\Gamma_1\Gamma_2}$ and the textile system

$$T_1T_2 = (p, q : \overline{\Gamma_1\Gamma_2} \to G_1G_2)$$

as follows: $A_{\overline{\Gamma_1\Gamma_2}} = \{\overline{\alpha_1\alpha_2} \mid \alpha_1\alpha_2 \in A_{\Gamma_1\Gamma_2}\}, V_{\overline{\Gamma_1\Gamma_2}} = V_{\Gamma_1\Gamma_2}$, and for $\overline{\alpha_1\alpha_2} \in A_{\overline{\Gamma_1\Gamma_2}}$ with $\alpha_1\alpha_2 \in A_{\Gamma_1\Gamma_2}$,

$$i_{\overline{\Gamma_1\Gamma_2}}(\overline{\alpha_1\alpha_2}) = i_{\Gamma_1\Gamma_2}(\alpha_1\alpha_2) \quad , \quad t_{\overline{\Gamma_1\Gamma_2}}(\overline{\alpha_1\alpha_2}) = t_{\Gamma_1\Gamma_2}(\alpha_1\alpha_2)$$
$$p(\overline{\alpha_1\alpha_2}) = p_1(\alpha_1)p_2(\alpha_2) \quad , \quad q(\overline{\alpha_1\alpha_2}) = q_1(\alpha_1)q_2(\alpha_2).$$

Note that $p_1(\alpha_1)p_2(\alpha_2)$ and $q_1(\alpha_1)q_2(\alpha_2)$ are in $A_{G_1G_2}$ by our assumption.

For a textile system T and $n \in \mathbf{N}$, define

$$T^{(n)} = \overbrace{TT\cdots T}^{n}.$$

we call $T^{(n)}$ the *n-th product-power* of T. For two textile systems $T_1 = (p_1, q_1 : \Gamma_1 \to G_1)$ and $T_2 = (p_2, q_2 : \Gamma_2 \to G_2)$ with $G_1 = G_2$, we can define

$$T_1 \circ T_2 = (T_1^* T_2^*)^*.$$

We call $T_1 \circ T_2$ the *composition* of T_1 and T_2. For a textile system T and $n \in \mathbf{N}$, define

$$T^n = \overbrace{T \circ T \circ \cdots \circ T}^{n}.$$

We call T^n the *n-th composition-power* of T. Clearly $T^n = ((T^*)^{(n)})^*$. For a textile system $T = (p, q : \Gamma \to G)$, the *inverse* T^{-1} of T is defined by $T^{-1} = (q, p : \Gamma \to G)$.

For a textile system T, let U_T denote the set of all textiles weaved by T.

Let $T = (p, q : \Gamma \to G)$ be a textile system with $T^* = (p^T, q^T : \Gamma^T \to G^T)$. Then in our definition the graphs Γ, G, Γ^T and G^T are not assumed to be always nondegenerate, because even if all these graphs of T and T^* are nondegenerate, it is possible that for example, $(T^{[n]})^*$ has degenerate graphs for some $n \geq 2$. A textile system T will be said to be *in the standard form* if all the graphs of T and T^* are nondegenerate (i.e., G and Γ are nondegenerate and p and q are onto). For any textile system T, we can always construct, by taking the "trimming closure", a textile system T_0 in the standard form such that $U_{T_0} = U_T$. We next define the trimming closure.

An arc $\alpha \in A_\Gamma$ is called an *end* in T if at least one of the following holds:

$$i_\Gamma(\alpha) \notin t_\Gamma(A_\Gamma), \quad t_\Gamma(\alpha) \notin i_\Gamma(A_\Gamma), \quad p_A(\alpha) \notin q_A(A_\Gamma), \quad q_A(\alpha) \notin p_A(A_\Gamma).$$

Let Γ_1 be the possibly degenerate graph obtained from Γ by deleting all ends in T. Let G_1 be the possibly degenerate subgraph of G whose arc-set is equal to $p_A(A_{\Gamma_1}) \cup q_A(A_{\Gamma_1})$. Define a textile system

$$\mathrm{trim}(T) = (p_1, q_1 : \Gamma_1 \to G_1)$$

with $p_1 = p|A_{\Gamma_1}$ and $q_1 = q|A_{\Gamma_1}$. Let $\mathrm{trim}^0(T) = T$ and $\mathrm{trim}^i(T) = \mathrm{trim}^{i-1}(T)$ for $i \geq 1$. Let

$$\mathrm{trim}^i(T) = (p_i, q_i : \Gamma_i \to G_i), \quad i \geq 0.$$

Then $A_{\Gamma_i} \supset A_{\Gamma_{i+1}}$ for all $i \geq 0$. Since A_Γ is finite, there is $j \geq 0$ such that $A_{\Gamma_j} = A_{\Gamma_{j+1}}$. Clearly $\mathrm{trim}^j(T) = \mathrm{trim}^{i+j}(T)$ for all $i \geq 1$. The *trimming closure* of T is defined by

$$\overline{\mathrm{trim}}(T) = \mathrm{trim}^j(T).$$

Then, $\overline{\mathrm{trim}}(T)$ is in the standard form and since $U_{\mathrm{trim}(T)} = U_T$, we have $U_{\overline{\mathrm{trim}}(T)} = U_T$.

Standing hypothesis. Throughout the remainder of this memoir, we assume, unless otherwise stated, that a textile system is in the standard form unless it is given in the form in which an operation of taking a higher block system, a product, or a composition is used. We also assume, unless otherwise stated, that a graph is nondegenerate.

Let $T = (p, q : \Gamma \to G)$ be a textile system. Let $\xi : X_\Gamma \to X_G$ and $\eta : X_\Gamma \to X_G$ be the factor maps of $(X_\Gamma, \sigma_\Gamma)$ into (X_G, σ_G) given by p and q, respectively. Let X_k and Z_k be defined for all nonnegative integers k as follows:

$$X_0 = X_G, \quad Z_0 = X_\Gamma$$

$$X_k = \xi(Z_{k-1}) \cap \eta(Z_{k-1}), \qquad k = 1, 2, \cdots$$

$$Z_k = \xi^{-1}(X_k) \cap \eta^{-1}(X_k), \qquad k = 1, 2, \cdots.$$

Define

$$X_T = \bigcap_{k=0}^{\infty} X_k \quad \text{and} \quad Z_T = \bigcap_{k=0}^{\infty} Z_k.$$

Since X_Γ is compact and ξ and η are continuous maps, X_k, Z_k, X_T and Z_T are closed. They are also shift invariant. We have subshifts (X_T, σ_T) and (Z_T, ς_T) with $\sigma_T = \sigma_G | X_T$ and $\varsigma_T = \sigma_\Gamma | Z_T$. We shall call (X_T, σ_T) the *woof shift* defined by T, while (X_{T^*}, σ_{T^*}) will be called the *warp shift* defined by T. We say that T is *nondegenerate* if $X_T = X_0$, and that T is *finitely saturated*, if there is a nonnegative integer n such that $X_n = X_T$. Let $\xi_T : Z_T \to X_T$ and $\eta_T : Z_T \to X_T$ be the restrictions of ξ and η. Then they are factor maps of (Z_T, ς_T) onto (X_T, σ_T). We say that T is *one-sided 1-1* if ξ_T is 1-1. If T is one-sided 1-1, then we define

$$\varphi_T = \eta_T \xi_T^{-1},$$

which is an onto endomorphism of (X_T, σ_T). We say that T is *1-1* if both ξ_T and η_T are 1-1. If T is 1-1, then φ_T is an automorphism of (X_T, σ_T).

We shall explain some facts about X_k, Z_k, X_T and Z_T. For $k \in \mathbf{N}$, a subconfiguration over A_Γ of the form

$$\downarrow (\rightarrow z_i)_{1 \leq i \leq k}$$

with

$$z_i \in X_\Gamma, \quad i = 1, \cdots, k, \quad \text{and}$$

$$\eta(z_i) = \xi(z_{i+1}), \quad i = 1, \cdots, k-1,$$

will be called an *obi* of *width* k weaved by T. Let W_k be the set of all obis of width k weaved by T. Then it is proved by induction that for $k \in \mathbf{N}$,

$$X_k = \{\eta(z_k) \mid \downarrow(\rightarrow z_i)_{1 \le i \le 2k} \in W_{2k}\}$$

and

$$Z_k = \{z_{k+1} \mid \downarrow(\rightarrow z_i)_{1 \le i \le 2k+1} \in W_{2k+1}\}.$$

Therefore X_T is the set of all bisequences (threads) in X_G that appear on some textile weaved by T, and Z_T is the set of all bisequences in X_Γ that appear on some textile weaved by T (by the compactness of the metric space of all configurations over A_Γ which will be defined after Fact 2.2).

Let $k \in \mathbf{N}$. Let

$$w = (\alpha_{ij})_{1 \le i \le k, j \in \mathbf{Z}}$$

be an obi in W_k, where $\alpha_{ij} \in A_\Gamma$. Put

$$\beta_j = \downarrow \alpha_{1j} \cdots \alpha_{kj} = \begin{pmatrix} \alpha_{1j} \\ \vdots \\ \alpha_{kj} \end{pmatrix}, \quad j \in \mathbf{Z}.$$

Then w is described as $(\beta_j)_{j \in \mathbf{Z}}$ which is a bisequence over $\{\downarrow u \mid u \in L_k(\Gamma^T)\}$. Therefore we can consider W_k as a set of bisequences over $\{\downarrow u \mid u \in L_k(\Gamma^T)\}$. In this sense, W_k is the space of a topological Markov shift. In fact, $W_k = X_{\Gamma_k}$, where Γ_k is the graph defined as follows:

$$A_{\Gamma_k} = \{\begin{pmatrix} \alpha_1 \\ \vdots \\ \alpha_k \end{pmatrix} \mid \alpha_1 \cdots \alpha_k \in L_k(\Gamma^T), \alpha_i \in A_{\Gamma^T}\}$$

and for $\begin{pmatrix} \alpha_1 \\ \vdots \\ \alpha_k \end{pmatrix} \in A_{\Gamma_k}$, its initial and terminal vertices are

$$\begin{pmatrix} i_\Gamma(\alpha_1) \\ \vdots \\ i_\Gamma(\alpha_k) \end{pmatrix} \quad \text{and} \quad \begin{pmatrix} t_\Gamma(\alpha_1) \\ \vdots \\ t_\Gamma(\alpha_k) \end{pmatrix}.$$

Let $h_k : \Gamma_{2k} \to G$ be the graph-homomorphism such that

$$h_k(\begin{pmatrix} \alpha_1 \\ \vdots \\ \alpha_{2k} \end{pmatrix}) = q(\alpha_k), \quad \begin{pmatrix} \alpha_1 \\ \vdots \\ \alpha_{2k} \end{pmatrix} \in A_{\Gamma_{2k}}, \quad \alpha_i \in A_\Gamma.$$

Let $\theta_k : X_{\Gamma_{2k}} \to X_G$ be the factor map given by h_k. Then

$$X_k = \theta_k(X_{\Gamma_{2k}}).$$

Therefore X_k is the space of a sofic system. Similarly Z_k is the space of a sofic system.

Proposition 2.1. (1) If T is a 1-1 and finitely saturated textile system, then (X_T, σ_T) is topologically conjugate to a topological Markov shift.

(2) If $T = (p, q : \Gamma \to G)$ is a one-sided 1-1 and finitely saturated textile system such that the factor map η given by q is onto, then (X_T, σ_T) is topologically conjugate to a topological Markov shift.

Proof. The proposition will be proved by using the following fact.

If (X, σ) is a topological Markov shift, $(\bar{X}, \bar{\sigma})$ is a subsystem of (X, σ), and there is an onto factor map $\psi : (X, \sigma) \to (\bar{X}, \bar{\sigma})$ such that $\psi|\bar{X} = \mathrm{id}_{\bar{X}}$, then $(\bar{X}, \bar{\sigma})$ is topologically conjugate to a topological Markov shift.

To prove this fact, let H be the graph with $(X_H, \sigma_H) = (X, \sigma)$. Suppose that ψ is a block map of (l, l) type with $l \in \mathbf{N}$ and that it is given by a mapping f of $L_{2l+1}(H)$ onto A_H. Let $\rho : (X_H, \sigma_H) \to (X_{H^{[2l+1]}}, \sigma_{H^{[2l+1]}})$ be the conjugacy such that for $(\beta_i)_{i\in\mathbf{Z}} \in X_H$ with $\beta_i \in A_H$,

$$\rho((\beta_i)_{i\in\mathbf{Z}}) = (\beta_{i-l} \cdots \beta_{i+l})_{i\in\mathbf{Z}}.$$

Let $\bar{L}$ be the set of all paths in $L_{2l+1}(H)$ that appear on some bisequence in $\bar{X}$. Let K be the graph defined as follows: $A_k = \bar{L}$ and for $\beta_1 \cdots \beta_{2l+1} \in A_K$ with $\beta_i \in A_H$,

$$i_K(\beta_1 \cdots \beta_{2l+1}) = \beta_1 \cdots \beta_{2l} \quad \text{and} \quad t_K(\beta_1 \cdots \beta_{2l+1}) = \beta_2 \cdots \beta_{2l+1}.$$

Then we have

$$\bar{X} \subset \rho^{-1}(X_K).$$

On the other hand, since $\psi|\bar{X} = \mathrm{id}_{\bar{X}}$, it follows that for $\beta_1 \cdots \beta_{2l+1} \in \bar{L}$ with $\beta_i \in A_H$,

$$f(\beta_1 \cdots \beta_{2l+1}) = \beta_{l+1},$$

so that

$$\psi\rho^{-1}(X_K) = \rho^{-1}(X_K).$$

Hence

$$\bar{X} = \psi(X_H) \supset \psi(\rho^{-1}(X_K)) = \rho^{-1}(X_K).$$

Thus we have

$$\bar{X} = \rho^{-1}(X_K),$$

so that $(\bar{X}, \bar{\sigma})$ is topologically conjugate to the topological Markov shift (X_K, σ_K).

Throughout the remainder of the proof we assume that notation is the same as in the three paragraphs preceding Proposition 2.1.

Let us prove (1). Since T is finitely saturated, there is $n \in \mathbf{N}$ such that $X_n = X_T$. We have

$$W_{2n} = X_{\Gamma_{2n}} \quad \text{and} \quad \theta_n(W_{2n}) = X_n.$$

Let $\bar{W}$ be the set of all obis in W_{2n} that appear on some textile weaved by T. That is,

$$\bar{W} = \{\downarrow (\rightarrow z_i)_{1\le i\le 2n} \mid \downarrow (\rightarrow z_i)_{i\in\mathbf{Z}} \in U_T, z_i \in X_T\}.$$

Then $\bar{W}$ is considered as a subset of $X_{\Gamma_{2n}}$. Since $X_n = X_T$, we have

$$\begin{aligned} X_n &= \{\eta(z_n) \mid \downarrow (\rightarrow z_i)_{1\le i\le 2n} \in \bar{W}, z_i \in X_\Gamma\} \\ &= \theta_n(\bar{W}). \end{aligned}$$

Let $\bar{\theta} : \bar{W} \rightarrow X_n$ be the onto mapping obtained by restricting θ_n. Since T is 1-1, it follows that $\bar{\theta}$ is a conjugacy. Let

$$\psi = \bar{\theta}^{-1}\theta_n.$$

Then ψ is a factor map of the topological Markov shift $(X_{\Gamma_{2n}}, \sigma_{\Gamma_{2n}})$ onto $(\bar{W}, \bar{\sigma})$ with $\bar{\sigma} = \sigma_{\Gamma_{2n}}|\bar{W}$, and $\psi|\bar{W}$ is the identity mapping on $\bar{W}$. Therefore by the fact above, $(\bar{W}, \bar{\sigma})$ is topologically conjugate to a topological Markov shift. Since $X_T = X_n = \bar{\theta}(\bar{W})$ and $\bar{\theta}$ is a conjugacy, (X_T, σ_T) is topologically conjugate to a topological Markov shift.

Next, we prove (2). There is $n \in \mathbf{N}$ such that $X_n = X_T$. Let $\theta : W_n \rightarrow X_G$ be defined by

$$\theta(\downarrow (\rightarrow z_i)_{1\le i\le n}) = \xi(z_1), \quad \downarrow (\rightarrow z_i)_{1\le i\le n} \in W_n, \quad z_i \in X_\Gamma.$$

Then $X_n = \theta(W_n)$ because η is onto. Let

$$\bar{W} = \{\downarrow (\rightarrow z_i)_{1\le i\le n} \mid \downarrow (\rightarrow z_i)_{i\in\mathbf{Z}} \in U_T, z_i \in X_\Gamma\}.$$

Then we have $\theta(\bar{W}) = X_n$. Since T is one-sided 1-1, the onto mapping $\bar{\theta} : \bar{W} \rightarrow X_n$ obtained by restricting θ is a conjugacy. If we let $\psi = \bar{\theta}^{-1}\theta$, then ψ is a factor map of $(X_{\Gamma_n}, \sigma_{\Gamma_n})$ onto $(\bar{W}, \bar{\sigma})$ with $\bar{\sigma} = \sigma_{\Gamma_n}|\bar{W}$, and $\psi|\bar{W}$ is the identity mapping on $\bar{W}$. The remainder is the same as in the proof of (1). □

Fact 2.2. Let G be a graph. Let k and l be nonnegative integers. Let $\varphi : X_G \rightarrow X_G$ be a block map of (k, l) type and let $f : L_{k+l+1}(G) \rightarrow A_G$ give φ. Let $T_\varphi = (p, q : \Gamma \rightarrow G)$ be the textile system defined as follows: $\Gamma = G^{[k+l+1]}$; for $a_1 \cdots a_{k+l+1} \in A_\Gamma$ with $a_i \in A_G$,

$$p(a_1 \cdots a_{k+l+1}) = a_{k+1}$$

and

$$q(a_1 \cdots a_{k+l+1}) = f(a_1 \cdots a_{k+l+1}).$$

(Generally, T_φ is not in the standard form.) Let $E(\varphi) = \bigcap_{n=1}^{\infty} \varphi^n(X_G)$. Then T_φ is one-sided 1-1 and

$$\varphi_{T_\varphi} = \varphi | E(\varphi).$$

If φ is onto, then T_φ is nondegenerate and is in the standard form.

Proof. Clear. □

Let A be an alphabet. Let U_A be the set of all configurations over A. We define a metric δ on U_A as follows. Let $u = (\alpha_{ij})_{i,j\in\mathbf{Z}}$ and $u' = (\alpha'_{ij})_{i,j\in\mathbf{Z}}$ be elements of U_A with $\alpha_{ij}, \alpha'_{ij} \in A$. Then $\delta(u, u') = 0$ if $u = u'$ and $\delta(u, u') = 1/(1+k)$ if $u \neq u'$, where

$$k = \min\{|i| + |j| \mid i, j \in \mathbf{Z}, \alpha_{ij} \neq \alpha'_{ij}\}.$$

Then with this metric, U_A is compact. For $k, l \in \mathbf{Z}$, let $\boldsymbol{\sigma}_A^{(k,l)} : U_A \to U_A$ be the homeomorphism defined by

$$\boldsymbol{\sigma}_A^{(k,l)}((\alpha_{ij})_{i,j\in\mathbf{Z}}) = (\alpha_{i+k,j+l})_{i,j\in\mathbf{Z}}, \quad (\alpha_{ij})_{i,j\in\mathbf{Z}} \in U_A.$$

The dynamical system $(U_A, \boldsymbol{\sigma}_A^{(k,l)})$ is called the *full (k,l)-shift of 2-dimensional configurations* over A. A subset U of U_A is said to be *shift-invariant* if $\boldsymbol{\sigma}_A^{(k,l)}(U) = U$ for all $(k,l) \in \mathbf{Z}^2$. Let U be a closed, shift-invariant subset of U_A. Let $\boldsymbol{\sigma}^{(k,l)}$ be the restriction of $\boldsymbol{\sigma}_A^{(k,l)}$ on U for $(k,l) \in \mathbf{Z}^2$. Then the dynamical system $(U, \boldsymbol{\sigma}^{(k,l)})$ is called a *(k,l)-subshift of 2-dimensional configurations* over A or a *subshift of 2-dimensional configurations* over A.

Let $T = (p, q : \Gamma \to G)$ be a textile system. Then U_T is a closed, shift-invariant subset of U_{A_Γ}. We have (k,l)-subshifts $(U_T, \boldsymbol{\sigma}_T^{(k,l)})$, $k, l \in \mathbf{Z}$, of 2-dimensional configurations, which will be called the *textile shifts* defined by T.

Let $T = (p, q : \Gamma \to G)$ be a textile system and let $T^* = (p^T, q^T : \Gamma^T \to G^T)$ be its dual. Let $t = (\alpha_{ij}) \in U_T$ with $\alpha_{ij} \in A_\Gamma = A_{\Gamma^T}$. Let $(x_i)_{i\in\mathbf{Z}}$ and $(y_j)_{j\in\mathbf{Z}}$ be the text and dual text of t, respectively, with

$$x_i = (a_{ij})_{j\in\mathbf{Z}} \in X_G, \quad i \in \mathbf{Z}$$

and

$$y_j = (b_{ij})_{i\in\mathbf{Z}} \in X_{G^T}, \quad j \in \mathbf{Z}.$$

Then for $i, j \in \mathbf{Z}$,

$$a_{ij} = p(\alpha_{ij}) = q(\alpha_{i-1,j})$$

and

$$b_{ij} = p^T(\alpha_{ij}) = q^T(\alpha_{i,j-1}).$$

Let $k, l \in \mathbf{Z}$ with $(k,l) \neq (0,0)$. Define graphs $\hat{G}^{(k,l)}$ and $\check{G}^{(k,l)}$ by

$$\hat{G}^{(k,l)} = G^l(G^T)^k \text{ and } \check{G}^{(k,l)} = (G^T)^k G^l \quad \text{if } k \neq 0 \text{ and } l \neq 0,$$
$$\hat{G}^{(k,0)} = \check{G}^{(k,0)} = (G^T)^k$$
$$\hat{G}^{(0,l)} = \check{G}^{(0,l)} = G^l.$$

For $(i,j) \in \mathbf{Z}^2$, let $\hat{c}_t^{(k,l)}(i,j)$ and $\check{c}_t^{(k,l)}(i,j)$ be the arcs of $\hat{G}^{(k,l)}$ and $\check{G}^{(k,l)}$, respectively, defined as follows: putting

$$c_t^{(k,0)}(i,j) = \begin{cases} b_{ij} \cdots b_{i+k-1,j} & \text{if } k > 0 \\ b_{i-1,j} \cdots b_{i+k,j} & \text{if } k < 0 \end{cases}$$

and

$$c_t^{(0,l)}(i,j) = \begin{cases} a_{ij} \cdots a_{i,j+l-1} & \text{if } l > 0 \\ a_{i,j-1} \cdots a_{i,j+l} & \text{if } l < 0, \end{cases}$$

we define

$$\begin{aligned} \hat{c}_t^{(k,0)}(i,j) &= \check{c}_t^{(k,0)}(i,j) = c_t^{(k,0)}(i,j) \\ \hat{c}_t^{(0,l)}(i,j) &= \check{c}_t^{(0,l)}(i,j) = c_t^{(0,l)}(i,j), \end{aligned}$$

and if $k \neq 0$ and $l \neq 0$, then we define

$$\begin{aligned} \hat{c}_t^{(k,l)}(i,j) &= c_t^{(0,l)}(i,j)c_t^{(k,0)}(i,j+l) \\ \check{c}_t^{(k,l)}(i,j) &= c_t^{(k,0)}(i,j)c_t^{(0,l)}(i+k,j). \end{aligned}$$

Let $\hat{\theta}_T^{(k,l)} : U_T \to X_{\hat{G}^{(k,l)}}$ and $\check{\theta}_T^{(k,l)} : U_T \to X_{\check{G}^{(k,l)}}$ be defined by

$$\begin{aligned} \hat{\theta}_T^{(k,l)}(t) &= (\hat{c}_t^{(k,l)}(mk,ml))_{m\in\mathbf{Z}}, \quad t \in U_T \\ \check{\theta}_T^{(k,l)}(t) &= (\check{c}_t^{(k,l)}(mk,ml))_{m\in\mathbf{Z}}, \quad t \in U_T. \end{aligned}$$

Then $\hat{\theta}_T^{(k,l)}$ and $\check{\theta}_T^{(k,l)}$ are continuous. Define

$$\hat{X}_T^{(k,l)} = \hat{\theta}_T^{(k,l)}(U_T) \quad \text{and} \quad \check{X}_T^{(k,l)} = \check{\theta}_T^{(k,l)}(U_T).$$

Then we have two subshifts $(\hat{X}_T^{(k,l)}, \hat{\sigma}_T^{(k,l)})$ and $(\check{X}_T^{(k,l)}, \check{\sigma}_T^{(k,l)})$, which are topologically conjugate. For if $\psi : \hat{X}_T^{(k,l)} \to \check{X}_T^{(k,l)}$ is defined by

$$\psi(\hat{\theta}_T^{(k,l)}(t)) = \check{\theta}_T^{(k,l)}(\sigma_T^{(0,l)}(t)), \quad t \in U_T,$$

then ψ is a topological conjugacy of $(\hat{X}_T^{(k,l)}, \hat{\sigma}_T^{(k,l)})$ onto $(\check{X}_T^{(k,l)}, \check{\sigma}_T^{(k,l)})$. Clearly we have

$$(X_T, \sigma_T) = (\hat{X}_T^{(0,1)}, \hat{\sigma}_T^{(0,1)}) = (\check{X}_T^{(0,1)}, \check{\sigma}_T^{(0,1)})$$

and

$$(X_{T^*}, \sigma_{T^*}) = (\hat{X}_T^{(1,0)}, \hat{\sigma}_T^{(1,0)}) = (\check{X}_T^{(1,0)}, \check{X}_T^{(1,0)}).$$

Noting that $\hat{\theta}_T^{(0,1)} = \check{\theta}_T^{(0,1)}$ and $\hat{\theta}_T^{(1,0)} = \check{\theta}_T^{(1,0)}$, we define

$$\theta_T = \check{\theta}_T^{(0,1)} \quad \text{and} \quad \theta_T^* = \check{\theta}_T^{(1,0)}.$$

Then for $t \in U_T, \theta_T(t)$ is the 0th thread of t and $\theta_T^*(t)$ is the 0th dual thread of t. We call $(\check{X}_T^{(k,l)}, \check{\sigma}_T^{(k,l)})$ the (k,l)-*bias shift* defined by T or a *bias shift* defined by T, if $kl \neq 0$.

Let (X, d) be a metric space. Let $\tau : X \to X$ be a continuous map. A sequence $(x_i)_{i\in\mathbf{Z}}$ with $x_i \in X$ and $\tau(x_i) = x_{i+1}$ for all $i \in \mathbf{Z}$, is called an *orbit* of τ. We say that τ is *expansive* if there is a real number $\epsilon > 0$ such that for any pair of orbits $(x_i)_{i\in\mathbf{Z}}$ and $(y_i)_{i\in\mathbf{Z}}$, if $d(x_i, y_i) \leq \epsilon$ for all $i \in \mathbf{Z}$, then $(x_i)_{i\in\mathbf{Z}} = (y_i)_{i\in\mathbf{Z}}$. Of course, if τ is a homeomorphism, then τ is expansive if and only if there is $\epsilon > 0$ such that for any $x, y \in X$, if $d(\tau^i(x), \tau^i(y)) \leq \epsilon$ for all $i \in \mathbf{Z}$, then $x = y$.

The following two fundamental propositions are easily proved by observing the figure of a textile (see Fig. 2.2).

Proposition 2.3. Let T be a textile system. Let $(k,l) \in \mathbf{Z}^2$ with $(k,l) \neq (0,0)$. Then the following statements are equivalent

(1) $\boldsymbol{\sigma}_T^{(k,l)} : U_T \to U_T$ is expansive.
(2) $\hat{\theta}_T^{(k,l)} : U_T \to \hat{X}_T^{(k,l)}$ is 1-1.
(3) $\check{\theta}_T^{(k,l)} : U_T \to \check{X}_T^{(k,l)}$ is 1-1.
(4) $(U_T, \boldsymbol{\sigma}_T^{(k,l)})$ is topologically conjugate to $(\check{X}_T^{(k,l)}, \check{\sigma}_T^{(k,l)})$.

In particular, if $k \neq 0$ and $l \neq 0$, then $\boldsymbol{\sigma}_T^{(k,l)}$ is expansive if and only if the correspondence

$$(*) \qquad \hat{\theta}_T^{(k,l)}(t) \mapsto \check{\theta}_T^{(k,l)}(t), \quad t \in U_T$$

is 1-1.

Proof. By the above, the equivalence of (2) and (3) is clear. It is also clear that (3) $\Rightarrow$ (4) $\Rightarrow$ (1).

To prove that (1) $\Rightarrow$ (3), assume that $\check{\theta}_T^{(k,l)}$ is not 1-1. Then there are two distinct textiles $t = (\alpha_{ij})_{i,j\in\mathbf{Z}}$ and $t' = (\alpha'_{ij})_{i,j\in\mathbf{Z}}$ with $\alpha_{ij}, \alpha'_{ij} \in A_\Gamma = A_{\Gamma^T}$ such that $\check{\theta}_T^{(k,l)}(t) = \check{\theta}_T^{(k,l)}(t')$. It is observed that $\mathbf{Z}^2$ is partitioned into the two disjoint subsets, say $E^{(k,l)}$ and $\bar{E}^{(k,l)}$, such that $\check{\theta}_T^{(k,l)}(t)$ divides t into two subconfigurations $(\alpha_{ij})_{(i,j)\in E^{(k,l)}}$ and $(\alpha_{ij})_{(i,j)\in\bar{E}^{(k,l)}}$. Clearly $\check{\theta}_T^{(k,l)}(t)$ also divides t' into $(\alpha'_{ij})_{(i,j)\in E^{(k,l)}}$ and $(\alpha'_{ij})_{(i,j)\in\bar{E}^{(k,l)}}$. Since

$t \neq t'$, we may assume without loss of generality $(\alpha_{ij})_{(i,j)\in E^{(k,l)}}$ and $(\alpha'_{ij})_{(i,j)\in E^{(k,l)}}$ are distinct. Let $u = (\beta_{ij})_{i,j\in \mathbf{Z}}$ be the configuration such that

$$\beta_{ij} = \begin{cases} \alpha'_{ij}, & \text{if } (i,j) \in E^{(k,l)} \\ \alpha_{ij}, & \text{if } (i,j) \in \bar{E}^{(k,l)}. \end{cases}$$

Then $u \in U_T$ because $\check{\theta}_T^{(k,l)}(t) = \check{\theta}_T^{(k,l)}(t')$. Clearly $u \neq t$ and for any $\epsilon > 0$ there is $(m,n) \in \mathbf{Z}^2$ such that

$$\delta((\boldsymbol{\sigma}_T^{(k,l)})^N \boldsymbol{\sigma}_T^{(m,n)}(t),\ (\boldsymbol{\sigma}_T^{(k,l)})^N \boldsymbol{\sigma}_T^{(m,n)}(u)) < \epsilon$$

for all $N \in \mathbf{Z}$. Therefore $\boldsymbol{\sigma}_T^{(k,l)}$ is not expansive. Thus we have shown that (1) $\Rightarrow$ (3).

Assume that $k \neq 0$ and $l \neq 0$. If $\boldsymbol{\sigma}_T^{(k,l)}$ is expansive, then the correspondence $(*)$ is 1-1 because (2) and (3) hold. If $\boldsymbol{\sigma}_T^{(k,l)}$ is not expansive, then $\check{\theta}_T^{(k,l)}$ is not 1-1, so that we may assume that there are t and u as above. It is easily observed that there is $(m,n) \in \mathbf{Z}^2$ such that

$$\check{\theta}_T^{(k,l)}(\boldsymbol{\sigma}_T^{(m,n)}(t)) = \check{\theta}_T^{(k,l)}(\boldsymbol{\sigma}_T^{(m,n)}(u)) \quad \text{and} \quad \hat{\theta}_T^{(k,l)}(\boldsymbol{\sigma}_T^{(m,n)}(t)) \neq \hat{\theta}_T^{(k,l)}(\boldsymbol{\sigma}_T^{(m,n)}(u))$$

or

$$\check{\theta}_T^{(k,l)}(\boldsymbol{\sigma}_T^{(m,n)}(t)) \neq \check{\theta}_T^{(k,l)}(\boldsymbol{\sigma}_T^{(m,n)}(u)) \quad \text{and} \quad \hat{\theta}_T^{(k,l)}(\boldsymbol{\sigma}_T^{(m,n)}(t)) = \hat{\theta}_T^{(k,l)}(\boldsymbol{\sigma}_T^{(m,n)}(u)),$$

so that the correspondence $(*)$ is not 1-1. □

Proposition 2.4. Let T be a textile system. Let $(k,l),(m,n) \in \mathbf{Z}^2$ with $(k,l) \neq (0,0)$ and $(m,n) \neq (0,0)$. Then the following statements are valid.

(1) If $\check{\theta}_T^{(k,l)}$ is 1-1, then the mapping $\psi_{T,(k,l)}^{(m,n)} : \check{X}_T^{(k,l)} \to \check{X}_T^{(k,l)}$ defined by

$$\psi_{T,(k,l)}^{(m,n)} = \check{\theta}_T^{(k,l)} \boldsymbol{\sigma}_T^{(m,n)} (\check{\theta}_T^{(k,l)})^{-1}$$

is an automorphism of $(\check{X}_T^{(k,l)}, \check{\sigma}_T^{(k,l)})$, and $(\check{X}_T^{(k,l)}, \psi_{T,(k,l)}^{(m,n)})$ is topologically conjugate to $(U_T, \boldsymbol{\sigma}_T^{(m,n)})$.

(2) If $\check{\theta}_T^{(k,l)}$ and $\check{\theta}_T^{(m,n)}$ are 1-1, then $(\check{X}_T^{(k,l)}, \psi_{T,(k,l)}^{(m,n)})$ is topologically conjugate to $(\check{X}_T^{(m,n)}, \check{\sigma}_T^{(m,n)})$ by the topological conjugacy that maps $\check{\theta}_T^{(k,l)}(t)$ to $\check{\theta}_T^{(m,n)}(t)$ for $t \in U_T$.

(3) If T is 1-1, then $(X_T, \varphi_T^m \sigma_T^n)$ is topologically conjugate to $(U_T, \boldsymbol{\sigma}_T^{(m,n)})$.

Proof. If $\check{\theta}_T^{(k,l)}$ is 1-1, then $\check{\theta}_T^{(k,l)} : (U_T, \boldsymbol{\sigma}_T^{(k,l)}) \to (\check{X}_T^{(k,l)}, \check{\sigma}_T^{(k,l)})$ is a topological conjugacy. It is easily seen that $\psi_{T,(k,l)}^{(m,n)}$ is an automorphism of $(\check{X}_T^{(k,l)}, \check{\sigma}_T^{(k,l)})$. Clearly $\check{\theta}_T^{(k,l)} : (U_T, \boldsymbol{\sigma}_T^{(m,n)}) \to (\check{X}_T^{(k,l)}, \psi_{T,(k,l)}^{(m,n)})$ is a topological conjugacy. If $\check{\theta}_T^{(m,n)}$ is 1-1 in addition, then $\check{\theta}_T^{(m,n)} : (U_T, \boldsymbol{\sigma}_T^{(m,n)}) \to (\check{X}_T^{(m,n)}, \check{\sigma}_T^{(m,n)})$ is a topological conjugacy. Thus we have proved (1) and (2).

Assume that T is 1-1. Then $\theta_T = \check{\theta}_T^{(0,1)}$ is 1-1. We have

$$\begin{aligned} \psi_{T,(0,1)}^{(m,n)} &= \theta_T \boldsymbol{\sigma}_T^{(m,n)} \theta_T^{-1} \\ &= (\theta_T \boldsymbol{\sigma}_T^{(1,0)} \theta_T^{-1})^m (\theta_T \boldsymbol{\sigma}_T^{(0,1)} \theta_T^{-1})^n \\ &= \varphi_T^m \sigma_T^n. \end{aligned}$$

Therefore (3) follows from (1). □

Theorem 2.5. Let T be a one-sided 1-1 textile system. Then φ_T is expansive if and only if T^* is 1-1. If both T and T^* are 1-1, then the mapping $\chi_T : X_T \to X_{T^*}$ defined by

$$\chi_T(\theta_T(t)) = \theta_T^*(t), \quad t \in U_T$$

is a homeomorphism and the following diagrams commute:

$$\begin{array}{ccc} X_T & \xrightarrow{\varphi_T} & X_T \\ {\scriptstyle \chi_T}\downarrow & & \downarrow{\scriptstyle \chi_T} \\ X_{T^*} & \xrightarrow[\sigma_{T^*}]{} & X_{T^*} \end{array} \qquad \begin{array}{ccc} X_T & \xrightarrow{\sigma_T} & X_T \\ {\scriptstyle \chi_T}\downarrow & & \downarrow{\scriptstyle \chi_T} \\ X_{T^*} & \xrightarrow[\varphi_{T^*}]{} & X_{T^*} \end{array} .$$

Proof. Since T^* is 1-1 if and only if $\theta_T^* = \theta_T^{(1,0)}$ is 1-1, it follows from Proposition 2.3 that T^* is 1-1 if and only if $\boldsymbol{\sigma}_T^{(1,0)}$ is expansive. Therefore, to prove the first claim, we shall show that $\boldsymbol{\sigma}_T^{(1,0)}$ is expansive if and only if φ_T is expansive.

Assume that φ_T is not expansive. Let $\epsilon > 0$. Since ξ_T^{-1} is continuous, there is $\epsilon' > 0$ such that for $x, y \in X_T$ if $d_{X_T}(x, y) < \epsilon'$, then $d_{Z_T}(\xi_T^{-1}(x), \xi_T^{-1}(y)) < \epsilon$. Since φ_T is not expansive, there are distinct two orbits $(x_i)_{i\in\mathbf{Z}}$ and $(y_i)_{i\in\mathbf{Z}}$ of (X_T, φ_T) such that $d_{X_T}(x_i, y_i) \le \epsilon'$ for all $i \in \mathbf{Z}$. Let

$$t = \downarrow (\xi_T^{-1}(x_i))_{i\in\mathbf{Z}} \quad \text{and} \quad t' = \downarrow (\xi_T^{-1}(y_i))_{i\in\mathbf{Z}}.$$

Then $t, t' \in U_T, t \neq t'$, and

$$\delta(\boldsymbol{\sigma}_T^{(i,0)}(t), \boldsymbol{\sigma}_T^{(i,0)}(t')) < \epsilon$$

for all $i \in \mathbf{Z}$. Thus $\boldsymbol{\sigma}_T^{(1,0)}$ is not expansive.

Assume that φ_T is expansive. Then there is $\epsilon_0 > 0$ such that for any orbits $(x_i)_{i\in\mathbf{Z}}$ and $(y_i)_{i\in\mathbf{Z}}$ of (X_T, φ_T), if $d_{X_T}(x_i, y_i) < \epsilon_0$ for all $i \in \mathbf{Z}$, then $(x_i)_{i\in\mathbf{Z}} = (y_i)_{i\in\mathbf{Z}}$. Let $t, t' \in U_T$ and let $(x_i)_{i\in\mathbf{Z}}$ and $(y_i)_{i\in\mathbf{Z}}$ be the texts of t and t', respectively. Suppose that

$$\delta(\boldsymbol{\sigma}_T^{(i,0)}(t), \boldsymbol{\sigma}_T^{(i,0)}(t')) < \epsilon_0$$

for all $i \in \mathbf{Z}$. Then $d_{X_T}(x_i, y_i) < \epsilon_0$ for all $i \in \mathbf{Z}$. Therefore since $(x_i)_{i\in\mathbf{Z}}$ and $(y_i)_{i\in\mathbf{Z}}$ are orbits of (X_T, φ_T), we have $(x_i)_{i\in\mathbf{Z}} = (y_i)_{i\in\mathbf{Z}}$. This implies that $t = t'$ because ξ_T is 1-1. Thus $\boldsymbol{\sigma}_T^{(1,0)}$ is expansive.

Assume that T and T^* are 1-1. Then both $\theta_T = \check{\theta}_T^{(0,1)}$ and $\theta_T^* = \check{\theta}_T^{(1,0)}$ are homeomorphisms and $\chi_T = \check{\theta}_T^{(1,0)}(\check{\theta}_T^{(0,1)})^{-1}$. Therefore χ_T is a homeomorphism. It follows as a special case of (2) of Proposition 2.4 that the first diagram commutes. This implies by duality that the second diagram commutes. □

Proposition 2.6. Let T be a finitely saturated textile system. If (X_{T^*}, σ_{T^*}) is topologically conjugate to a topological Markov shift, then T^* is finitely saturated.

Proof. Let $T = (p, q : \Gamma \to G)$ and let $T^* = (p^T, q^T : \Gamma^T \to G^T)$. Since (X_{T^*}, σ_{T^*}) is topologically conjugate to a topological Markov shift, it is a subshift of a finite type (see e.g., [DGS]). Hence by definition, there is a finite set F of words over A_{G^T} such that for each bisequence $(b_i)_{i\in\mathbf{Z}} \in X_{A_{G^T}}$ with $b_i \in A_{G^T}$, there are $i', i'' \in \mathbf{Z}$ with $i' \le i''$ and $b_{i'} \cdots b_{i''} \in F$ if and only if $(b_i)_{i\in\mathbf{Z}} \notin X_{T^*}$. Let m be a positive integer such that $2m$ is not less than the maximum length of the words in F. Since T is finitely saturated, there is a nonnegative integer n such that T is *n-saturated*, that is, $X_n = X_T$, where X_n is given as follows: $X_0 = X_G$ and for $n \in \mathbf{N}$, X_n is the set of all bisequences in X_G written in the form $\eta(z_n)$ for some obi $\downarrow (\to z_i)_{1\le i\le 2n}$ of width $2n$ weaved by T with $z_i \in X_\Gamma$, where $\eta : X_\Gamma \to X_G$ is the factor map given by q. Let

$$N = (\# A_{G^T})^{2m+2n}.$$

We claim that T^* is N-saturated.

Assume that T^* is not N-saturated. Then there is an obi

$$\downarrow (\to z_j^*)_{1\le j\le 2N}$$

weaved by T^* with $z_j^* \in X_{\Gamma^T}$ such that

$$\eta^*(z_N^*) \notin X_{T^*},$$

where $\eta^* : X_{\Gamma^T} \to X_{G^T}$ is the factor map given by q^T. For $j = 1, \cdots, 2N$, put

$$z_j^* = (\alpha_{ij})_{i\in\mathbf{Z}}$$

with $\alpha_{ij} \in A_{\Gamma^T} = A_\Gamma$. Consider the subconfiguration

$$(\alpha_{ij})_{i\in\mathbf{Z}, 1\le j\le 2N}.$$

Then this is well weaved by T and we have

$$\eta^*(z_N^*) = (q^T(\alpha_{iN}))_{i\in\mathbf{Z}}.$$

Since $\eta^*(z_N^*) \notin X_{T^*}$, there are $i', i'' \in \mathbf{Z}$ with $i' \leq i''$ such that

$$q^T(\alpha_{i'N}) \cdots q^T(\alpha_{i''N}) \in F.$$

We may assume that

$$n+1 \leq i' \quad \text{and} \quad i'' \leq n+2m.$$

Since $N = (\#A_{GT})^{2m+2n}$, there are integers j_1, j_2 with $1 \leq j_1 < j_2 \leq N$ such that

$$(p^T(\alpha_{ij_1}))_{1 \leq i \leq 2m+2n} = (q^T(\alpha_{ij_2}))_{1 \leq i \leq 2m+2n}$$

and there are integers j_3, j_4 with $N+1 \leq j_3 < j_4 \leq 2N$ such that

$$(p^T(\alpha_{ij_3}))_{1 \leq i \leq 2m+2n} = (q^T(\alpha_{ij_4}))_{1 \leq i \leq 2m+2n}$$

For $j \in \mathbf{Z}$, let $r(j)$ be the integer such that $r(j) \equiv j \bmod j_2 - j_1 + 1$ with $0 \leq r(j) < j_2 - j_1 + 1$ and let $s(j)$ be the integer such that $s(j) \equiv j \bmod (j_4 - j_3 + 1)$ with $0 \leq s(j) < j_4 - j_3 + 1$. Let

$$(\alpha'_{ij})_{1 \leq i \leq 2m+2n, j \in \mathbf{Z}}$$

be the subconfiguration such that for $i = 1, \cdots, 2m+2n$

$$\alpha'_{ij} = \begin{cases} \alpha_{ij}, & \text{if } \ j_1 \leq j \leq j_4 \\ \alpha_{i,j_1+r(j-j_1)} & \text{if } \ j < j_1 \\ \alpha_{i,j_3+s(j-j_3)} & \text{if } \ j > j_4. \end{cases}$$

Then $(\alpha'_{ij})_{1 \leq i \leq 2m+2n, j \in \mathbf{Z}}$ is an obi of width $2m+2n$ weaved by T. Since T is n-saturated, there is a textile $t = (\beta_{ij})_{i,j \in \mathbf{Z}}$ weaved by T with $\beta_{ij} \in A_\Gamma$ such that

$$\eta((\alpha'_{nj})_{j \in \mathbf{Z}}) = \eta((\beta_{nj})_{j \in \mathbf{Z}})$$

and there is a textile $t' = (\beta'_{ij})_{i,j \in \mathbf{Z}}$ weaved by T with $\beta'_{ij} \in A_\Gamma$ such that

$$\eta((\alpha'_{n+2m,j})_{j \in \mathbf{Z}}) = \eta((\beta'_{n+2m,j})_{j \in \mathbf{Z}}).$$

Let $u = (\gamma_{ij})_{i,j \in \mathbf{Z}}$ be the configuration defined by

$$\begin{aligned} (\gamma_{ij})_{i \leq n, j \in \mathbf{Z}} &= (\beta_{ij})_{i \leq n, j \in \mathbf{Z}} \\ (\gamma_{ij})_{n < i \leq n+2m, j \in \mathbf{Z}} &= (\alpha'_{ij})_{n < i \leq n+2m, j \in \mathbf{Z}} \\ (\gamma_{ij})_{i > n+2m, j \in \mathbf{Z}} &= (\beta'_{ij})_{i > n+2m, j \in \mathbf{Z}} \end{aligned}$$

Then clearly $u \in U_T$. Therefore

$$(q^T(\gamma_{iN}))_{i \in \mathbf{Z}} \in X_{T^*}$$

but

$$q^T(\gamma_{i'N})\cdots q^T(\gamma_{i''N}) = q^T(\alpha_{i'N})\cdots q^T(\alpha_{i''N}) \in F,$$

which is a contradiction. $\square$

Theorem 2.7. Let T be a 1-1 and finitely saturated textile system. Then (X_T, φ_T) is topologically conjugate to a topological Markov shift if and only if T^* is 1-1 and finitely saturated.

Proof. If T^* is 1-1 and finitely saturated, then (X_{T^*}, σ_{T^*}) is topologically conjugate to a topological Markov shift, by Proposition 2.1. Moreover, since T^* is 1-1, (X_T, φ_T) is topologically conjugate to (X_{T^*}, σ_{T^*}), by Theorem 2.5. Therefore if T^* is 1-1 and finitely saturated, then (X_T, φ_T) is topologically conjugate to a topological Markov shift.

Assume that (X_T, φ_T) is topologically conjugate to a topological Markov shift. Then by Theorem 2.5, T^* is 1-1 and (X_T, φ_T) is topologically conjugate to (X_{T^*}, σ_{T^*}) so that (X_{T^*}, σ_{T^*}) is topologically conjugate to a topological Markov shift. Thus, by Proposition 2.6, T^* is finitely saturated. $\square$

Let $T = (p, q : \Gamma \to G)$ and $T' = (p', q' : \Gamma' \to G')$ be two textile systems. We say that T and T' are *topologically conjugate* if there are topological conjugacies ψ : $(X_T, \upsilon_T) \to (X_{T'}, \sigma_{T'})$ and $\Psi : (Z_T, \varsigma_T) \to (Z_{T'}, \varsigma_{T'})$ such that

$$\psi\xi_T = \xi_{T'}\Psi \quad \text{and} \quad \psi\eta_T = \eta_{T'}\Psi.$$

We say that T and T' are *strongly topologically conjugate* if there are topological conjugacies $\psi : (X_G, \sigma_G) \to (X_{G'}, \sigma_{G'})$ and $\Psi : (X_\Gamma, \sigma_\Gamma) \to (X_{\Gamma'}, \sigma_{\Gamma'})$ such that

$$\psi\xi = \xi'\Psi \quad \text{and} \quad \psi\eta = \eta'\Psi,$$

where ξ, η, ξ', and η' are the factor maps given by p, q, p', and q', respectively.

Lemma 2.8. Let T be a textile system and let $n \in \mathbf{N}$. Then the following statements are valid.

(1) T and $T^{[n]}$ are strongly topologically conjugate.

(2) If T is one-sided 1-1, then T and $((T^*)^{[n]})^*$ are topologically conjugate.

(3) T is one-sided 1-1 if and only if $((T^*)^{[n]})^*$ is one-sided 1-1.

(4) T is 1-1 if and only if $((T^*)^{[n]})^*$ is 1-1.

(5) If T is nondegenerate, then so is $((T^*)^{[n]})^*$.

Proof. It is clear that (1) is valid.

Since the lemma is trivially valid for $n = 1$, we assume that $n \geq 2$. Put $((T^*)^{[n]})^* = T'$. For any integer $m \geq 2$ and any $t = (\alpha_{ij})_{i,j\in\mathbf{Z}} \in U_T$, let $w_m(t)$ be defined by

$$w_m(t) = (\alpha_{ij})_{0\leq i\leq m-1, j\in\mathbf{Z}}.$$

Then we can write

$$X_{T'} = \{w_{n-1}(t) \mid t \in U_T\},$$

and

$$Z_{T'} = \{w_n(t) \mid t \in U_T\},$$

and $\xi_{T'}$ and $\eta_{T'}$ are given by

$$\xi_{T'}(w_n(t)) = w_{n-1}(t), \quad t \in U_T$$

and

$$\eta_{T'}(w_n(t)) = w_{n-1}(\sigma_T^{(1,0)}(t)), \quad t \in U_T$$

Let $\psi : X_{T'} \to X_T$ and $\Psi : Z_{T'} \to Z_T$ be defined by

$$\psi(w_{n-1}(t)) = \theta_T(t), \quad t \in U_T$$

and

$$\Psi(w_n(t)) = w_1(t), \quad t \in U_T.$$

Clearly ψ and Ψ are well defined. Since $\xi_T(w_1(t)) = \theta_T(t)$ and $\eta_T(w_1(t)) = \theta_T(\sigma_T^{(1,0)}(t))$ for $t \in U_T$, it easily follows that

$$\xi_T \Psi = \psi \xi_{T'} \quad \text{and} \quad \eta_T \Psi = \psi \eta_{T'}.$$

If ξ_T is 1-1, then $\psi : (X_{T'}, \sigma_{T'}) \to (X_T, \sigma_T)$ and $\Psi : (Z_{T'}, \varsigma_{T'}) \to (Z_T, \varsigma_T)$ are topological conjugacies. Thus (2) is proved.

It is clear that ξ_T is 1-1 if and only if $\xi_{T'}$ is 1-1. It is also clear that η_T is 1-1 if and only if $\eta_{T'}$ is 1-1. Thus (3) and (4) are proved.

It T is nondegenerate, then for any obi w of width $n-1$ weaved by T, T can weave a textile t such that $w_{n-1}(t) = w$. Therefore (5) follows. □

Proposition 2.9. Let T be a one-sided 1-1 and nondegenerate textile system such that (X_{T^*}, σ_{T^*}) is topologically conjugate to a topological Markov shift. Then there is $n \in \mathbf{N}$ such that if

$$T_0 = \overline{\text{trim}}\ ((T^*)^{[n]})^*,$$

then T_0 is one-sided 1-1 and nondegenerate, T_0^* is nondegenerate, T and T_0 are topologically conjugate, and T^* and T_0^* are topologically conjugate, and if T is 1-1 in addition, then T_0 is 1-1.

Proof. Let $T = (p, q : \Gamma \to G)$ with $T^* = (p^T, q^T : \Gamma^T \to G^T)$. Since (X_{T^*}, σ_{T^*}) is topologically conjugate to a topological Markov shift, there is a finite set F of words of the same length, say n, over A_{G^T} such that for $(b_i)_{i \in \mathbf{Z}}$ with $b_i \in A_{G^T}$, there is $i' \in \mathbf{Z}$

with $b_{i'} \cdots b_{i'+n-1} \in F$ if and only if $(b_i)_{i \in \mathbf{Z}} \notin X_{T^*}$. Let $T_0 = \overline{\text{trim}}((T^*)^{[n]})^*$. Put $((T^*)^{[n]})^* = T_1$. Then we have

$$U_{T_0} = U_{T_1}.$$

Since T is one-sided 1-1 and nondegenerate, it follows from (3) and (5) of Lemma 2.8 that T_1 is one-sided 1-1 and nondegenerate. Therefore T_0 is one-sided 1-1 and nondegenerate. By (2) of Lemma 2.8, T and T_1 are topologically conjugate so that T and T_0 are topologically conjugate. By (1) of Lemma 2.8, T^* and $(T^*)^{[n]} = T_1^*$ are topologically conjugate so that T^* and T_0^* are topologically conjugate.

Put $T_0 = (p_0, q_0 : \Gamma_0 \to G_0)$. Then we note that $V_{\Gamma_0} \subset A_{(G^T)^{[n]}}$. But no element in F is contained in V_{Γ_0}. For otherwise, since Γ_0 is nondegenerate, there would be a bisequence in X_{Γ_0} which passes through a vertex which is an element in F. Then, since T_0 is nondegenerate, there would be a textile t in U_{T_0} such that the element in F appears on a dual thread of t. But this cannot be the case, because $U_{T_0} = U_{T_1}$ and no dual thread having an element in F as an arc appearing on it can appear on a textile in U_{T_1}. Put $T_0^* = (p_0^{T_0}, q_0^{T_0} : \Gamma_0^{T_0} \to G_0^{T_0})$. Then

$$X_{G_0^{T_0}} \subset X_{(G^T)^{[n]}}.$$

Since $V_{\Gamma_0} \cap F = \emptyset$, $A_{G_0^{T_0}} \cap F = \emptyset$ so that

$$X_{G_0^{T_0}} \subset X_{(T^*)^{[n]}} = X_{T_1^*}.$$

Since $U_{T_0^*} = U_{T_1^*}$, $X_{T_0^*} = X_{T_1^*}$. Thus

$$X_{G_0^{T_0}} \subset X_{T_0^*}$$

so that T_0^* is nondegenerate.

If T is 1-1 in addition, then T_1 is 1-1 by (4) of Lemma 2.8 so that T_0 is 1-1. □

Theorem 2.10. Let (X, σ) be a topological Markov shift and φ an automorphism of (X, σ) such that (X, φ) is topologically conjugate to a topological Markov shift. Then there is (we can construct) a 1-1 nondegenerate textile system $T = (p, q : \Gamma \to G)$ such that there is a topological conjugacy $\psi : (X, \sigma) \to (X_G, \sigma_G)$ with $\varphi_T = \psi\varphi\psi^{-1}$ and $T^* = (p^T, q^T : \Gamma^T \to G^T)$ is 1-1 and nondegenerate, so that (X, φ) is topologically conjugate to (X_{G^T}, σ_{G^T}).

Proof. This follows from Fact 2.2, Proposition 2.9 and Theorem 2.5. □

The theorem above asserts that every automorphism of a topological Markov shift which defines a dynamical system which is topologically conjugate to a topological Markov shift, is associated, through a topological conjugacy, with a 1-1 and nondegenerate textile system whose dual is also 1-1 and nondegenerate. It also asserts that for any automorphism of a topological Markov shift, if the dynamical system defined by it is topologically conjugate to a topological Markov shift, then we can obtain the defining matrix of a topological Markov shift to which the dynamical system is topologically conjugate (see Section 10).

Question 2.a. Is there an expansive automorphism φ of a topological Markov shift (X, σ) such that (X, φ) is not topologically conjugate to any topological Markov shift?

Let A be an alphabet. Let $\tilde{X}_A$ be the set of all sequences $(a_i)_{i\in\mathbf{N}}$ with $a_i \in A$. A metric $\tilde{d}$ on $\tilde{X}_A$ is defined as follows. For $x = (a_i)_{i\in\mathbf{N}}$ and $y = (b_i)_{i\in\mathbf{N}}$ in $\tilde{X}_A$, $\tilde{d}(x, y) = 0$ if $x = y$, and $\tilde{d}(x, y) = 1/k$ if $x \neq y$, where

$$k = \min\{i \in \mathbf{N} \mid a_i \neq b_i\}.$$

Let $\tilde{\sigma}_A : \tilde{X}_A \to \tilde{X}_A$ be defined by

$$\tilde{\sigma}_A((a_i)_{i\in\mathbf{N}}) = (a_{i+1})_{i\in\mathbf{N}}, \quad (a_i)_{i\in\mathbf{N}} \in \tilde{X}_A.$$

The dynamical system $(\tilde{X}_A, \tilde{\sigma}_A)$ is called the *one-sided full shift* over A. Let $\tilde{X}$ be a closed $\tilde{\sigma}_A$-invariant subset of $\tilde{X}_A$. Let $\tilde{\sigma} = \tilde{\sigma}_A|\tilde{X}$. Then we have a dynamical system $(\tilde{X}, \tilde{\sigma})$ which is called a *one-sided subshift* over A. Let G be a graph with $M_G = M$. Then the definition of the *one-sided topological Markov shift*

$$(\tilde{X}_G, \tilde{\sigma}_G) = (\tilde{X}_M, \tilde{\sigma}_M)$$

should be clear.

For $x = (a_i)_{i\in\mathbf{Z}} \in X_A$, let $\tilde{x}$ denote $(a_i)_{i\in\mathbf{N}}$. For a subshift (X, σ), let $\tilde{X}$ denote the set $\{\tilde{x} \mid x \in X\}$. Then $(\tilde{X}, \tilde{\sigma})$ is a one-sided subshift. (Note that the notation is consistent with the definitions above.) Let $s_X : X \to \tilde{X}$ be the mapping defined by

$$s_X(x) = \tilde{x}, \quad x \in X.$$

For a configuration $u = (\alpha_{ij})_{i,j\in\mathbf{Z}}$ with $\alpha_{ij} \in A$, let $\bar{u}$ denote the subconfiguration $(\alpha_{ij})_{i\in\mathbf{N},j\in\mathbf{Z}}$. Let

$$\bar{U}_A = \{\bar{u} \mid u \in U_A\}.$$

We define a metric $\bar{\delta}$ on $\bar{U}_A$ as follows. Let $v = (\alpha_{ij})_{i\in\mathbf{N},j\in\mathbf{Z}}$ and $v' = (\alpha'_{ij})_{i\in\mathbf{N},j\in\mathbf{Z}}$ be points of $\bar{U}_A$ with $\alpha_{ij}, \alpha'_{ij} \in A$. Then $\bar{\delta}(v, v') = 0$ if $v = v'$ and $\bar{\delta}(v, v') = 1/k$ if $v \neq v'$, where

$$k = \min\{i + |j| \mid i \in \mathbf{N}, j \in \mathbf{Z}, \alpha_{ij} \neq \alpha'_{ij}\}.$$

Let T be a textile system. Then we have one-sided subshifts $(\tilde{X}_T, \tilde{\sigma}_T)$ and $(\tilde{Z}_T, \tilde{\varsigma}_T)$ with

$$\tilde{X}_T = \{\tilde{x} \mid x \in X_T\} \quad \text{and} \quad \tilde{Z}_T = \{\tilde{z} \mid z \in Z_T\}.$$

Let $\tilde{\xi}_T : \tilde{Z}_T \to \tilde{X}_T$ and $\tilde{\eta}_T : \tilde{Z}_T \to \tilde{X}_T$ be defined by

$$\tilde{\xi}_T(s_{Z_T}(z)) = s_{X_T}(\xi_T(z)), \quad z \in Z_T$$

and

$$\tilde{\eta}_T(s_{Z_T}(z)) = s_{X_T}(\eta_T(z)), \quad z \in Z_T,$$

respectively. Clearly $\tilde{\xi}_T$ and $\tilde{\eta}_T$ are well defined and they are factor maps of $(\tilde{Z}_T, \tilde{\varsigma}_T)$ onto $(\tilde{X}_T, \tilde{\sigma}_T)$. If $\tilde{\xi}_T$ is 1-1, then we can define an onto endomorphism $\tilde{\varphi}_T$ of $(\tilde{X}_T, \tilde{\sigma}_T)$ by

$$\tilde{\varphi}_T = \tilde{\eta}_T \tilde{\xi}_T^{-1}.$$

Let

$$\bar{U}_T = \{\bar{t} \mid t \in U_T\}.$$

Let $\bar{\theta}_T : \bar{U}_T \to X_T$ and $\tilde{\theta}_T : \bar{U}_T \to \tilde{X}_{T^*}$ as follows: for $s = (\alpha_{ij})_{i \in \mathbf{N}, j \in \mathbf{Z}}$ with $\alpha_{ij} \in A_\Gamma$

$$\bar{\theta}_T(s) = (p(\alpha_{1j}))_{j \in \mathbf{Z}} \quad \text{and} \quad \tilde{\theta}_T(s) = (p^T(\alpha_{i0}))_{i \in \mathbf{N}},$$

where $T = (p, q : \Gamma \to G)$ with $T^* = (p^T, q^T : \Gamma^T \to G^T)$. Clearly $\bar{\theta}_T$ and $\tilde{\theta}_T$ are continuous.

For $k \in \mathbf{N}$ and $l \in \mathbf{Z}$, define $\bar{\boldsymbol{\sigma}}_T^{(k,l)} : \bar{U}_T \to \bar{U}_T$ by

$$\bar{\boldsymbol{\sigma}}_T^{(k,l)}((\alpha_{ij})_{i \in \mathbf{N}, j \in \mathbf{Z}}) = (\alpha_{i+k, j+l})_{i \in \mathbf{N}, j \in \mathbf{Z}}, \quad (\alpha_{ij})_{i \in \mathbf{N}, j \in \mathbf{Z}} \in \bar{U}_T.$$

Then $\bar{\boldsymbol{\sigma}}_T^{(k,l)}$ is an onto continuous map. The dynamical systems $(\bar{U}_T, \bar{\boldsymbol{\sigma}}_T^{(k,l)})$, $k \in \mathbf{N}$, $j \in \mathbf{Z}$, are called the *half-textile shifts* defined by T.

Let (X, d) be a metric space. A continuous map $\tau : X \to X$ is said to be *positively expansive* if there is a real number $\delta > 0$ such that for $x, y \in X$ if $x \neq y$, then there is $n \in \mathbf{N}$ such that

$$d(\tau^n(x), \tau^n(y)) > \delta.$$

Theorem 2.11. Let T be a textile system. Then the following statements are valid.

(1) If T is one-sided 1-1, then φ_T is positively expansive if and only if both $\tilde{\xi}_{T^*}$ and $\tilde{\eta}_{T^*}$ are 1-1.

(2) If $\tilde{\xi}_T$ is 1-1, then $\tilde{\varphi}_T$ is expansive if and only if T^* is one-sided 1-1.

(3) If T is one-sided 1-1 and both $\tilde{\xi}_{T^*}$ and $\tilde{\eta}_{T^*}$ are 1-1, then the mapping $\bar{\chi}_T : X_T \to \tilde{X}_{T^*}$ defined by

$$\bar{\chi}_T(\bar{\theta}_T(s)) = \tilde{\theta}_T(s), \quad s \in \bar{U}_T$$

is a homeomorphism and the following diagrams commute:

$$\begin{array}{ccc} X_T & \xrightarrow{\varphi_T} & X_T \\ {\scriptstyle \bar{\chi}_T}\downarrow & & \downarrow{\scriptstyle \bar{\chi}_T} \\ \tilde{X}_{T^*} & \xrightarrow[\tilde{\sigma}_{T^*}]{} & \tilde{X}_{T^*} \end{array} \qquad \begin{array}{ccc} X_T & \xrightarrow{\sigma_T} & X_T \\ {\scriptstyle \bar{\chi}_T}\downarrow & & \downarrow{\scriptstyle \bar{\chi}_T} \\ \tilde{X}_{T^*} & \xrightarrow[\tilde{\varphi}_{T^*}]{} & \tilde{X}_{T^*} \end{array} .$$

Proof. First we prove (3). Assume that T is one-sided 1-1 and both $\tilde{\xi}_{T^*}$ and $\tilde{\eta}_{T^*}$ are 1-1. Since T is one-sided 1-1, $\bar{\theta}_T : \bar{U}_T \to X_T$ is a homeomorphism. Since $\tilde{\xi}_{T^*}$ and $\tilde{\eta}_{T^*}$ are 1-1, $\tilde{\xi}_{T^*}$ and $\tilde{\eta}_{T^*}$ are homeomorphisms. Therefore it follows that $\tilde{\theta}_T : \bar{U}_T \to \tilde{X}_{T^*}$ is a homeomorphism. Thus $\bar{\chi}_T = \tilde{\theta}_T(\bar{\theta}_T)^{-1}$ is a homeomorphism. It is clear that the diagrams commute.

(1) Suppose that T is one-sided 1-1. Assume that at least one of $\tilde{\xi}_{T^*}$ and $\tilde{\eta}_{T^*}$ is not 1-1. If $\tilde{\xi}_{T^*}$ is not 1-1, then for any $l \in \mathbf{N}$ we have $s_l = (\alpha_{ij})_{i\in\mathbf{N},j\in\mathbf{Z}}$ and $s_l' = (\alpha_{ij}')_{i\in\mathbf{N},j\in\mathbf{Z}}$ in $\bar{U}_T$ such that $s_l \neq s_l'$ but $\alpha_{ij} = \alpha_{ij}'$ for all $i \in \mathbf{N}$ and $j \leq l$. This implies that φ_T is not positively expansive because for all $n \in \mathbf{N}$,

$$d(\varphi_T^n(\bar{\theta}_T(s_l)), \varphi_T^n(\bar{\theta}_T(s_l'))) \leq \frac{1}{1+l}.$$

Similarly if η_{T^*} is not 1-1, then φ_T is not positively expansive.

It follows from (3) that if both $\tilde{\xi}_{T^*}$ and $\tilde{\eta}_{T^*}$ are 1-1, then (X_T, φ_T) is topologically conjugate to the one-sided subshift $(\tilde{X}_{T^*}, \tilde{\sigma}_{T^*})$ so that φ_T is positively expansive.

(2) Suppose that $\tilde{\xi}_T$ is 1-1. Assume that T^* is not one-sided 1-1. Then there are two textiles $t = (\alpha_{ij})_{i,j\in\mathbf{Z}}$ and $t' = (\alpha_{ij}')_{i,j\in\mathbf{Z}}$ such that $t \neq t'$ but $\alpha_{ij} = \alpha_{ij}'$ for all $i \in \mathbf{Z}$ and $j < 0$. Let

$$x_{n,l} = s_{X_T}\theta_T\boldsymbol{\sigma}_T^{(n,-l)}(t) \quad \text{and} \quad x_{n,l}' = s_{X_T}\theta_T\boldsymbol{\sigma}_T^{(n,-l)}(t')$$

for any $n \in \mathbf{Z}$ and $l \in \mathbf{N}$. Then $(x_{n,l})_{n\in\mathbf{Z}}$ and $(x_{n,l}')_{n\in\mathbf{Z}}$ are distinct orbits of $(\tilde{X}_T, \tilde{\varphi}_T)$ and we have

$$\tilde{d}_{\tilde{X}_T}(x_{n,l}, x_{n,l}') \leq \frac{1}{l}$$

for all $n \in \mathbf{Z}$ and $l \in \mathbf{N}$. Thus $\tilde{\varphi}_T$ is not expansive.

Conversely assume that $\tilde{\varphi}_T$ is not expansive. Since $\tilde{\xi}_T^{-1}$ is continuous, there is $\epsilon > 0$ such that for $x, x' \in \tilde{X}_T$, if $\tilde{d}_{\tilde{X}_T}(x, x') \leq \epsilon$, then $\tilde{d}_{\tilde{Z}_T}(\tilde{\xi}_T^{-1}(x), \tilde{\xi}_T^{-1}(x')) \leq \frac{1}{2}$. Since $\tilde{\varphi}_T$ is not expansive, there are two textiles $t = (\alpha_{ij})_{i,j\in\mathbf{Z}}$ and $t' = (\alpha_{ij}')_{i,j\in\mathbf{Z}}$ in U_T such that

$$s_{X_T}\theta_T(t) \neq s_{X_T}\theta_T(t')$$

and

$$\tilde{d}_{\tilde{X}_T}(s_{X_T}\theta_T\sigma_T^{(i,0)}(t), s_{X_T}\theta_T\sigma_T^{(i,0)}(t')) \le \epsilon \quad \text{for all } i \in \mathbf{Z}.$$

Therefore

$$\tilde{d}_{\tilde{Z}_T}(\tilde{\xi}_T^{-1}s_{X_T}\theta_T\sigma_T^{(i,0)}(t), \tilde{\xi}_T^{-1}s_{X_T}\theta_T\sigma_T^{(i,0)}(t')) \le \frac{1}{2} \quad \text{for all } i \in \mathbf{Z}$$

so that $\alpha_{i0} = \alpha'_{i0}$ for all $i \in \mathbf{Z}$, but there is $n \in \mathbf{N}$ with $\alpha_{0n} \neq \alpha'_{0n}$. This implies that T^* is not one-sided 1-1. □

For a configuration $u = (\alpha_{ij})_{i,j\in\mathbf{Z}} \in U_A$, where A is an alphabet, let $\hat{u}$ denote the subconfiguration $(\alpha_{ij})_{i,j\in\mathbf{N}}$. Let

$$\hat{U}_A = \{\hat{u} \mid u \in U_A\}.$$

We define a metric $\hat{\delta}$ on $\hat{U}_A$ as follows. Let $v = (\alpha_{ij})_{i,j\in\mathbf{N}}$ and $v' = (\alpha'_{ij})_{i,j\in\mathbf{N}}$ be elements of $\hat{U}_A$ with $\alpha_{ij}, \alpha'_{ij} \in A$. Then $\hat{\delta}(v, v') = 0$ if $v = v'$, and $\hat{\delta}(v, v') = 1/(k-1)$ if $v \neq v'$, where

$$k = \min\{i + j \mid i, j \in \mathbf{N}, \alpha_{ij} \neq \alpha'_{ij}\}$$

Let $T = (p, q : \Gamma \to G)$ with $T^* = (p^T, q^T : \Gamma^T \to G^T)$. Let

$$\hat{U}_T = \{\hat{t} \mid t \in U_T\}.$$

Let $\hat{\theta}_T : \hat{U}_T \to \tilde{X}_T$ and $\hat{\theta}^*_T : \hat{U}_T \to \tilde{X}_{T^*}$ be defined as follows: for $s = (\alpha_{ij})_{i,j\in\mathbf{N}} \in \hat{U}_T$ with $\alpha_{ij} \in A_\Gamma$,

$$\hat{\theta}_T(s) = (p(\alpha_{1j}))_{j\in\mathbf{N}} \quad \text{and} \quad \hat{\theta}^*_T(s) = (p^T(\alpha_{i1}))_{i\in\mathbf{N}}.$$

Clearly $\hat{\theta}_T$ and $\hat{\theta}^*_T$ are continuous.

For $k, l \in \mathbf{N}$, define $\hat{\sigma}_T^{(k,l)} : \hat{U}_T \to \hat{U}_T$ by

$$\hat{\sigma}_T^{(k,l)}((\alpha_{ij})_{i,j\in\mathbf{N}}) = (\alpha_{i+k,j+l})_{i,j\in\mathbf{N}}, \quad (\alpha_{ij})_{i,j\in\mathbf{N}} \in \hat{U}_T.$$

Then $\hat{\sigma}_T^{(k,l)}$ is an onto continuous map. The dynamical systems $(\hat{U}_T, \hat{\sigma}_T^{(k,l)})$, $k, l \in \mathbf{N}$, are called the *quarter-textile shifts* defined by T.

Theorem 2.12. Let T be a textile system such that $\tilde{\xi}_T$ is 1-1. Then $\tilde{\varphi}_T = \tilde{\eta}_T\tilde{\xi}_T^{-1}$ is positively expansive if and only if $\tilde{\xi}_{T^*}$ is 1-1. If $\tilde{\xi}_{T^*}$ is 1-1, then the mapping $\hat{\chi}_T : \tilde{X}_T \to \tilde{X}_{T^*}$ defined by

$$\hat{\chi}_T(\hat{\theta}_T(s)) = \hat{\theta}^*_T(s), \quad s \in \hat{U}_T,$$

is a homeomorphism and the following diagrams commute:

$$
\begin{array}{ccc}
\tilde{X}_T & \xrightarrow{\tilde{\varphi}_T} & \tilde{X}_T \\
\hat{\chi}_T \downarrow & & \downarrow \hat{\chi}_T \\
\tilde{X}_{T^*} & \xrightarrow[\tilde{\sigma}_{T^*}]{} & \tilde{X}_{T^*}
\end{array}
\qquad
\begin{array}{ccc}
\tilde{X}_T & \xrightarrow{\tilde{\sigma}_T} & \tilde{X}_T \\
\hat{\chi}_T \downarrow & & \downarrow \hat{\chi}_T \\
\tilde{X}_{T^*} & \xrightarrow[\tilde{\varphi}_{T^*}]{} & \tilde{X}_{T^*}
\end{array} .
$$

Proof. Assume that $\tilde{\xi}_{T^*}$ is not 1-1. Then for any $l \in \mathbf{N}$ there are $s_l = (\alpha_{ij})_{i,j\in\mathbf{N}}$ and $s'_l = (\alpha'_{ij})_{i,j\in\mathbf{N}}$ in $\hat{U}_T$ such that $s_l \neq s'_l$ but $\alpha_{ij} = \alpha'_{ij}$ for all $i \in \mathbf{N}$ and $1 \leq j \leq l$. This implies that $\tilde{\varphi}_T$ is not positively expansive because for all $n \in \mathbf{N}$,

$$\tilde{d}(\tilde{\varphi}_T^n(\hat{\theta}_T(s_l)), \tilde{\varphi}_T^n(\hat{\theta}_T(s'_l))) \leq \frac{1}{1+l}.$$

Assume that $\tilde{\xi}_{T^*}$ is 1-1. Since $\tilde{\xi}_T$ is 1-1, $\tilde{\xi}_T$ is a homeomorphism. Therefore it follows that $\hat{\theta}_T$ is a homeomorphism. Since $\tilde{\xi}_{T^*}$ is 1-1, it also follows that $\hat{\theta}^*_T$ is a homeomorphism. Therefore $\hat{\chi}_T = \hat{\theta}^*_T \hat{\theta}_T^{-1}$ is a homeomorphism. Clearly the diagrams commute. Therefore $(\tilde{X}_T, \tilde{\varphi}_T)$ is topologically conjugate to the one-sided subshift $(\tilde{X}_{T^*}, \tilde{\sigma}_{T^*})$ so that $\tilde{\varphi}_T$ is positively expansive. $\square$

For a continuous map φ of a compact space into itself, we let $h(\varphi)$ denote the topological entropy of φ. For a textile system T, we are interested in $h(\boldsymbol{\sigma}_T^{(k,l)}), k, l \in \mathbf{Z}$. Especially we define

$$h(T) = h(\boldsymbol{\sigma}_T^{(1,0)}).$$

Theorem 2.13. Let T be a textile system. Then

$$h(T) = \lim_{k\to\infty} h(\sigma_{(T^{[k]})^*}).$$

If T is one-sided 1-1, then

$$h(\varphi_T) = h(T).$$

If $\tilde{\xi}_T$ is 1-1, then

$$h(\tilde{\varphi}_T) = h(T).$$

Proof. For $t = (\alpha_{ij})_{i,j\in\mathbf{Z}} \in U_T$ with $\alpha_{ij} \in A_\Gamma$ and for $k, n \in \mathbf{N}$, let

$$w_t(n,k) = (\alpha_{ij})_{0\leq i\leq n-1, -k\leq j\leq k}.$$

Let $s(n,k)$ be the largest cardinality of any subset of U_T such that for any two distinct members t_1 and t_2 of the subset,

$$d((\sigma_T^{(1,0)})^i(t_1),(\sigma_T^{(1,0)})^i(t_2)) \geq \frac{1}{1+k} \quad \text{for some } i, 0 \leq i \leq n-1.$$

Let $N(n,k)$ be the cardinality of the set

$$\{w_t(n,k) \mid t \in U_T\}.$$

Then it follows that

$$N(n,k) \leq s(n,k) \leq N(n+2k,k).$$

Therefore

$$\limsup_{n\to\infty} \frac{1}{n}\log s(n,k) = \limsup_{n\to\infty} \frac{1}{n}\log N(n,k) = h(\sigma_{(T^{[2k+1]})^*}).$$

Thus

$$h(T) = h(\sigma_T^{(1,0)}) = \lim_{k\to\infty} h(\sigma_{(T^{[k]})^*})$$

(cf. [Wal],[M]).

By similar arguments, we also see that

$$h(\bar{\sigma}_T^{(1,0)}) = \lim_{k\to\infty} h(\sigma_{(T^{[k]})^*})$$

and

$$h(\hat{\sigma}_T^{(1,0)}) = \lim_{k\to\infty} h(\sigma_{(T^{[k]})^*}).$$

If ξ_T is 1-1, then (X_T, φ_T) is topologically conjugate to $(\bar{U}_T, \bar{\sigma}_T^{(1,0)})$ so that $h(\varphi_T) = h(T)$ by the above. If $\tilde{\xi}_T$ is 1-1, then $(\tilde{X}_T, \tilde{\varphi}_T)$ is topologically conjugate to $(\hat{U}_T, \hat{\sigma}_T^{(1,0)})$ so that $h(\tilde{\varphi}_T) = h(T)$ by the above. $\square$

Let $T = (p, q : \Gamma \to G)$ be a textile system. Let $L(X_T)$ be the set of all paths in $L(G)$ that appear on some bisequence in X_T and let $L(Z_T)$ be the set of all paths in $L(\Gamma)$ that appear on some bisequence in Z_T. Let $\xi_T^L : L(Z_T) \to L(X_T)$ and $\eta_T^L : L(Z_T) \to L(X_T)$ be defined as follows: for $\alpha_1 \cdots \alpha_k \in L(Z_T)$ with $k \in \mathbf{N}, \alpha_i \in A_\Gamma$,

$$\xi_T^L(\alpha_1 \cdots \alpha_k) = p(\alpha_1)\cdots p(\alpha_k) \quad \text{and} \quad \eta_T^L(\alpha_1 \cdots \alpha_k) = q(\alpha_1)\cdots q(\alpha_k).$$

Corollary 2.14. Let T be a textile system. If at least one of $\xi_{T^*}^L$ and $\eta_{T^*}^L$ is bounded-to-one, then

$$h(T) = h(\sigma_{T^*})$$

Proof. Assume that $\xi_{T^*}^L$ is bounded-to-one. Then $\xi_{(T^{[n]})^*}^L : L(X_{(T^{[n+1]})^*}) \to L(X_{(T^{[n]})^*})$ is bounded-to-one for all $n \in \mathbf{N}$. Therefore for all $n \in \mathbf{N}$,

$$h(\sigma_{(T^{[n]})^*}) = h(\sigma_{(T^{[n+1]})^*})$$

(cf. [CP]). Thus by the theorem above

$$h(T) = h(\sigma_{T^*}).$$

If $\eta^L_{T^*}$ is bounded-to-one, then similarly we have the same conclusion. □

Proposition 2.15. If T is a nondegenerate textile system such that (X_{T^*}, σ_{T^*}) is topologically conjugate to a topological Markov shift, then we can obtain the defining matrix of a topological Markov shift to which (X_{T^*}, σ_{T^*}) is topologically conjugate, and hence we can compute $h(\sigma_{T^*})$.

Proof. This follows from the proof of Proposition 2.9. □

3. RESOLVING TEXTILE SYSTEMS

A graph-homomorphism $h : \Gamma \to G$ is said to be *right resolving* if for each vertex $u \in V_\Gamma$ and for each arc $a \in A_G$ with $i_G(a) = h_V(u)$ there is a unique arc $\alpha \in A_\Gamma$ such that $i_\Gamma(\alpha) = u$ and $h_A(\alpha) = a$. The graph-homomorphism h is said to be *left resolving* if for each vertex $u \in V_\Gamma$ and for each arc $a \in A_G$ with $t_G(a) = h_V(u)$, there is a unique arc $\alpha \in A_\Gamma$ such that $t_\Gamma(\alpha) = u$ and $h_A(\alpha) = a$. We say that h is *biresolving* if h is both right resolving and left resolving.

By direct observation we have the following proposition.

Proposition 3.1. Let $T = (p, q : \Gamma \to G)$ be a textile system with $T^* = (p^T, q^T : \Gamma^T \to G^T)$. Then

(1) p is right resolving if and only if p^T is right resolving;

(2) p is left resolving if and only if q^T is right resolving;

(3) q is right resolving if and only if p^T is left resolving;

(4) q is left resolving if and only if q^T is left resolving.

A textile system $T = (p, q : \Gamma \to G)$ is said to be *LR* if p is left resolving and q is right resolving. We say that T is *RL* if $T^{-1} = (q, p : \Gamma \to G)$ is LR. The textile system T is said to be *LL* if both p and q are left resolving. An *RR* textile system is defined similarly.

Corollary 3.2. If T is an LR textile system, then T^* is also LR.

For an onto graph-homomorphism $h : \Gamma \to G$ which is either right resolving or left resolving, the factor map of $(X_\Gamma, \sigma_\Gamma)$ into (X_G, σ_G) given by h, is onto. This fact together with Proposition 3.1 (and our standing hypothesis) gives the following fact.

Fact 3.3. If a textile system T is either LL, LR, RL or RR, then T is nondegenerate.

An endomorphism φ of a topological Markov shift is said to be *LL* if there is a one-sided 1-1 LL textile system such that $\varphi_T = \varphi$. Similarly an *LR endomorphism*, an *RL endomorphism* and an *RR endomorphism* are defined. The definitions of an *LL automorphism* and an *LR automorphism* should be clear.

Proposition 3.4. Let $T = (p, q : \Gamma \to G)$ be a textile system. If at least one of p and q is either right resolving or left resolving, then

$$h(T) = h(\sigma_{T^*}).$$

Proof. By Proposition 3.1, at least one of p^T and q^T is either right resolving or left resolving. Therefore at least one of $\xi^L_{T^*}$ and $\eta^L_{T^*}$ is bounded-to-one. Thus the result follows from Corollary 2.14. □

By the proposition above and for the same reason stated before Fact 3.3, we have:

Proposition 3.5. Let $T = (p, q : \Gamma \to G)$ be a textile system with $T^* = (p^T, q^T : \Gamma^T \to G^T)$. Then

$$h(T) = h(\sigma_{G^T})$$

if any one of the following conditions holds:
(1) T is LR;
(2) T is RL;
(3) q is biresolving;
(4) p is biresolving.

Lemma 3.6. Let $T = (p, q : \Gamma \to G)$ be a textile system. Then the following two statements are valid.

(1) T is nondegenerate and $\tilde{\xi}_T$ is 1-1 if and only if T is one-sided 1-1 and nondegenerate and p is left resolving.

(2) T is nondegenerate and both $\tilde{\xi}_T$ and $\tilde{\eta}_T$ are 1-1 if and only if T is 1-1 and LL.

Proof. We note that by our standing hypothesis, G and Γ are nondegenerate and p and q are onto.

(1) Assume that T is nondegenerate and $\tilde{\xi}_T$ is 1-1. Since T is nondegenerate, it follows that $\tilde{X}_G = \tilde{X}_T$ and $\tilde{X}_\Gamma = \tilde{Z}_T$. Since $\tilde{\xi}_T$ is 1-1, it follows that ξ_T is 1-1 so that T is one-sided 1-1. Let $u \in V_\Gamma$. Let $a \in A_G$ with $t_G(a) = p_V(u)$. There is $(\alpha_{i+1})_{i\in\mathbf{N}} \in \tilde{X}_\Gamma, \alpha_{i+1} \in A_\Gamma$, such that $i_\Gamma(\alpha_2) = u$. Let $(a_i)_{i\in\mathbf{N}}$ be a sequence such that $a_1 = a$ and $a_i = p(\alpha_i)$ for $i \geq 2$. Then $(a_i)_{i\in\mathbf{N}} \in \tilde{X}_G$. Since $\tilde{\xi}_T : \tilde{X}_\Gamma \to \tilde{X}_G$ is 1-1 and onto, there is a unique element $(\beta_i)_{i\in\mathbf{N}} \in \tilde{X}_\Gamma, \beta_i \in A_\Gamma$, such that $\tilde{\xi}((\beta_i)_{i\in\mathbf{N}}) = (a_i)_{i\in\mathbf{N}}$. Hence $p(\beta_1) = a$. Since

$$\tilde{\xi}_T((\beta_{i+1})_{i\in\mathbf{N}}) = (a_{i+1})_{i\in\mathbf{N}} = \tilde{\xi}_T((\alpha_{i+1})_{i\in\mathbf{N}})$$

and $\tilde{\xi}_T$ is 1-1, we have $\beta_2 = \alpha_2$ so that

$$t_\Gamma(\beta_1) = i_\Gamma(\beta_2) = i_\Gamma(\alpha_2) = u.$$

Therefore we have $\beta_1 \in A_\Gamma$ such that $p(\beta_1) = a$ and $t_\Gamma(\beta_1) = u$. Such β_1 is unique because $(\beta_i)_{i\in\mathbf{N}} \in \tilde{X}_\Gamma$ with $\tilde{\xi}_T((\beta_i)_{i\in\mathbf{N}}) = (a_i)_{i\in\mathbf{N}}$, is unique. Thus p is left resolving.

Conversely assume that T is one-sided 1-1 and nondegenerate and p is left resolving. Then, since p is left resolving, for any $x \in X_T$ and $(\alpha_i)_{i\in\mathbf{N}} \in \tilde{Z}_T, \alpha_i \in A_\Gamma$, such that $\tilde{\xi}_T((\alpha_i)_{i\in\mathbf{N}}) = s_{X_T}(x), (\alpha_i)_{i\in\mathbf{N}}$ can be extended to $(\alpha_i)_{i\in\mathbf{Z}} \in Z_T$ such that $\xi_T((\alpha_i)_{i\in\mathbf{Z}}) = x$. Therefore the injectivity of ξ_T implies that of $\tilde{\xi}_T$. Thus T is nondegenerate and $\tilde{\xi}_T$ is 1-1.

(2) Assume that T is nondegenerate and both $\tilde{\xi}_T$ and $\tilde{\eta}_T$ are 1-1. Then it follows from (1) that T is one-sided 1-1, p is left resolving, T^{-1} is one-sided 1-1 and q is left resolving. Therefore T is 1-1 and LL.

Conversely assume that T is 1-1 and LL. Then since T is LL, it follows that T is nondegenerate. By (1), both $\tilde{\xi}_T$ and $\tilde{\eta}_T$ are 1-1. □

Proposition 3.7. (1) If T is a 1-1 LL textile system over a graph G, then we can define the automorphism $\tilde{\varphi}_T$ of the one-sided topological Markov shift $(\tilde{X}_T, \tilde{\sigma}_T) = (\tilde{X}_G, \tilde{\sigma}_G)$. If $\tilde{\varphi}$ is an automorphism of a one-sided topological Markov shift, then there is a 1-1 LL textile system T with $\tilde{\varphi}_T = \tilde{\varphi}$.

(2) If $T = (p, q : \Gamma \to G)$ is a one-sided 1-1 and nondegenerate textile system with p left resolving, then we can define the onto endomorphism $\tilde{\varphi}_T$ of the one-sided topological Markov shift $(\tilde{X}_T, \tilde{\sigma}_T) = (\tilde{X}_G, \tilde{\sigma}_G)$. If $\tilde{\varphi}$ is an onto endomorphism of a one-sided topological Markov shift, then there is a one-sided 1-1 and nondegenerate textile system $T = (p, q : \Gamma \to G)$ with p left resolving such that $\tilde{\varphi}_T = \tilde{\varphi}$.

Proof. (1) The first claim of (1) is proved by (2) of Lemma 3.6.

To prove the second claim, let $\tilde{\varphi}$ be an automorphism of a one-sided topological Markov shift $(\tilde{X}_G, \tilde{\sigma}_G)$. The automorphism $\tilde{\varphi}$ naturally induces the automorphism φ of the topological Markov shift (X_G, σ_G). It follows that φ and φ^{-1} are block maps of $(0, m)$ type and $(0, n)$ type, respectively, for some $m, n \in \mathbf{N}$. Let φ and φ^{-1} be given by $f : L_{m+1}(G) \to A_G$ and $g : L_{n+1}(G) \to A_G$, respectively. Let $l = m + n$. Considering φ as a block map of $(0, l)$ type, let $T_\varphi = (p, q : G^{[l+1]} \to G)$ be as in Fact 2.2. Then for $a_1 \cdots a_{l+1} \in L_{l+1}(G)$ with $a_i \in A_G$, $p(a_1 \cdots a_{l+1}) = a_1$ and $q(a_1 \cdots a_{l+1}) = f(a_1 \cdots a_{m+1})$. Clearly T_φ is 1-1 and $\tilde{\varphi}_{T_\varphi} = \tilde{\varphi}$. It is clear that p is left resolving. We see that q is also left resolving. For if $a_1 \cdots a_l \in L_l(G)$ with $a_i \in A_G$ and if $b \in A_G$ with $bf(a_1 \cdots a_{m+1}) \in L_2(G)$, then there is unique $a \in A_G$ such that $aa_1 \cdots a_l \in L_{l+1}(G)$ and $f(aa_1 \cdots a_m) = b$. In fact, it is easily seen that a is given as

$$a = g(bf(a_1 \cdots a_{m+1}) \cdots f(a_n a_{n+1} \cdots a_{m+n})).$$

Thus T_φ is LL.

(2) By using (1) of Lemma 3.6, (2) of the proposition is easily proved. □

Theorem 3.8. (1) Let T be a 1-1 LL textile system. Then $(\tilde{X}_T, \tilde{\varphi}_T)$ is topologically conjugate to a topological Markov shift if and only if T^* is one-sided 1-1 and finitely saturated.

(2) Let $T = (p, q : \Gamma \to G)$ be a one-sided 1-1 and nondegenerate textile system with p left resolving. Then $(\tilde{X}_T, \tilde{\varphi}_T)$ is topologically conjugate to a one-sided topological Markov shift if and only if $\tilde{\xi}_{T^*}$ is 1-1 and T^* is finitely saturated.

Proof. (1) Let $T^* = (p^T, q^T : \Gamma^T \to G^T)$. Since T is LL, q^T is biresolving by Proposition 3.1. Hence the factor map η^* given by q^T is onto. Assume that T^* is one-sided 1-1 and finitely saturated. Then, by Proposition 2.1, (X_{T^*}, σ_{T^*}) is topologically conjugate to a topological Markov shift. Moreover, since T^* is one-sided 1-1 and both $\tilde{\xi}_T$ and $\tilde{\eta}_T$ are 1-1, it follows from Theorem 2.11(3) that (X_{T^*}, σ_{T^*}) is topologically conjugate to $(\tilde{X}_T, \tilde{\varphi}_T)$. Thus $(\tilde{X}_T, \tilde{\varphi}_T)$ is topologically conjugate to a topological Markov shift.

Conversely assume that $(\tilde{X}_T, \tilde{\varphi}_T)$ is topologically conjugate to a topological Markov shift. Then it follows from Theorem 2.11 (2)(3) that T^* is one-sided 1-1 and (X_{T^*}, σ_{T^*}) is topologically conjugate to $(\tilde{X}_T, \tilde{\varphi}_T)$ so that (X_{T^*}, σ_{T^*}) is topologically conjugate to a topological Markov shift. Thus, by Proposition 2.6, T^* is finitely saturated.

(2) Let $T^* = (p^T, q^T : \Gamma^T \to G^T)$. Since p is left resolving, q^T is right resolving so that the factor map η^* given by q^T is onto. Assume that $\tilde{\xi}_{T^*}$ is 1-1 and T^* is finitely saturated. Then since $\tilde{\xi}_{T^*}$ is 1-1, ξ_{T^*} is 1-1 so that T^* is one-sided 1-1. Hence (X_{T^*}, σ_{T^*}) is topologically conjugate to a topological Markov shift, by Proposition 2.1. Since both $\tilde{\xi}_T$ and $\tilde{\xi}_{T^*}$ are 1-1, it follows from Theorem 2.12 that $(\tilde{X}_T, \tilde{\varphi}_T)$ is topologically conjugate to $(\tilde{X}_{T^*}, \tilde{\sigma}_{T^*})$. Thus $(\tilde{X}_T, \tilde{\varphi}_T)$ is topologically conjugate to a one-sided topological Markov shift.

Conversely assume that $(\tilde{X}_T, \tilde{\varphi}_T)$ is topologically conjugate to a one-sided topological Markov shift. Then by Theorem 2.12, $\tilde{\xi}_{T^*}$ is 1-1 and $(\tilde{X}_T, \tilde{\varphi}_T)$ is topologically conjugate to $(\tilde{X}_{T^*}, \tilde{\sigma}_{T^*})$. Hence (X_{T^*}, σ_{T^*}) is topologically conjugate to a topological Markov shift. Thus, by Proposition 2.6, T^* is finitely saturated. □

Let $\phi : (X, \sigma) \to (X', \sigma')$ be a factor map between subshifts. We say that ϕ is *right closing* if for any two distinct points $x, y \in X$ with $\lim_{n\to\infty} d(\sigma^{-n}(x), \sigma^{-n}(y)) = 0$, $\phi(x)$ and $\phi(y)$ are distinct, and we say that ϕ is *left closing* if for any two distinct points $x, y \in X$ with $\lim_{n\to\infty} d(\sigma^n(x), \sigma^n(y)) = 0$, $\phi(x)$ and $\phi(y)$ are distinct [Ki]. We say that ϕ is *biclosing* if ϕ is both right closing and left closing.

For a factor map $\phi : (X, \sigma) \to (X', \sigma')$ between topological Markov shifts, ϕ is biclosing and onto if and only if ϕ is topologically conjugate to a factor map given by a biresolving graph-homomorphism, that is, there is a biresolving graph homomorphism $h : \Gamma \to G$ and topological conjugacies $v : (X, \sigma) \to (X_\Gamma, \sigma_\Gamma)$ and $v' : (X', \sigma') \to (X_G, \sigma_G)$ such that $\psi v = v' \phi$, where ψ is the factor map given by h. This result for the case where both (X, σ) and (X', σ') are irreducible topological Markov shifts, was given by [N1] together with the result that in this case, $\phi : (X, \sigma) \to (X', \sigma')$ is biclosing and onto if and only if ϕ is constant-to-one and onto. Here an *irreducible topological Markov shift* means a topological Markov shift whose defining matrix is irreducible. The general result above follows from Boyle's observation [B1] of Kitchen's result [Ki].

Theorem 3.9. (1) Let (X, σ) be a topological Markov shift and φ an onto endomorphism of (X, σ). If (X, φ) is topologically conjugate to a one-sided topological Markov shift, then there is (we can construct) a one-sided 1-1 and nondegenerate textile system $T = (p, q : \Gamma \to G)$ such that q is biresolving, T^* is 1-1 (and LL), and there is a topological conjugacy $\psi : (X, \sigma) \to (X_G, \sigma_G)$ such that $\varphi_T = \psi\varphi\psi^{-1}$.

(2) Let $(\tilde{X}, \tilde{\sigma})$ be a one-sided topological Markov shift and $\tilde{\varphi}$ an automorphism of $(\tilde{X}, \tilde{\sigma})$. If $(\tilde{X}, \tilde{\varphi})$ is topologically conjugate to a topological Markov shift, then there is (we can construct) a 1-1 and LL textile system $T = (p, q : \Gamma \to G)$ such that T^* is one-sided 1-1 and nondegenerate (and q^T is biresolving if $T^* = (p^T, q^T : \Gamma^T \to G^T)$) and there is a topological conjugacy $\tilde{\psi} : (\tilde{X}, \tilde{\sigma}) \to (\tilde{X}_G, \tilde{\sigma}_G)$ with $\tilde{\varphi}_T = \tilde{\psi}\tilde{\varphi}\tilde{\psi}^{-1}$.

(3) If $T = (p, q : \Gamma \to G)$ is a one-sided 1-1 and nondegenerate textile system such that q is biresolving and $T^* = (p^T, q^T : \Gamma^T \to G^T)$ is 1-1 (and LL), then the mapping $\bar{\chi}_T : X_G \to \tilde{X}_{G^T}$ defined by

$$\bar{\chi}_T(\bar{\theta}_T(s)) = \tilde{\theta}_T(s), \quad s \in \bar{U}_T$$

is a homeomorphism and the following diagrams commute:

$$\begin{array}{ccc} X_G & \xrightarrow{\varphi_T} & X_G \\ {\scriptstyle \bar{\chi}_T}\downarrow & & \downarrow{\scriptstyle \bar{\chi}_T} \\ \tilde{X}_{G^T} & \xrightarrow[\tilde{\sigma}_{G^T}]{} & \tilde{X}_{G^T} \end{array} \qquad \begin{array}{ccc} X_G & \xrightarrow{\sigma_G} & X_G \\ {\scriptstyle \bar{\chi}_T}\downarrow & & \downarrow{\scriptstyle \bar{\chi}_T} \\ \tilde{X}_{G^T} & \xrightarrow[\tilde{\varphi}_{T^*}]{} & \tilde{X}_{G^T} \end{array} .$$

Proof. (1) Let (X, σ) be a topological Markov shift and φ an onto endomorphism of (X, σ) such that (X, φ) is topologically conjugate to a one-sided topological Markov shift. By Fact 2.2, there is a one-sided 1-1 nondegenerate textile system T_0 such that $(X_{T_0}, \sigma_{T_0}) = (X, \sigma)$ and $\varphi_{T_0} = \varphi$. Since (X_{T_0}, φ_{T_0}) is topologically conjugate to a one-sided topological Markov shift, it follows from Theorem 2.11 that (X_{T_0}, φ_{T_0}) is topologically conjugate to $(\tilde{X}_{T_0^*}, \tilde{\sigma}_{T_0^*})$, so that this is topologically conjugate to a one-sided topological Markov shift. Therefore $(X_{T_0^*}, \sigma_{T_0^*})$ is topologically conjugate to a topological Markov shift. Thus by Proposition 2.9, there is (we can construct) a one-sided 1-1 and nondegenerate textile system $T = (p, q : \Gamma \to G)$ such that T^* is nondegenerate and

T and T_0 are topologically conjugate. Since T and T_0 are topologically conjugate, there is a topological conjugacy $\psi : (X_{T_0}, \sigma_{T_0}) \to (X_T, \sigma_T)$ such that $\varphi_T = \psi\varphi_{T_0}\psi^{-1}$. Hence (X_T, φ_T) is topologically conjugate to (X, φ), so that φ_T is positively expansive. Therefore, by Theorem 2.11, $\tilde{\xi}_{T^*}$ and $\tilde{\eta}_{T^*}$ are 1-1. Hence, since T^* is nondegenerate, it follows from Lemma 3.6 that T^* is 1-1 and LL. Since T^* is LL, q is biresolving, by Proposition 3.1. Since T is nondegenerate, $(X_T, \sigma_T) = (X_G, \sigma_G)$. Thus $T = (p, q : \Gamma \to G)$ is a one-sided 1-1 nondegenerate textile system such that q is biresolving, T^* is 1-1 and LL, and $\psi : (X, \sigma) \to (X_G, \sigma_G)$ is a topological conjugacy with $\varphi_T = \psi\varphi\psi^{-1}$.

(2) Let $(\tilde{X}, \tilde{\sigma})$ be a one-sided topological Markov shift and $\tilde{\varphi}$ an automorphism of $(\tilde{X}, \tilde{\sigma})$ such that $(\tilde{X}, \tilde{\varphi})$ is topologically conjugate to a topological Markov shift. Let (X, σ) be the topological Markov shift such that $\tilde{X} = \{s_X(x) | x \in X\}$ and $\tilde{\sigma}s_X = s_X\sigma$. The automorphism $\tilde{\varphi}$ of $(\tilde{X}, \tilde{\sigma})$ uniquely induces the automorphism φ of (X, σ) with $\tilde{\varphi}s_X = s_X\varphi$. Let T_φ be as in the proof of Proposition 3.7. Then T_φ is 1-1 and LL with $\varphi_{T_\varphi} = \varphi$. Since $(\tilde{X}_{T_\varphi}, \tilde{\varphi}_{T_\varphi})$ is topologically conjugate to a topological Markov shift, it follows from Theorem 2.11 (2) that T_φ^* is one-sided 1-1. Since T_φ^* is one-sided 1-1 and $\tilde{\xi}_{T_\varphi}$ and $\tilde{\eta}_{T_\varphi}$ are 1-1, it follows from Theorem 2.11 (3) that $(\tilde{X}_{T_\varphi}, \tilde{\varphi}_{T_\varphi})$ is topologically conjugate to $(X_{T_\varphi^*}, \sigma_{T_\varphi^*})$. Therefore $(X_{T_\varphi^*}, \sigma_{T_\varphi^*})$ is topologically conjugate to a topological Markov shift. Thus T_φ is a 1-1 and nondegenerate textile system such that $(X_{T_\varphi^*}, \sigma_{T_\varphi^*})$ is topologically conjugate to a topological Markov shift. By Proposition 2.9 there is (we can construct) a 1-1 nondegenerate textile system $T = (p, q : \Gamma \to G)$ such that T^* is nondegenerate, T and T_φ are topologically conjugate, and T^* and T_φ^* are also topologically conjugate. Since T^* and T_φ^* are topologically conjugate and T_φ^* is one-sided 1-1, T^* is one-sided 1-1.

Put $T_\varphi = (p_\varphi, q_\varphi : \Gamma_\varphi \to G_\varphi)$. Since T and T_φ are topologically conjugate and T and T_φ are nondegenerate, there are topological conjugacies $\psi : (X_{G_\varphi}, \sigma_{G_\varphi}) \to (X_G, \sigma_G)$ and $\Psi : (X_{\Gamma_\varphi}, \sigma_{\Gamma_\varphi}) \to (X_\Gamma, \sigma_\Gamma)$ such that $\psi\xi_{T_\varphi} = \xi_T\Psi$ and $\psi\eta_{T_\varphi} = \eta_T\Psi$. By construction, ξ_{T_φ} is a topological conjugacy which is a block map of $(0,0)$ type and $\xi_{T_\varphi}^{-1}$ is a block map of $(0, l)$ type for some $l \geq 0$ (see the proof of Proposition 3.7). Therefore it follows from the proofs of Proposition 2.9 and Lemma 2.8 (2) that ψ and Ψ can be given as block maps of $(0, nl)$ type for some $n \in \mathbf{N}$ such that ψ^{-1} and Ψ^{-1} are block maps of $(0,0)$ type. Thus we can define topological conjugacies $\tilde{\psi} : (\tilde{X}_{G_\varphi}, \tilde{\sigma}_{G_\varphi}) \to (\tilde{X}_G, \tilde{\sigma}_G)$ and $\tilde{\Psi} : (\tilde{X}_{\Gamma_\varphi}, \tilde{\sigma}_{\Gamma_\varphi}) \to (\tilde{X}_\Gamma, \tilde{\sigma}_\Gamma)$ such that $\tilde{\psi}s_{X_{G_\varphi}} = s_{X_G}\psi$ and $\tilde{\Psi}s_{X_{\Gamma_\varphi}} = s_{X_\Gamma}\Psi$, and we have

$$\tilde{\xi}_T = \tilde{\psi}\tilde{\xi}_{T_\varphi}\tilde{\Psi}^{-1} \quad \text{and} \quad \tilde{\eta}_T = \tilde{\psi}\tilde{\eta}_{T_\varphi}\tilde{\Psi}^{-1}.$$

Therefore, since $\tilde{\xi}_{T_\varphi}$ and $\tilde{\eta}_{T_\varphi}$ are 1-1, $\tilde{\xi}_T$ and $\tilde{\eta}_T$ are 1-1, so that T is 1-1 and LL by Lemma 3.6 (because T is nondegenerate). We also have

$$\tilde{\varphi}_T = \tilde{\eta}_T\tilde{\xi}_T^{-1} = \tilde{\psi}\tilde{\eta}_{T_\varphi}\tilde{\xi}_{T_\varphi}^{-1}\tilde{\psi}^{-1} = \tilde{\psi}\tilde{\varphi}_{T_\varphi}\tilde{\psi}^{-1} = \tilde{\psi}\tilde{\varphi}\tilde{\psi}^{-1}.$$

Thus T has all the required properties.

(3) Since $(X_T, \sigma_T) = (X_G, \sigma_G)$ for a nondegenerate textile system T over a graph G, (3) of the theorem directly follows from (3) of Theorem 2.11. □

Question 3.a. (1) Is there a positively expansive onto endomorphism φ of a topological Markov shift (X, σ) such that (X, φ) is not topologically conjugate to any one-sided topological Markov shift ?

(2) Is there an expansive automorphism $\tilde{\varphi}$ of a one-sided topological Markov shift $(\tilde{X}, \tilde{\sigma})$ such that $(\tilde{X}, \tilde{\varphi})$ is not topologically conjugate to any topological Markov shift ?

A graph G is said to be *column-reduced* if all columns of M_G are different.

Lemma 3.10. Let G be a column-reduced graph. Let $h : \Gamma \to G$ and $h' : \Gamma \to G$ be left-resolving graph-homomorphisms which give topological conjugacies of $(X_\Gamma, \sigma_\Gamma)$ onto (X_G, σ_G). Then, for $u, v \in V_\Gamma$, $h_V(u) = h_V(v)$ if and only if $h'_V(u) = h'_V(v)$.

Proof. Assume that there are $u, v \in V_\Gamma$ such that

$$h_V(u) = h_V(v) \quad \text{and} \quad h'_V(u) \neq h'_V(v).$$

Let E be the set of all arcs going to $h_V(u)$ in G. Since h is left resolving, for each $a \in E$, there is a unique arc $\alpha(a) \in A_\Gamma$ such that $t_\Gamma(\alpha(a)) = u$ and $h_A(\alpha(a)) = a$, and also there is a unique arc $\beta(a) \in A_\Gamma$ such that $t_\Gamma(\beta(a)) = v$ and $h_A(\beta(a)) = a$. Clearly $\{\alpha(a) \mid a \in E\}$ is the set of all arcs going to u in Γ, and $\{\beta(a) \mid a \in E\}$ is the set of all arcs going to v in Γ. Since h' is left resolving, $\{h'_A(\alpha(a)) \mid a \in E\}$ is the set of all arcs going to $h'_V(u)$ in G and $\{h'_A(\beta(a)) \mid a \in E\}$ is the set of all arcs going to $h'_V(v)$ in G. If

$$i_G(h'_A(\alpha(a))) = i_G(h'_A(\beta(a)))$$

for all $a \in E$, then the two columns of M_G corresponding to $h'_V(u)$ and $h'_V(v)$ would be the same, contrary to the hypothesis that G is column-reduced. Therefore there is $a_0 \in E$ such that

$$i_G(h'_A(\alpha(a_0))) \neq i_G(h'_A(\beta(a_0))).$$

Let $\alpha_0 = \alpha(a_0)$ and let $\beta_0 = \beta(a_0)$. Then

$$h_A(\alpha_0) = h_A(\beta_0), \quad t_\Gamma(\alpha_0) = u \quad \text{and} \quad t_\Gamma(\beta_0) = v.$$

Put

$$i_\Gamma(\alpha_0) = u_1 \quad \text{and} \quad i_\Gamma(\beta_0) = v_1.$$

Then

$$h_V(u_1) = h_V(v_1) \quad \text{but} \quad h'_V(u_1) \neq h'_V(v_1).$$

By the same argument as above, we see that there are $\alpha_1, \beta_1 \in A_\Gamma$ such that

$$h_A(\alpha_1) = h_A(\beta_1), \quad t_\Gamma(\alpha_1) = u_1 \quad \text{and} \quad t_\Gamma(\beta_1) = v_1$$

and if we put

$$i_\Gamma(\alpha_1) = u_2 \quad \text{and} \quad i_\Gamma(\beta_1) = v_2,$$

then

$$h_V(v_2) = h_V(u_2) \quad \text{but} \quad h'_V(u_2) \neq h'_V(v_2).$$

By continuation of this process, we have for all $l \in \mathbf{N}$, two paths

$$\alpha_l \alpha_{l-1} \cdots \alpha_0 \quad \text{and} \quad \beta_l \beta_{l-1} \cdots \beta_0, \quad \alpha_i, \beta_i \in A_\Gamma$$

such that

$$h_A(\alpha_l) h_A(\alpha_{l-1}) \cdots h_A(\alpha_0) = h_A(\beta_l) h_A(\beta_{l-1}) \cdots h_A(\beta_0)$$

but

$$i_\Gamma(\alpha_i) \neq i_\Gamma(\beta_i) \quad \text{for all } i = 0, \cdots, l.$$

This contradicts the assumption that h gives a topological conjugacy of $(X_\Gamma, \sigma_\Gamma)$ onto (X_G, σ_G).

Therefore we conclude that for $u, v \in A_\Gamma$ if $h_V(u) = h_V(v)$, then $h'_V(u) = h'_V(v)$. The converse also holds by symmetry. □

A continuous transformation $\tau : X \to X$ is said to be *topologically transitive* if there is some $x \in X$ with $\{\tau^n(x) \mid n \in \mathbf{Z}\}$ dense in X.

Lemma 3.11. Let $T = (p, q : \Gamma \to G)$ be a 1-1 and LL textile system such that $\sigma^n_{T^*}$ is topologically transitive for all $n \in \mathbf{N}$. Then there is a 1-1 and LL textile system $\bar{T} = (\bar{p}, \bar{q} : \Gamma \to G_0)$ such that (X_{G_0}, σ_{G_0}) is a full shift (i.e., G_0 has exactly one vertex), there is a topological conjugacy $\tilde{\psi} : (\tilde{X}_G, \tilde{\sigma}_G) \to (\tilde{X}_{G_0}, \tilde{\sigma}_{G_0})$ between one-sided subshifts with $\tilde{\varphi}_{\bar{T}} = \tilde{\psi} \tilde{\varphi}_T \tilde{\psi}^{-1}$, and $X_{\bar{T}^*} = X_{T^*}$.

Proof. As is well known, there are a column-reduced graph G_0 and a left-resolving graph-homomorphism $k : G \to G_0$ which gives a topological conjugacy, say ψ_k, of (X_G, σ_G) onto (X_{G_0}, σ_{G_0}). Define a textile system

$$\bar{T} = (\bar{p}, \bar{q} : \Gamma \to G_0)$$

with $\bar{p} = kp$ and $\bar{q} = kq$. Since T and $\bar{T}$ are LL, we have $(X_T, \sigma_T) = (X_G, \sigma_G)$ and $(X_{\bar{T}}, \sigma_{\bar{T}}) = (X_{G_0}, \sigma_{G_0})$ and we have

$$\xi_{\bar{T}} = \psi_k \xi_T \quad \text{and} \quad \eta_{\bar{T}} = \psi_k \eta_T.$$

Since T and $\bar{T}$ are 1-1 and LL, the automorphism $\tilde{\varphi}_T$ of the one-sided topological Markov shift $(\tilde{X}_G, \tilde{\sigma}_G)$ and the automorphism $\tilde{\varphi}_{\bar{T}}$ of the one-sided topological Markov shift $(\tilde{X}_{G_0}, \tilde{\sigma}_{G_0})$ can be defined by $\tilde{\varphi}_T = \tilde{\eta}_T \tilde{\xi}_T^{-1}$ and $\tilde{\varphi}_{\bar{T}} = \tilde{\eta}_{\bar{T}} \tilde{\xi}_{\bar{T}}^{-1}$. Since k is left resolving, ψ_k^{-1} is of $(0, l)$-type for some $l \geq 0$ so that ψ_k naturally induces the topological conjugacy

$\tilde{\psi}_k : (\tilde{X}_G, \tilde{\sigma}_G) \to (\tilde{X}_{G_0}, \tilde{\sigma}_{G_0})$ with $\tilde{\psi}_k s_{X_G} = s_{X_{G_0}} \psi_k$. We have $\tilde{\xi}_{\bar{T}} = \tilde{\psi}_k \tilde{\xi}_T$ and $\tilde{\eta}_{\bar{T}} = \tilde{\psi}_k \tilde{\eta}_T$ so that

$$\tilde{\varphi}_{\bar{T}} = \tilde{\psi} \tilde{\varphi}_T \tilde{\psi}^{-1},$$

where $\tilde{\psi} = \tilde{\psi}_k$.

We see that X_{T^*} and $X_{\bar{T}^*}$ give the same subset of the set X_{V_Γ} of bisequences over V_Γ. For we can show that for any two dimensional configuration $(\alpha_{ij})_{i,j\in\mathbf{Z}}$ with $\alpha_{i,j} \in A_\Gamma$, $(\alpha_{ij})_{i,j\in\mathbf{Z}} \in U_T$ if and only if $(\alpha_{ij})_{i,j\in\mathbf{Z}} \in U_{\bar{T}}$. In fact, suppose that $(\alpha_{ij})_{i,j\in\mathbf{Z}} \in U_T$. Then

$$t_\Gamma(\alpha_{i,j-1}) = i_\Gamma(\alpha_{ij}) \quad \text{and} \quad q(\alpha_{i-1,j}) = p(\alpha_{ij}) \quad \text{for all } i,j \in \mathbf{Z}$$

so that

$$t_\Gamma(\alpha_{i,j-1}) = i_\Gamma(\alpha_{ij}) \quad \text{and} \quad \bar{q}(\alpha_{i-1,j}) = \bar{p}(\alpha_{ij}) \quad \text{for all } i,j \in \mathbf{Z}.$$

This implies that $(\alpha_{ij})_{i,j\in\mathbf{Z}} \in U_{\bar{T}}$. Conversely suppose that $(\alpha_{ij})_{i,j\in\mathbf{Z}} \in U_{\bar{T}}$. Then, we have

$$t_\Gamma(\alpha_{i,j-1}) = i_\Gamma(\alpha_{ij}) \quad \text{for all } \ i,j \in \mathbf{Z}$$

and

$$\eta_{\bar{T}}((\alpha_{i-1,j})_{j\in\mathbf{Z}}) = \xi_{\bar{T}}((\alpha_{ij})_{j\in\mathbf{Z}}) \quad \text{for all } \ i \in \mathbf{Z},$$

so that we have

$$\eta_T((\alpha_{i-1,j})_{j\in\mathbf{Z}}) = \xi_T((\alpha_{ij}))_{j\in\mathbf{Z}} \quad \text{for all } \ i \in \mathbf{Z},$$

because ψ_k is 1-1. Therefore $(\alpha_{ij})_{j\in\mathbf{Z}} \in U_T$.

Since $\sigma^n_{T^*}$ is topologically transitive for all $n \in \mathbf{N}$, $\sigma^n_{\bar{T}^*}$ is topologically transitive for all $n \in \mathbf{N}$. Let $\bar{T}^* = (\bar{p}^{\bar{T}}, \bar{q}^{\bar{T}} : \Gamma^{\bar{T}} \to G_0^{\bar{T}})$. Since G_0 is column-reduced and $\bar{p}$ and $\bar{q}$ are left resolving, it follows from Lemma 3.10 that $G_0^{\bar{T}}$ has the property that for any $a, b \in A_{G_0^{\bar{T}}}$,

$$i_{G_0^{\bar{T}}}(a) = i_{G_0^{\bar{T}}}(b) \quad \Leftrightarrow \quad t_{G_0^{\bar{T}}}(a) = t_{G_0^{\bar{T}}}(b).$$

Therefore the nondegenerate component (the maximal nondegenerate subgraph) of $G_0^{\bar{T}}$ is the union of the irreducible components, say $H_1, \cdots, H_m$, of $G_0^{\bar{T}}$ with $m \geq 1$. Hence $X_{G_0^{\bar{T}}} = X_{H_1} \cup \cdots \cup X_{H_m}$. If $X_{\bar{T}^*}$ had nonempty intersections with different X_{H_i} and X_{H_j}, then $\sigma_{\bar{T}^*}$ could not be topologically transitive. Hence we may assume that $X_{\bar{T}^*} \subset X_{H_1}$. If H_1 had period $n > 1$, then $\sigma^n_{\bar{T}^*}$ would not be topologically transitive. Therefore H_1 must be irreducible and aperiodic. By the property stated above, for any $a, b \in A_{H_1}$,

$$i_{H_1}(a) = i_{H_1}(b) \quad \Leftrightarrow \quad t_{H_1}(a) = t_{H_1}(b).$$

Thus H_1 must have exactly one vertex, say v. This implies that for any textile weaved by $\bar{T}^*$, v is the only vertex that appears on it as a vertex in $V_{G_0^{\bar{T}}}$. Alternatively, for any textile weaved by $\bar{T}$, v is the only vertex that appears on it as a vertex in V_{G_0}. Hence for every $x \in X_{\bar{T}}$, v is the only vertex that appears on x as a vertex in V_{G_0}. Since $(X_{\bar{T}}, \sigma_{\bar{T}}) = (X_{G_0}, \sigma_{G_0})$, $V_{G_0} = \{v\}$ so that (X_{G_0}, σ_{G_0}) is a full shift. $\square$

We say that a topological Markov shift is *aperiodic* if its defining matrix is aperiodic.

Theorem 3.12. (1) If (X,σ) is an irreducible and aperiodic topological Markov shift having more than one point and if φ is an onto endomorphism of (X,σ) such that (X,φ) is topologically conjugate to a one-sided topological Markov shift, then φ is N-to-one for some $N \geq 3$ and (X,φ) is topologically conjugate to the one-sided full N-shift.

(2) If $(\tilde{X},\tilde{\sigma})$ is a one-sided topological Markov shift having more than one point and having an expansive automorphism $\tilde{\varphi}$ such that $\tilde{\varphi}^n$ is topologically transitive for all $n \in \mathbf{N}$, then $(\tilde{X},\tilde{\sigma})$ is topologically conjugate to the one-sided full N-shift for some $N \geq 3$.

Proof. First we prove (2). Assume that $(\tilde{X},\tilde{\sigma})$ is a one-sided topological Markov shift with more than one point and assume that $\tilde{\varphi}$ is an expansive automorphism of $(\tilde{X},\tilde{\sigma})$ such that $\tilde{\varphi}^n$ is topologically transitive for all $n \in \mathbf{N}$. Let (X,σ) be the topological Markov shift such that $\tilde{X} = \{s_X(x) \mid x \in X\}$ and $\tilde{\sigma}s_X = s_X\sigma$. The automorphism $\tilde{\varphi}$ of $(\tilde{X},\tilde{\sigma})$ naturally induces the automorphism φ of (X,σ) with $\tilde{\varphi}s_X = s_X\varphi$. Let $T = T_\varphi$, where T_φ is as in the proof of Proposition 3.7. Then T is 1-1 and LL with $\tilde{\varphi}_T = \tilde{\varphi}$. Since $\tilde{\varphi}$ is expansive, it follows from Theorem 2.11 that T^* is one-sided 1-1 and $(\tilde{X},\tilde{\varphi})$ is topologically conjugate to (X_{T^*},σ_{T^*}). Since $\tilde{\varphi}^n$ is topologically transitive for all $n \in \mathbf{N}$, $\sigma_{T^*}^n$ is topologically transitive for all $n \in \mathbf{N}$. Therefore, by Lemma 3.11, there are a 1-1 and LL textile system $\bar{T} = (\bar{p},\bar{q} : \Gamma \to G_0)$ such that (X_{G_0},σ_{G_0}) is the full N-shift for some $N \in \mathbf{N}$, and a topological conjugacy $\tilde{\psi} : (\tilde{X},\tilde{\sigma}) \to (\tilde{X}_{G_0},\tilde{\sigma}_{G_0})$ with $\tilde{\varphi}_{\bar{T}} = \tilde{\psi}\tilde{\varphi}\tilde{\psi}^{-1}$. Thus $(\tilde{X},\tilde{\sigma})$ is topologically conjugate to the one-sided full N-shift.

Clearly $N > 1$. Assume that $N = 2$. Then since an LL endomorphism is a "leftmost-permutive" block-map (see the paragraph following Question 3.c below), it follows from Theorem 6.9 of [H] that $\varphi_{\bar{T}}^2$ is the identity map on X_{G_0}, which contradicts our hypothesis. Thus we conclude that $N \geq 3$.

The proof of (2) is completed.

To prove (1), assume that (X,σ) is an irreducible and aperiodic topological Markov shift with $h(\sigma) > 0$ and that φ is an onto endomorphism of (X,σ) such that (X,φ) is topologically conjugate to a one-sided topological Markov shift. By Theorem 3.9 (1), there is a one-sided 1-1 and nondegenerate textile system $T = (p,q : \Gamma \to G)$ such that q is biresolving, T^* is 1-1 and LL and there is a topological conjugacy $\psi : (X,\sigma) \to (X_G,\sigma_G)$ such that $\varphi_T = \psi\varphi\psi^{-1}$. Since (X,σ) is an irreducible topological Markov shift, so is (X_G,σ_G). Since q is biresolving, φ_T is N-to-one for some $N \in \mathbf{N}$, and q_V is also N-to-one (by Proposition 2.2 of [N1]). Let $T^* = (p^T,q^T : \Gamma^T \to G^T)$. Since q_V is N-to-one, it follows that every vertex of G^T has exactly N arcs ending in itself so that $h(\sigma_{G^T}) = \log N$.

It also follows from Theorem 3.9 (3) that (X_T,φ_T) is topologically conjugate to $(\tilde{X}_{G^T},\tilde{\sigma}_{G^T})$, $\tilde{\varphi}_{T^*}$ is an automorphism of $(\tilde{X}_{G^T},\tilde{\sigma}_{G^T})$, and $(\tilde{X}_{G^T},\tilde{\varphi}_{T^*})$ is topologically conjugate to (X_G, σ_G), which is an irreducible and aperiodic with $h(\sigma_G) > 0$ because (X, σ) is an irreducible and aperiodic topological Markov shift with $h(\sigma) > 0$. Hence $\tilde{\varphi}^n_{T^*}$ is topologically transitive for all $n \in \mathbf{N}$ and $h(\tilde{\varphi}_{T^*}) > 0$. Since $h(\sigma_{G^T}) = \log N$, it follows from (2) proved above that $(\tilde{X}_{G^T}, \tilde{\sigma}_{G^T})$ is topologically conjugate to the one-sided full N-shift with $N \geq 3$. Since (X, φ) is topologically conjugate to $(\tilde{X}_{G^T}, \tilde{\sigma}_{G^T})$, (1) is proved. □

Trow showed in [T1](see also [T2]) that there is a strong restriction for an irreducible topological Markov shift having a constant-to-one endomorphism. But we have not yet had a clear grasp on the irreducible topological Markov shifts (X, σ) having an onto endomorphism φ such that (X, φ) is topologically conjugate to a one-sided topological Markov shift.

Question 3.b. If (X, σ) is an irreducible and aperiodic topological Markov shift having an onto endomorphism φ such that (X, φ) is topologically conjugate to a one-sided full shift, then is (X, σ) topologically conjugate to a full shift ? (If $\tilde{\varphi}$ is an automorphism of a one-sided full shift $(\tilde{X}, \tilde{\sigma})$ such that $(\tilde{X}, \tilde{\varphi})$ is topologically conjugate to an irreducible and aperiodic topological Markov shift, then is $(\tilde{X}, \tilde{\varphi})$ topologically conjugate to a full-shift ?)

Theorem 3.13. Let $(\tilde{X}, \tilde{\sigma})$ be a one-sided topological Markov shift and $\tilde{\varphi}$ an onto endomorphism of $(\tilde{X}, \tilde{\sigma})$ such that $(\tilde{X}, \tilde{\varphi})$ is topologically conjugate to a one-sided topological Markov shift. Then there is (we can construct) a one-sided 1-1 and LR textile system $T = (p, q : \Gamma \to G)$ such that there is a topological conjugacy $\tilde{\psi} : (\tilde{X}, \tilde{\sigma}) \to (\tilde{X}_G, \tilde{\sigma}_G)$ with $\tilde{\varphi}_T = \tilde{\psi}\tilde{\varphi}\tilde{\psi}^{-1}$, $T^* = (p^T, q^T : \Gamma^T \to G^T)$ is one-sided 1-1, and $(\tilde{X}, \tilde{\varphi})$ is topologically conjugate to $(\tilde{X}_{G^T}, \tilde{\sigma}_{G^T})$.

Proof. Let (X, σ) be the topological Markov shift such that $\{s_X(x) \mid x \in X\} = \tilde{X}$ and $s_X\sigma = \tilde{\sigma}s_X$. Then the onto endomorphism $\tilde{\varphi}$ of $(\tilde{X}, \tilde{\sigma})$ induces the onto endomorphism φ of (X, σ) with $\tilde{\varphi}s_X = s_X\varphi$, and φ is a block map of $(0, l)$ type for some nonnegative integer l. Let $T_\varphi = (p_\varphi, q_\varphi : \Gamma_\varphi \to G_\varphi)$ be as in Fact 2.2. Then T_φ is one-sided 1-1 and nondegenerate, $\tilde{\xi}_{T_\varphi}$ is 1-1, and $\tilde{\varphi}_{T_\varphi} = \tilde{\varphi}$. Since $(\tilde{X}_{T_\varphi}, \tilde{\varphi}_{T_\varphi})$ is topologically conjugate to a one-sided topological Markov shift, it follows from Theorem 2.12 that $\tilde{\xi}_{T^*_\varphi}$ is 1-1 and $(\tilde{X}_{T_\varphi}, \tilde{\varphi}_{T_\varphi})$ is topologically conjugate to $(\tilde{X}_{T^*_\varphi}, \tilde{\sigma}_{T^*_\varphi})$, so that this is topologically conjugate to a one-sided topological Markov shift. Since T_φ is one-sided 1-1 and nondegenerate and $(X_{T^*_\varphi}, \sigma_{T^*_\varphi})$ is topologically conjugate to a topological Markov shift, it follows from Proposition 2.9 that there is (we can construct) a one-sided 1-1 nondegenerate textile system $T = (p, q : \Gamma \to G)$ with $T^* = (p^T, q^T : \Gamma^T \to G^T)$ such that T^* is nondegenerate and T and T_φ are topologically conjugate. By a similar argument to that in (2) of the proof of Theorem 3.9, we see that $\tilde{\xi}_T$ is 1-1 and there is a topological conjugacy $\tilde{\psi} : (\tilde{X}_{G_\varphi}, \tilde{\sigma}_{G_\varphi}) \to (\tilde{X}_G, \tilde{\sigma}_G)$ such that $\tilde{\varphi}_T = \tilde{\psi}\tilde{\varphi}\tilde{\psi}^{-1}$. Since $\tilde{\varphi}$ is positively expansive, so is

$\tilde{\varphi}_T$. Therefore it follows from Theorem 2.12 that $\tilde{\xi}_{T^*}$ is 1-1 and $(\tilde{X}_T, \tilde{\varphi}_T)$ is topologically conjugate to $(\tilde{X}_{T^*}, \tilde{\sigma}_{T^*})$, which is equal to $(\tilde{X}_{G^T}, \tilde{\sigma}_{G^T})$ because T^* is nondegenerate. Thus $(\tilde{X}, \tilde{\varphi})$ is topologically conjugate to $(\tilde{X}_{G^T}, \tilde{\sigma}_{G^T})$. Since T is nondegenerate and $\tilde{\xi}_T$ is 1-1, it follows from Lemma 3.6 that T is one-sided 1-1 and p is left resolving. Since T^* is nondegenerate and $\tilde{\xi}_{T^*}$ is 1-1, it also follows from Lemma 3.6 that T^* is one-sided 1-1 and p^T is left resolving. Since p^T is left resolving, q is right resolving, by Proposition 3.1. Thus T is one-sided 1-1 and LR and T^* is one-sided 1-1. □

Question 3.c. Is there a positively expansive onto endomorphism $\tilde{\varphi}$ of a one-sided topological Markov shift $(\tilde{X}, \tilde{\sigma})$ such that $(\tilde{X}, \tilde{\varphi})$ is not topologically conjugate to any one-sided topological Markov shift ?

Let (X_G, σ_G) be a topological Markov shift, where G is a graph. Let k and l be nonnegative integers. Let $\varphi : X_G \to X_G$ be a block map of (k, l) type and assume that it is given by a mapping $f : L_{k+l+1}(G) \to A_G$. Let $T_\varphi = (p, q : \Gamma \to G)$ be the textile system as in Fact 2.2. We say that φ is *leftmost-permutive* if q is left resolving, we say that φ is *rightmost-permutive* if q is right resolving, and we say that φ is *bipermutive* if q is biresolving. In particular, if G is a graph with only one vertex, then (X_G, σ_G) is the full shift (X_A, σ_A) over the alphabet $A = A_G$. In this case, φ is leftmost-permutive if $f(aa_1 \cdots a_{k+l}) \neq f(a'a_1 \cdots a_{k+l})$ for any $a, a', a_1, \cdots, a_{k+l} \in A$ with $a \neq a'$, φ is rightmost-permutive if $f(a_1 \cdots a_{k+l}a) \neq f(a_1 \cdots a_{k+l}a')$ for any $a_1, \cdots, a_{k+l}, a, a' \in A$ with $a \neq a'$, and φ is bipermutive if φ is both leftmost-permutive and rightmost-permutive. If φ is a block map of $(0, l)$ type, then $\tilde{\varphi} : \tilde{X}_G \to \tilde{X}_G$ such that $\tilde{\varphi} s_{X_G} = s_{X_G} \varphi$, can be defined.

Proposition 3.14. Let (X, σ) be a topological Markov shift with $h(\sigma) > 0$. Let k and l be nonnegative integers. Let $\varphi : X \to X$ be a bipermutive block map of (k, l) type. Then if $k > 0$ and $l > 0$, then (X, φ) is topologically conjugate to a one-sided topological Markov shift, and if either $k = 0$ or $l = 0$, then (X, φ) is not expansive. In particular, if (X, σ) is a full-shift, $k > 0$, and $l > 0$, then (X, φ) is topologically conjugate to the one-sided full shift $(\tilde{X}, \tilde{\sigma}^{k+l})$.

Proof. Let G be a graph such that $(X_G, \sigma_G) = (X, \sigma)$. Let T_φ be as in Fact 2.2 and put $T_\varphi = T = (p, q : \Gamma \to G)$. Then T is one-sided 1-1 and nondegenerate, $\varphi_T = \varphi$, and q is biresolving. Hence T^* is LL so that T^* is nondegenerate.

Assume that $k > 0$ and $l > 0$. Then for $\alpha \in A_\Gamma$, $i_\Gamma(\alpha)$ uniquely determines $p(\alpha)$. Therefore, using the fact that q is right resolving, we see that for $\tilde{x} \in \tilde{X}_{T^*}$, the element $\tilde{z} \in \tilde{Z}_{T^*}$ such that $\tilde{\xi}_{T^*}(\tilde{z}) = \tilde{x}$ is uniquely determined by $\tilde{x}$. Thus, $\tilde{\xi}_{T^*}$ is 1-1. Similarly for $\alpha \in A_\Gamma$, $t_\Gamma(\alpha)$ uniquely determines $p(\alpha)$. Using this and the fact that q is left resolving, we see that $\tilde{\eta}_{T^*}$ is 1-1. Therefore by Theorem 2.11, $(X, \varphi) = (X_T, \varphi_T)$ is topologically conjugate to $(\tilde{X}_{T^*}, \tilde{\sigma}_{T^*})$. Since T^* is nondegenerate, $(\tilde{X}_{T^*}, \tilde{\sigma}_{T^*}) = (\tilde{X}_{G^T}, \tilde{\sigma}_{G^T})$. Thus (X, φ) is topologically conjugate to a one-sided topological Markov shift. If $(X, \sigma) = (X_A, \sigma_A)$,where A is an alphabet, then it follows that $(\tilde{X}_{G^T}, \tilde{\sigma}_{G^T}) = (\tilde{X}_{A^{k+l}}, \tilde{\sigma}_{A^{k+l}})$, which is topologically conjugate to $(\tilde{X}, \tilde{\sigma}^{k+l})$.

Assume that $k = 0$. Then p is left resolving. Therefore T is LL so that q^T is biresolving. Since $h(\sigma_G) > 0$, G has distinct two arcs b and b' with $t_G(b) = t_G(b')$. Since q^T is biresolving, it follows that there are two bisequences z and z' in X_{Γ^T} such that $\eta_{T^*}(z) = \eta_{T^*}(z')$, z passes through $b \in V_{\Gamma_T}$ and z' passes through $b' \in V_{\Gamma_T}$. Therefore η_{T^*} is not 1-1 so that $\varphi = \varphi_T$ is not expansive, by Theorem 2.5.

Assume that $l = 0$. Then p is right resolving. Therefore T is RR so that p^T is biresolving. This and the fact that G has distinct two arcs a and a' with $i_G(a) = i_G(a')$ (because $h(\sigma_G) > 0$), similarly show that ξ_{T^*} is not 1-1, so that φ is not expansive, by Theorem 2.5. □

The last statement of the above proposition has been independently proved by M.A. Shereshevsky and V.S. Afraimovich [SA]. Although the above proposition contains a statement about bipermutive endomorphisms of general topological Markov shifts, no example is known for a bipermutive endomorphism of an irreducible and aperiodic topological Markov shift which is not topologically conjugate to any full-shift. (Cf. Question 3.b.)

Proposition 3.15. Let (X, σ) be a topological Markov shift. Let $l \in \mathbf{N}$. Let $\varphi : X \to X$ be a rightmost permutive block map of $(0, l)$ type, then $(\tilde{X}, \tilde{\varphi})$ is topologically conjugate to a one-sided topological Markov shift. In particular, if (X, σ) is a full shift, then $(\tilde{X}, \tilde{\varphi})$ is topologically conjugate to $(\tilde{X}, \tilde{\sigma}^l)$ (Coven [C]).

Proof. Let G be a graph such that $(X_G, \sigma_G) = (X, \sigma)$. Let T_φ be as in Fact 2.2, and put $T_\varphi = T = (p, q : \Gamma \to G)$. Then T is one-sided 1-1 and LR, $\tilde{\xi}_T$ is 1-1 and $\tilde{\varphi}_T = \tilde{\varphi}$. Since T^* is LR, T^* is nondegenerate. It follows that for $\alpha \in A_\Gamma$, $i_\Gamma(\alpha)$ uniquely determines $p(\alpha)$. Therefore, since q is right resolving, it follows that $\tilde{\xi}_{T^*}$ is 1-1. Since T^* is nondegenerate, it follows from Theorem 2.12 that $(\tilde{X}, \tilde{\varphi}) = (\tilde{X}_T, \tilde{\varphi}_T)$ is topologically conjugate to $(\tilde{X}_{G^T}, \tilde{\sigma}_{G^T})$, where $T^* = (p^T, q^T : \Gamma^T \to G^T)$. In particular, if $(X, \sigma) = (X_A, \sigma_A)$, where A is an alphabet, then $(\tilde{X}_{G^T}, \tilde{\sigma}_{G^T}) = (\tilde{X}_{A^l}, \tilde{\sigma}_{A^l})$. □

Let (X, σ) and (X', σ') be irreducible topological Markov shifts with $h(\sigma) = h(\sigma')$. Let $\phi : (X, \sigma) \to (X', \sigma')$ be an onto factor map. Then there is $d \in \mathbf{N}$ such that $\#\phi^{-1}(y) = d$ for any bilaterally transitive point y in X' ([H],[CP]). This number d is called the *degree* of ϕ and denoted by $\deg(\phi)$ (see [B2]).

For an onto endomorphism φ of an irreducible topological Markov shift (X, σ) with $h(\sigma) = \log \lambda$, Boyle considered in [B2] positive numbers $L(\varphi)$ and $R(\varphi)$ defined by using the natural measures on unstable and stable sets, which are in the ring generated by $1/\lambda$ and the algebraic integers of $\boldsymbol{Q}[\lambda]$ and have the following properties:

$$L(\varphi)R(\varphi)\deg(\varphi) = 1;$$

$L(\varphi)$ and $R(\varphi)$ are invariants of topological conjugacies for endomorphisms (that is, $L(\varphi) = L(\psi\varphi\psi^{-1})$ and $R(\varphi) = R(\psi\varphi\psi^{-1})$ for any topological conjugacy $\psi : (X,\sigma) \to (X',\sigma'))$; for any onto endomorphisms φ_1 and φ_2 of (X,σ)

$$L(\varphi_2\varphi_1) = L(\varphi_1)L(\varphi_2) \quad \text{and} \quad R(\varphi_2\varphi_1) = R(\varphi_1)R(\varphi_2).$$

(Actually, Boyle considered $L(\varphi)$ and $R(\varphi)$ more generally for endomorphisms φ of irreducible sofic systems in [B2]). Let us call $L(\varphi)$ and $R(\varphi)$ the *L-number* of φ and the *R-number* of φ, respectively. We shall define similar numbers for textile systems. For our purpose, we shall adopt a different approach.

Let $h : \Gamma \to G$ be a graph-homomorphism. For $U \subset V_\Gamma$ and $w \in L(G)$, define

$$S_+(U,w) = \{t_\Gamma(\alpha_n) | \exists\alpha_1\cdots\alpha_n \in L(\Gamma), \alpha_i \in A_\Gamma, i_\Gamma(\alpha_1) \in U \text{ and } h_A(\alpha_1)\cdots h_A(\alpha_n) = w\}$$

if $w \in L_n(G)$ with $n \in \mathbf{N}$; if $w \in V_G$, then

$$S_+(U,w) = h_V^{-1}(w) \cap U.$$

We also define $S_-(w,U)$ for $U \subset V_\Gamma$ and $w \in L(G)$ by

$$S_-(w,U) = \{i_\Gamma(\alpha_1) | \exists\alpha_1\cdots\alpha_n \in L(\Gamma), \alpha_i \in A_\Gamma, t_\Gamma(\alpha_n) \in U \text{ and } h_A(\alpha_1)\cdots h_A(\alpha_n) = w\}$$

if $w \in L_n(G)$ with $n \in \mathbf{N}$; if $w \in V_G$, then $S_-(w,U)$ is the same as $S_+(U,w)$. We call $S_+(U,w)$ the *w-successor* of U or a *successor* of U and $S_-(U,w)$ the *w-predecessor* of U or a *predecessor* of U. A *right-compatible set* (for h) is defined to be a nonempty successor of a set consisting of exactly one vertex in V_Γ. A *left-compatible set* is defined to be a nonempty predecessor of a set consisting of exactly one vertex in V_Γ. Let $\phi : (X_\Gamma, \sigma_\Gamma) \to (X_G, \sigma_G)$ be the factor map given by h. Assume that M_Γ and M_G are irreducible, they have the same maximal eigenvalue, say λ, and ϕ is onto. Then any successor of a maximal right-compatible set is also a maximal right-compatible set and any predecessor of a maximal left-compatible set is also a maximal left-compatible set [N1]. Let l and r be respectively left and right eigenvectors of M_Γ corresponding to λ. Let l_0 and r_0 be respectively left and right eigenvectors of M_G corresponding to λ. For $v \in V_\Gamma$, let $l(v)$ and $r(v)$ denote the components of l and r corresponding to v. For a subset U of V_Γ, let

$$l(U) = \sum_{u \in U} l(u)$$

and let $r(U)$ be defined similarly. For a subset U_0 of V_G, let $l_0(U_0)$ and $r_0(U_0)$ be defined similarly. By a discussion analogous to that in [B2], we see that

$$\frac{l(U_1)}{l_0(h_V(U_1))} = \frac{l(U_2)}{l_0(h_V(U_2))}$$

for any two maximal left-compatible sets U_1 and U_2 for h and

$$\frac{r(V_1)}{r_0(h_V(V_1))} = \frac{r(V_2)}{r_0(h_V(V_2))}$$

for any two maximal right-compatible sets V_1 and V_2 for h. Hence we define

$$L(h, l, l_0) = \frac{l(U)}{l_0(h_V(U))} \quad \text{and} \quad R(h, r, r_0) = \frac{r(V)}{r_0(h_V(V))},$$

where U is a maximal left-compatible set for h and V is a maximal right-compatible set for h. Corresponding to the theorem of L.R. Welch (Theorem 14.9 of [H]), we have

$$\frac{lr}{l_0 r_0} = L(h, l, l_0)R(h, r, r_0) \deg \phi$$

by discussions analogous to those in Section 14 of [H].

Let $T = (p, q : \Gamma \to G)$ be a nondegenerate textile system such that M_Γ and M_G are irreducible and have the same maximal eigenvalue, say λ. We define the *L-number* of T, the *R-number* of T, and the *degree* of T by

$$L(T) = \frac{L(q, l, l_0)}{L(p, l, l_0)}, \quad R(T) = \frac{R(q, r, r_0)}{R(p, r, r_0)},$$

and

$$\deg(T) = \frac{\deg(q)}{\deg(p)},$$

respectively, where l and r are left and right eigenvectors of M_Γ corresponding to λ and l_0 and r_0 are left and right eigenvectors of M_G corresponding to λ. Clearly, $L(T)$ and $R(T)$ as well as $\deg(T)$ are independent of choice of l and r as well as l_0 and r_0. We have

$$L(T)R(T) \deg(T) = 1.$$

We shall see in Proposition 5.4 that $L(T)$ and $R(T)$ as well as $\deg(T)$ are invariants of topological conjugacy of textile systems.

If T is one-sided 1-1, then

$$L(T) = L(\varphi_T), \quad R(T) = R(\varphi_T)$$

and

$$\deg(T) = \deg(\varphi_T),$$

where L and R of endomorphisms are used in Boyle's sense; the reader is referred to [B2] for the definitions. In fact, generally for an onto endomorphism φ of (X_G, σ_G), we see, by direct computation, that

$$L(\varphi) = L(T_\varphi) \quad \text{and} \quad R(\varphi) = R(T_\varphi),$$

where T_φ is defined as in Fact 2.2. Therefore $L(\varphi_T) = L(T_{\varphi_T})$ and $R(\varphi_T) = R(T_{\varphi_T})$. It is easy to see that T and T_{φ_T} are strongly topologically conjugate. Therefore $L(T) = L(T_{\varphi_T})$ and $R(T) = R(T_{\varphi_T})$ so that $L(T) = L(\varphi_T)$ and $R(T) = R(\varphi_T)$. It is clear that $\deg(T) = \deg(\varphi_T)$.

Moreover, as we shall see in Theorem 6.31 that if T is one-sided 1-1 and LR, then

$$h(\varphi_T) = -\log R(T)$$

and if T is one-sided 1-1 and RL, then

$$h(\varphi_T) = -\log L(T).$$

Therefore, if φ is a rightmost-permutive endomorphism of $(0, n)$ type of an irreducible topological Markov shift with $n \geq 0$, then

$$h(\tilde{\varphi}) = h(\varphi) = -\log R(\varphi),$$

and if φ is a leftmost-permutive endomorphism of $(m, 0)$ type of an irreducible topological Markov shift with $m \geq 0$, then

$$h(\varphi) = -\log L(\varphi).$$

In Section 6, we shall study LR textile systems in more detail. In closing this section, we give some remarks on the properties of resolving textile systems concerning the operations introduced in Section 2.

Let $T = (p, q : \Gamma \to G)$ be a textile system. Let $n \in \mathbf{N}$. Let $T^{[n]} = (p^{[n]}, q^{[n]} : \Gamma^{[n]} \to G^{[n]})$ and let $T^{(n)} = (p^{(n)}, q^{(n)} : \Gamma^n \to G^n)$. As is straightforwardly seen, if p is left resolving, then so are $p^{[n]}$ and $p^{(n)}$, and if p is right resolving, then so are $p^{[n]}$ and $p^{(n)}$. Of course, the "q-versions" of these properties also hold. It is also straightforward to see:

Fact 3.16. Let $T_1 = (p_1, q_1 : \Gamma_1 \to G)$ and $T_2 = (p_2, q_2 : \Gamma_2 \to G)$ be textile systems and let $T_1 \circ T_2 = (p_{12}, q_{12} : \Gamma_{12} \to G)$. If both p_1 and p_2 are left resolving, then so is p_{12}, and if both p_1 and p_2 are right resolving, then so is p_{12}. The "q-versions" of these statements also hold.

Hence, we have:

Corollary 3.17. Let $T_i = (p_i, q_i : \Gamma_i \to G)$ be a textile system for $i = 1, 2$. Let $T_1 \circ T_2 = (p_{12}, q_{12} : \Gamma_{12} \to G)$. Then the following statements are valid.

(1) If both T_1 and T_2 are LR, then so is $T_1 \circ T_2$.

(2) If both T_1 and T_2 are LL, then so is $T_1 \circ T_2$.

(3) If both q_1 and q_2 are biresolving, then so is q_{12}.

If T_1 and T_2 are one-sided 1-1 and nondegenerate textile systems over the same graph, then

$$\varphi_{T_1 \circ T_2} = \varphi_{T_2} \varphi_{T_1},$$

so that we have:

Corollary 3.18. Let φ_1 and φ_2 be endomorphisms of a topological Markov shift (X_G, σ_G). Then the following statements are valid.

(1) If both φ_1 and φ_2 are LR, then so is $\varphi_2 \varphi_1$.

(2) If both φ_1 and φ_2 are LL, then so is $\varphi_2 \varphi_1$.

(3) If there is a one-sided 1-1 and nondegenerate textile system $T_i = (p_i, q_i : \Gamma_i \to G)$ with q_i biresolving and $\varphi_{T_i} = \varphi_i$, for $i = 1, 2$, then there is a one-sided 1-1 and nondegenerate textile system $T = (p, q : \Gamma \to G)$ with q biresolving and $\varphi_T = \varphi_2 \varphi_1$.

4. SOFIC TEXTILE SYSTEMS

Let $\mathcal{G} = (G, \lambda)$ be a λ-graph over an alphabet A. The sofic cover defined by $\mathcal{G}$ will be denoted by $\pi_{\mathcal{G}}$. Let $n \in \mathbf{N}$. Let $\lambda^{[n]}$ denote the mapping of $A_{G^{[n]}} = L_n(G)$ onto the set

$$\{\lambda(a_1)\cdots\lambda(a_n) \mid a_1\cdots a_n \in L_n(G), a_i \in A_G\}$$

with $\lambda^{[n]}(a_1\cdots a_n) = \lambda(a_1)\cdots\lambda(a_n)$. The *higher block λ-graph* $\mathcal{G}^{[n]}$ *of order* n of $\mathcal{G}$ is defined by

$$\mathcal{G}^{[n]} = (G^{[n]}, \lambda^{[n]}).$$

A *sofic textile system* $\mathcal{T}$ is a textile system $T = (p, q : \Gamma \to G)$ combined with two onto mappings (labelings) $\lambda : A_G \to A$ and $\mu : V_\Gamma \to B$, where A and B are alphabets. The textile system T is called the *support* of $\mathcal{T}$. Write

$$\mathcal{T} = (T, \lambda : A_G \to A, \mu : V_\Gamma \to B)$$

with $T = (p, q : \Gamma \to G)$. For each $\alpha \in A_\Gamma$, let

$$\nu(\alpha) = (\mu(i_\Gamma(\alpha)), \mu(t_\Gamma(\alpha)), \lambda(p(\alpha)), \lambda(q(\alpha)))$$

and let

$$E_{\mathcal{T}} = \{\nu(\alpha) \mid \alpha \in A_\Gamma\}.$$

Let $j_{left} : E_{\mathcal{T}} \to B, j_{right} : E_{\mathcal{T}} \to B, j_{upper} : E_{\mathcal{T}} \to A$, and $j_{lower} : E_{\mathcal{T}} \to A$ be the projections with

$$j_{left}(\nu(\alpha)) = \mu(i_\Gamma(\alpha)), \quad j_{right}(\nu(\alpha)) = \mu(t_\Gamma(\alpha))$$
$$j_{upper}(\nu(\alpha)) = \lambda(p(\alpha)), \quad j_{lower}(\nu(\alpha)) = \lambda(q(\alpha))$$

for $\alpha \in A_\Gamma$. Let $\mathcal{G}$ and $\mathcal{H}$ be the λ-graphs defined by

$$\mathcal{G} = (G, \lambda) \quad \text{and} \quad \mathcal{H} = (\Gamma, \nu).$$

Let

$$X_{\mathcal{T}} = \pi_{\mathcal{G}}(X_T) \quad \text{and} \quad Z_{\mathcal{T}} = \pi_{\mathcal{H}}(Z_T).$$

Then we have a subshift $(X_{\mathcal{T}}, \sigma_{\mathcal{T}})$ which is a subsystem of $(X_{\mathcal{G}}, \sigma_{\mathcal{G}})$, and a subshift $(Z_{\mathcal{T}}, \varsigma_{\mathcal{T}})$ which is a subsystem of $(X_{\mathcal{H}}, \sigma_{\mathcal{H}})$. Let $\xi_{\mathcal{T}} : (Z_{\mathcal{T}}, \varsigma_{\mathcal{T}}) \to (X_{\mathcal{T}}, \sigma_{\mathcal{T}})$ and $\eta_{\mathcal{T}} : (Z_{\mathcal{T}}, \varsigma_{\mathcal{T}}) \to (X_{\mathcal{T}}, \sigma_{\mathcal{T}})$ be the 1-block factor maps given by j_{upper} and j_{lower}, respectively.

We say that $\mathcal{T}$ is *1-1* if both $\xi_{\mathcal{T}}$ and $\eta_{\mathcal{T}}$ are 1-1. We say that $\mathcal{T}$ is *nondegenerate* if T is nondegenerate. When $\mathcal{T}$ is 1-1, we define an automorphism $\varphi_{\mathcal{T}}$ of the subshift $(X_{\mathcal{T}}, \sigma_{\mathcal{T}})$ by

$$\varphi_{\mathcal{T}} = \eta_{\mathcal{T}} \xi_{\mathcal{T}}^{-1}.$$

The *dual* $\mathcal{T}^*$ of $\mathcal{T}$ is defined as the sofic textile system

$$\mathcal{T}^* = (T^*, \mu, \lambda).$$

Clearly we have

$$(\mathcal{T}^*)^* = \mathcal{T}$$

For $n \in \mathbf{N}$, we define the *higher block system* $\mathcal{T}^{[n]}$ *of order* n of $\mathcal{T}$ by

$$\mathcal{T}^{[n]} = (T^{[n]}, \lambda^{[n]}, \nu^{[n-1]}).$$

Recall that $T^{[n]} = (p^{[n]}, q^{[n]}; \Gamma^{[n]} \to G^{[n]})$ is the higher block system of order n of T and $\nu^{[n]}$ is the labeling such that $\mathcal{H}^{[n]} = (\Gamma^{[n]}, \nu^{[n]})$. We define $\nu^{[0]} = \mu$.

For $t \in U_T$ with $t = (\alpha_{ij})_{i,j \in \mathbf{Z}}, \alpha_{ij} \in A_\Gamma = A_{\Gamma^T}$, let

$$\pi_{\mathcal{T}}(t) = (\nu(\alpha_{ij}))_{i,j \in \mathbf{Z}}.$$

Let

$$\mathcal{U}_{\mathcal{T}} = \{\pi_{\mathcal{T}}(t) \mid t \in U_T\}.$$

Then $\pi_{\mathcal{T}} : U_T \to \mathcal{U}_{\mathcal{T}}$ is a continuous map and hence $\mathcal{U}_{\mathcal{T}}$ is compact. An element of $\mathcal{U}_{\mathcal{T}}$ is called a *sofic textile weaved* by $\mathcal{T}$. Let $\theta_{\mathcal{T}} : \mathcal{U}_{\mathcal{T}} \to X_{\mathcal{T}}$ and $\theta^*_{\mathcal{T}} : \mathcal{U}_{\mathcal{T}} \to X_{\mathcal{T}^*}$ be defined as follows: for $s = (\beta_{ij})_{i,j \in \mathbf{Z}}$ in $\mathcal{U}_{\mathcal{T}}$ with $\beta_{ij} \in E_{\mathcal{T}}$,

$$\theta_{\mathcal{T}}(s) = (j_{upper}(\beta_{0j}))_{j \in \mathbf{Z}} \quad \text{and} \quad \theta^*_{\mathcal{T}}(s) = (j_{left}(\beta_{i0}))_{i \in \mathbf{Z}}.$$

Then $\mathcal{T}$ is 1-1 if and only if $\theta_{\mathcal{T}}$ is a homeomorphism, and $\mathcal{T}^*$ is 1-1 if and only if $\theta^*_{\mathcal{T}}$ is a homeomorphism.

Theorem 4.1. Let $\mathcal{T}$ be a 1-1 sofic textile system. If there is no $n \in \mathbf{N}$ such that $(\mathcal{T}^{[n]})^*$ is 1-1, then $\varphi_{\mathcal{T}}$ is not expansive. If there is $n \in \mathbf{N}$ such that $(\mathcal{T}^{[n]})^*$ is 1-1, then $(X_{\mathcal{T}}, \varphi_{\mathcal{T}})$ is topologically conjugate to $(X_{(\mathcal{T}^{[n]})^*}, \sigma_{(\mathcal{T}^{[n]})^*})$, and $(X_{(\mathcal{T}^{[n]})^*}, \varphi_{(\mathcal{T}^{[n]})^*})$ is topologically conjugate to $(X_{\mathcal{T}}, \sigma_{\mathcal{T}})$.

Proof. If $(\mathcal{T}^{[n+1]})^*$ is not 1-1 with $n \in \mathbf{N}$, then there are two distinct sofic textiles $s = (\beta_{ij})_{i,j \in \mathbf{Z}}$ and $s' = (\beta'_{ij})_{i,j \in \mathbf{Z}}$ in $\mathcal{U}_{\mathcal{T}}$ with $\beta_{ij}, \beta'_{ij} \in E_{\mathcal{T}}$ such that

$$(\beta_{ij})_{i \in \mathbf{Z}, -m \le j \le m} = (\beta'_{ij})_{i \in \mathbf{Z}, -m \le j \le m}$$

where m is the largest nonnegative integer with $2m+1 \leq n$. Let $y = \theta_{\mathcal{T}}(s)$ and $y' = \theta_{\mathcal{T}}(s')$. Then $y, y' \in X_{\mathcal{T}}, y \neq y'$, and

$$d(\varphi_{\mathcal{T}}^i(y), \varphi_{\mathcal{T}}^i(y')) < \frac{1}{m+1} \quad \text{for all } i \in \mathbf{Z}.$$

Therefore if $(\mathcal{T}^{[n]})^*$ is not 1-1 for all $n \in \mathbf{N}$, then $(X_{\mathcal{T}}, \varphi_{\mathcal{T}})$ is not expansive.

To prove the remainder of the theorem, it suffices to show that if $\mathcal{T}^*$ is 1-1, then $(X_{\mathcal{T}}, \varphi_{\mathcal{T}})$ is topologically conjugate to $(X_{\mathcal{T}^*}, \sigma_{\mathcal{T}^*})$ and $(X_{\mathcal{T}^*}, \varphi_{\mathcal{T}^*})$ is topologically conjugate to $(X_{\mathcal{T}}, \sigma_{\mathcal{T}})$, because $(X_{\mathcal{T}^{[n]}}, \varphi_{\mathcal{T}^{[n]}})$ is topologically conjugate to $(X_{\mathcal{T}}, \varphi_{\mathcal{T}})$ and $(X_{\mathcal{T}^{[n]}}, \sigma_{\mathcal{T}^{[n]}})$ is topologically conjugate $(X_{\mathcal{T}}, \sigma_{\mathcal{T}})$ for all $n \in \mathbf{N}$. Assume that $\mathcal{T}^*$ is 1-1. Define

$$\chi_{\mathcal{T}} = \theta_{\mathcal{T}}^* \theta_{\mathcal{T}}^{-1}.$$

Then $\chi_{\mathcal{T}}$ is a homeomorphism and we have

$$\chi_{\mathcal{T}} \varphi_{\mathcal{T}} = \sigma_{\mathcal{T}^*} \chi_{\mathcal{T}} \quad \text{and} \quad \chi_{\mathcal{T}} \sigma_{\mathcal{T}} = \varphi_{\mathcal{T}^*} \chi_{\mathcal{T}}.$$

□

Here, we introduce the definition of the canonical sofic covers and their properties which were given by W. Krieger [Kr2], [Kr3].

Let $\mathcal{G} = (G, \lambda)$ be a λ-graph. We say that $\mathcal{G}$ is *irreducible* if G is *irreducible* (i.e., M_G is irreducible). We say that $\mathcal{G}$ is *right resolving* if there are no distinct $a, a' \in A_G$ such that $i_G(a) = i_G(a')$ and $\lambda(a) = \lambda(a')$. We also say that $\mathcal{G}$ is *left resolving* if $\mathcal{G}^{-1}$ is left resolving. For $v \in V_G$, let

$$L(\mathcal{G}, v) = \{\lambda(a_1) \cdots \lambda(a_n) \mid n \in \mathbf{N}, a_1 \cdots a_n \in L(G), a_i \in A_G, i_G(a_1) = v\}.$$

When $\mathcal{G}$ is right resolving, we say that $\mathcal{G}$ is *reduced* if $L(\mathcal{G}, v) \neq L(\mathcal{G}, v')$ for any two distinct $v, v' \in V_G$. When $\mathcal{G}$ is left resolving, we say that $\mathcal{G}$ is *reduced* if $\mathcal{G}^{-1}$ is reduced.

Recall that for a subshift (X, σ) over an alphabet A and $n \in \mathbf{N}$,

$$L_n(X) = \{a_0 \cdots a_{n-1} \mid (a_i)_{i \in \mathbf{N}} \in X, a_i \in A\}.$$

We introduce the null-word ϵ and let $L_0(X) = \{\epsilon\}$. Let

$$L(X) = \bigcup_{n \geq 0} L_n(X).$$

For an alphabet A, a sequence $(a_i)_{i \leq 0}$ with $a_i \in A$, will be called a *left-infinite sequence* over A. For a left-infinite sequence $x = (a_i)_{i \leq 0}$ over A and a word $w = b_1 \cdots b_n$ with $n \in \mathbf{N}$ and $b_i \in A$, let xw denote the left-infinite sequence $(c_i)_{i \leq 0}$ such $c_i = b_{n+i}$ for $-n < i \leq 0$ and $c_i = a_{i+n}$ for $i \leq -n$. We define $x\epsilon = x$. Let (Y, σ) be a sofic system over A. We define

$$Y_- = \{(a_i)_{i \leq 0} \mid (a_i)_{i \in \mathbf{Z}} \in Y, a_i \in A\}.$$

For $x, y \in Y_-$, let $x \sim y$ mean that for any word w in $L(Y)$, $xw \in Y_-$ if and only if $yw \in Y_-$. Then $\sim$ is an equivalence relation on Y_- such that for $x, y \in Y_-$ and $w \in L(Y)$, if $x \sim y$ and $xw \in Y_-$, then $xw \sim yw$. The set of equivalence classes of $\sim$, is finite [Kr2]. For $x \in Y_-$, let $C(x)$ denote the equivalence class containing x. Let $\mathcal{K}_Y^+ = (K^+, \lambda^+)$ be the λ-graph such that

$$\begin{aligned} V_{K^+} &= \{C(x) \mid x \in Y_-\} \\ A_{K^+} &= \{(C(x), a) \mid x \in Y_-, a \in A \text{ with } xa \in Y_-\} \end{aligned}$$

and for $(C(x), a) \in A_{K^+}$

$$i_{K^+}((C(x), a)) = C(x), \quad t_{K^+}((C(x), a)) = C(xa)$$

and

$$\lambda^+((C(x), a)) = a.$$

Then $\mathcal{K}_Y^+$ is right resolving and reduced. Let $(Y^{(-1)}, \sigma^{(-1)})$ denote the sofic system obtained from (Y, σ) by reversing the direction of time. We define a left-resolving and reduced λ-graph $\mathcal{K}_Y^-$ by $\mathcal{K}_Y^- = (\mathcal{K}_{Y^{(-1)}}^+)^{-1}$. R. Fischer [F2] showed that if (Y, σ) is *one-sided topologically transitive* (i.e., there is a point of Y whose forward orbit is dense in Y), then there is a unique irreducible, reduced right resolving λ-graph that defines (Y, σ). Let it be denoted by $\mathcal{F}_Y^+$. Then $\mathcal{F}_Y^+$ is the unique ergodic component of $\mathcal{K}_Y^+$ [Kr2]. If we define $\mathcal{F}_Y^- = (\mathcal{F}_{Y^{(-1)}}^+)^{-1}$,then $\mathcal{F}_Y^-$ is a unique irreducible, reduced left-resolving λ-graph that defines (Y, σ).

The following theorem is the "right" version of Krieger's theorem in [Kr2] and [Kr3]. The "left" version should be clear from it.

Theorem 4.2. (Krieger [Kr2],[Kr3]). If $\phi : (Y, \sigma) \to (Y', \sigma')$ is a topological conjugacy between sofic systems and if $\mathcal{K}_Y^+ = (K^+, \lambda^+)$ and $\mathcal{K}_{Y'}^+ = (K'^+, \lambda'^+)$, then there is a unique topological conjugacy ϕ_0 of (X_{K^+}, σ_{K^+}) onto $(X_{K'^+}, \sigma_{K'^+})$ such that $\pi_{\mathcal{K}_{Y'}^+} \phi_0 = \phi \pi_{\mathcal{K}_Y^+}$. If $\phi : (Y, \sigma) \to (Y', \sigma')$ is a topological conjugacy between one-sided topologically transitive sofic systems and if $\mathcal{F}_Y^+ = (F^+, \lambda^+)$ and $\mathcal{F}_{Y'}^+ = (F'^+, \lambda'^+)$, then there is a unique topological conjugacy ϕ_0 of (X_{F^+}, σ_{F^+}) onto $(X_{F'^+}, \sigma_{F'^+})$ such that $\pi_{\mathcal{F}_{Y'}^+} \phi_0 = \phi \pi_{\mathcal{F}_Y^+}$.

If we want to have a canonical λ-graph for a general sofic system (Y, σ) which coincides $\mathcal{F}_Y^+$ when (Y, σ) is one-sided topologically transitive, we can use the "induced right resolving λ-graph" of $\mathcal{K}_Y^-$ [N3]. Let $\mathcal{G} = (G, \lambda)$ be a λ-graph over an alphabet A. Let $L(\mathcal{G})$ denote $L(X_\mathcal{G})$, and for an integer $n \geq 0$, let $L_n(\mathcal{G})$ denote $L_n(X_\mathcal{G})$. For $U \subset V_G$ and $w \in L_n(\mathcal{G})$ with $n \in \mathbf{N}$, let

$$S_+(U, w) = \{t_G(b_n) \mid \exists b_1 \cdots b_n \in L_n(G), b_i \in A_G, i_G(b_1) \in U \text{ and } \lambda(b_1) \cdots \lambda(b_n) = w\},$$

and for the null-word ϵ, let $S_+(U,\epsilon) = U$. We call $S_+(U,w)$ the *w-successor* of U or a *successor* of U for $U \subset V_G$ and $w \in L(\mathcal{G})$. A *right-compatible set* is defined to be a nonempty successor of a set consisting of one vertex. Let C_+ be the family of all subsets of V_G that are a maximal compatible set or a successor of a maximal compatible set. The *induced right resolving λ-graph* $\mathcal{G}_+$ of $\mathcal{G}$ is defined as the λ-graph $\mathcal{G}_+ = (G_+, \lambda_+)$ such that

$$\begin{aligned} V_{G_+} &= C_+ \\ A_{G_+} &= \{(U,a) \mid U \in C_+, \quad a \in A, \quad S_+(U,a) \neq \emptyset\} \end{aligned}$$

and for $(U,a) \in A_{G_+}$,

$$i_{G_+}((U,a)) = U, \quad t_{G_+}((U,a)) = S_+(U,a),$$

and

$$\lambda_+((U,a)) = a.$$

Clearly $\mathcal{G}_+$ is right resolving and $X_{\mathcal{G}_+} = X_{\mathcal{G}}$. We shall identify each arc $(U,a) \in A_{G_+}$ with the set of all arcs $b \in A_G$ such that $i_G(b) \in U$ and $\lambda(b) = a$. For $(\beta_i)_{i\in\mathbf{Z}} \in X_{G_+}$ with $\beta_i \in A_{G_+}$, let $\gamma_+^{\mathcal{G}}((\beta_i)_{i\in\mathbf{Z}})$ denote the set of all points $(b_i)_{i\in\mathbf{Z}} \in X_G$ such that $b_i \in \beta_i$ for all $i \in \mathbf{Z}$.

For $U \subset V_G$ and $w \in L_n(\mathcal{G})$ with $n \in \mathbf{N}$, let

$$S_-(w,U) = \{i_G(b_1) \mid \exists b_1 \cdots b_n \in L_n(G), b_i \in A_G, t_G(b_n) \in U, \lambda(b_1)\cdots\lambda(b_n) = w\},$$

and let $S_-(\epsilon, U) = U$. We call $S_-(w,U)$ the *w-predecessor* of U or a *predecessor* of U for $U \subset V_G$ and $w \in L(\mathcal{G})$. A *left-compatible set* is defined to be a nonempty predecessor of a set consisting of one vertex. The definition of the *induced left resolving λ-graph* $\mathcal{G}_-$ of $\mathcal{G}$ is analogous to that of $\mathcal{G}_+$ or it is given by $\mathcal{G}_- = ((\mathcal{G}^{-1})_+)^{-1}$.

The following theorem is a refinement of Theorem 3.1 of [N3].

Theorem 4.3. Let $\mathcal{G} = (G,\lambda)$ and $\mathcal{G}' = (G',\lambda')$ be λ-graphs. Let $\mathcal{G}_+ = (G_+,\lambda_+)$ and $\mathcal{G}'_+ = (G'_+,\lambda'_+)$ be their induced right resolving λ-graphs. Suppose that $\phi_0 : (X_G, \sigma_G) \to (X_{G'}, \sigma_{G'})$ and $\phi : (X_{\mathcal{G}}, \sigma_{\mathcal{G}}) \to (X_{\mathcal{G}'}, \sigma_{\mathcal{G}'})$ are topological conjugacies such that $\pi_{\mathcal{G}'}\phi_0 = \phi\pi_{\mathcal{G}}$. Then there is a unique topological conjugacy $(\phi_0)_+ : (X_{G_+}, \sigma_{G_+}) \to (X_{G'_+}, \sigma_{G'_+})$ such that $\pi_{\mathcal{G}'_+}(\phi_0)_+ = \phi\pi_{\mathcal{G}_+}$ and $\gamma_+^{\mathcal{G}'}((\phi_0)_+(y)) = \phi_0(\gamma_+^{\mathcal{G}}(y))$ for all $y \in X_{G_+}$.

Proof. By the proof of Theorem 3.1 of [N3], we readily know that there is a topological conjugacy $(\phi_0)_+ : (X_{G_+}, \sigma_{G_+}) \to (X_{G'_+}, \sigma_{G'_+})$ such that $\pi_{\mathcal{G}'_+}(\phi_0)_+ = \phi\pi_{\mathcal{G}_+}$ and $\gamma_+^{\mathcal{G}'}((\phi_0)_+(y)) = \phi_0(\gamma_+^{\mathcal{G}}(y))$ for all $y \in X_{G_+}$. We shall prove uniqueness in the following.

Let $y = (\beta_i)_{i\in\mathbf{Z}} \in X_{G_+}$ with

$$\beta_i = (U_i, a_i),$$

where U_i is a maximal compatible set or a successor of a maximal compatible set for $\mathcal{G}$ and $a_i \in L_1(\mathcal{G})$ for $i \in \mathbf{Z}$. We say that y is *complete* if the following holds: if $\gamma_+^{\mathcal{G}}(y) = \{(b_i^{(1)})_{i\in\mathbf{Z}}, \cdots, (b_i^{(l)})_{i\in\mathbf{Z}}\}$, $b_i^{(j)} \in A_G$, then for every $i \in \mathbf{Z}$, there is $j \in \mathbf{Z}$ with $j < i$ such that

$$S_+(\{i_G(b_j^{(1)}), \cdots, i_G(b_j^{(l)})\}, a_j \cdots a_{i-1}) = U_i.$$

We know that if y is complete, then it is uniquely determined by $\gamma_+^{\mathcal{G}}(y)$. As is easily seen, even if y is not complete, there are $k \in \mathbf{Z}$ and a set of left-infinite sequences

$$\{(\bar{b}_i^{(1)})_{i\le k}, \cdots, (\bar{b}_i^{(m)})_{i\le k}\}$$

such that $\bar{b}_i^{(j)} \in \beta_i$ and $t_G(\bar{b}_{i-1}^{(j)}) = i_G(\bar{b}_i^{(j)})$ for all $j = 1, \cdots, m$ and $i \le k$ and that for all $i \in \mathbf{Z}$ there is $j < \min\{i, k\}$ with

$$S_+(\{i_G(\bar{b}_j^{(1)}), \cdots, i_G(\bar{b}_j^{(m)}\}, a_j \cdots a_{i-1}) = U_i.$$

Therefore if $z = \lim_{j\to\infty} \sigma_{G_+}^{-n_j}(y)$ with $0 < n_1 < n_2 < \cdots$, then z is complete. Moreover it follows that y is uniquely determined by z, $\gamma_+^{\mathcal{G}}(y)$, and the sequence $n_1, n_2, \cdots$, so that y is uniquely determined by $\gamma_+^{\mathcal{G}}(z)$, $\gamma_+^{\mathcal{G}}(y)$, and the sequence $n_1, n_2, \cdots$.

Let $y \in X_{G_+}$. Since X_{G_+} is compact, there is a sequence $0 < n_1 < n_2 < \cdots$ such that $\lim_{j\to\infty} \sigma_{G_+}^{-n_j}(y)$ exists. Let z be this limit point. Then $\lim_{j\to\infty} \sigma_{G'_+}^{-n_j}((\phi_0)_+(y)) = (\phi_0)_+(z)$. Since $\gamma_+^{\mathcal{G}'}((\phi_0)_+(y)) = \phi_0(\gamma_+^{\mathcal{G}}(y))$ and $\gamma_+^{\mathcal{G}'}((\phi_0)_+(z)) = \phi_0(\gamma_+^{\mathcal{G}}(z))$, we conclude by the above that $(\phi_0)_+(y)$ is uniquely determined by ϕ_0. □

Furthermore it was proved in [N3] that for any one-sided topologically transitive sofic system (Y, σ),

$$(\mathcal{K}_Y^-)_+ = \mathcal{F}_Y^+.$$

For a sofic system (Y, σ), let the *right canonical* λ*-graph* or simply the *canonical* λ*-graph* of (Y, σ) mean $(\mathcal{K}_Y^-)_+$ and let the *left canonical* λ*-graph* of (Y, σ) mean $(\mathcal{K}_Y^+)_-$, while we shall call $\mathcal{K}_Y^+$ and $\mathcal{K}_Y^-$ the *right Krieger* λ*-graph* of (Y, σ) and the *left Krieger* λ*-graph* of (Y, σ), respectively. Let us say that a λ-graph $\mathcal{G}$ is *canonical* if $\mathcal{G}$ is the right canonical λ-graph of $(X_{\mathcal{G}}, \sigma_{\mathcal{G}})$.

Proposition 4.4. Let φ be an automorphism of a sofic system (Y, σ). Then there is a 1-1 sofic textile system $\mathcal{T} = (T, \lambda, \mu)$ such that T is 1-1 and nondegenerate and

$$\varphi_{\mathcal{T}} = \varphi.$$

Proof. Let $\mathcal{G} = (G, \lambda) = (\mathcal{K}_Y^-)_+$. Then by Theorems 4.2 and 4.3, there is an automorphism φ_0 of (X_G, σ_G) such that $\pi_{\mathcal{G}}\varphi_0 = \varphi\pi_{\mathcal{G}}$. By Fact 2.2, there is a 1-1 and nondegenerate textile system $T = (p, q : \Gamma \to G)$ such that $\varphi_T = \varphi_0$. Let b be a symbol and let μ be the mapping of V_Γ onto $\{b\}$. Define $\mathcal{T} = (T, \lambda, \mu)$. Then $\mathcal{T}$ has all of the required properties. □

Let $\mathcal{T} = (T, \lambda : A_G \to A, \mu : V_\Gamma \to B)$ be a sofic textile system with $T = (p, q : \Gamma \to G)$. Then $U_\mathcal{T}$ is a closed and shift invariant subset of $U_{E_\mathcal{T}}$. Therefore we have subshifts $(U_\mathcal{T}, \sigma_\mathcal{T}^{(k,l)})$, $k, l \in \mathbf{Z}$, of 2-dimensional configurations, which will be called the *sofic textile shifts* defined by $\mathcal{T}$. Associated with $\mathcal{T}$, we have two λ-graphs

$$\mathcal{G} = (G, \lambda) \quad \text{and} \quad \mathcal{G}^T = (G^T, \mu).$$

Let $(k, l) \in \mathbf{Z}^2$ with $(k, l) \neq (0, 0)$. We define λ-graphs $\hat{\mathcal{G}}^{(k,l)}$ and $\check{\mathcal{G}}^{(k,l)}$ as follows: if $k \neq 0$ and $l \neq 0$, then

$$\hat{\mathcal{G}}^{(k,l)} = \mathcal{G}^l(\mathcal{G}^T)^k \quad \text{and} \quad \check{\mathcal{G}}^{(k,l)} = (\mathcal{G}^T)^k\mathcal{G}^l;$$
$$\hat{\mathcal{G}}^{(k,0)} = \check{\mathcal{G}}^{(k,0)} = (\mathcal{G}^T)^k \quad \text{and} \quad \hat{\mathcal{G}}^{(0,l)} = \check{\mathcal{G}}^{(l,0)} = \mathcal{G}^l.$$

We define $\hat{\theta}_\mathcal{T}^{(k,l)} : U_\mathcal{T} \to X_{\hat{\mathcal{G}}^{(k,l)}}$ and $\check{\theta}_\mathcal{T}^{(k,l)} : U_\mathcal{T} \to X_{\check{\mathcal{G}}^{(k,l)}}$ naturally. For example, $\check{\theta}_\mathcal{T}^{(k,l)}$ with $k \neq 0$ and $l \neq 0$, is defined as follows: if $s \in U_\mathcal{T}$ and $s = \pi_\mathcal{T}(t)$ with $t \in U_T$ and if

$$\check{\theta}_T^{(k,l)}(t) = (b_1^{(m)} \cdots b_k^{(m)} a_1^{(m)} \cdots a_l^{(m)})_{m \in \mathbf{Z}}$$

with $b_i^{(m)} \in A_{G^T}$ and $a_j^{(m)} \in A_G$, then

$$\check{\theta}_\mathcal{T}^{(k,l)}(s) = (\mu(b_1^{(m)}) \cdots \mu(b_k^{(m)})\lambda(a_1^{(m)}) \cdots \lambda(a_l^{(m)}))_{m \in \mathbf{Z}}.$$

Clearly

$$\hat{\theta}_\mathcal{T}^{(0,1)} = \check{\theta}_\mathcal{T}^{(0,1)} = \theta_\mathcal{T} \quad \text{and} \quad \hat{\theta}_\mathcal{T}^{(1,0)} = \check{\theta}_\mathcal{T}^{(1,0)} = \theta_\mathcal{T}^*.$$

Define

$$\hat{X}_\mathcal{T}^{(k,l)} = \hat{\theta}_\mathcal{T}^{(k,l)}(U_\mathcal{T}) \quad \text{and} \quad \check{X}_\mathcal{T}^{(k,l)} = \check{\theta}_\mathcal{T}^{(k,l)}(U_\mathcal{T}).$$

Then we have two subshifts $(\hat{X}_\mathcal{T}^{(k,l)}, \hat{\sigma}_\mathcal{T}^{(k,l)})$ and $(\check{X}_\mathcal{T}^{(k,l)}, \check{\sigma}_\mathcal{T}^{(k,l)})$ which are topologically conjugate. Clearly we have

$$(\hat{X}_\mathcal{T}^{(k,0)}, \hat{\sigma}_\mathcal{T}^{(k,0)}) = (\check{X}_\mathcal{T}^{(k,0)}, \check{\sigma}_\mathcal{T}^{(k,0)}) = (X_{\mathcal{T}^*}^{(k)}, \sigma_{\mathcal{T}^*}^{(k)})$$

and

$$(\hat{X}_\mathcal{T}^{(0,l)}, \hat{\sigma}_\mathcal{T}^{(0,l)}) = (\check{X}_\mathcal{T}^{(0,l)}, \check{\sigma}_\mathcal{T}^{(0,l)}) = (X_\mathcal{T}^{(l)}, \sigma_\mathcal{T}^{(l)})$$

The following two facts which correspond to Propositions 2.3 and 2.4, respectively, are clear.

Fact 4.5. Let $\mathcal{T}$ be a sofic textile system. Let $(k,l) \in \mathbf{Z}^2$ with $(k,l) \neq (0,0)$. Then $\hat{\theta}_{\mathcal{T}}^{(k,l)}$ is 1-1 if and only if $\check{\theta}_{\mathcal{T}}^{(k,l)}$ is 1-1. If $\check{\theta}_{\mathcal{T}}^{(k,l)}$ is 1-1, then $(U_{\mathcal{T}}, \boldsymbol{\sigma}_{\mathcal{T}}^{(k,l)})$ is topologically conjugate to $(\check{X}_{\mathcal{T}}^{(k,l)}, \check{\sigma}_{\mathcal{T}}^{(k,l)})$.

Fact 4.6. Let $\mathcal{T}$ be a sofic textile system. Let $(k,l),(m,n) \in \mathbf{Z}^2$ with $(k,l) \neq (0,0)$ and $(m,n) \neq (0,0)$. Then the following statements are valid.

(1) If $\check{\theta}_{\mathcal{T}}^{(k,l)}$ is 1-1, then the mapping defined by

$$\psi_{\mathcal{T},(k,l)}^{(m,n)} = \check{\theta}_{\mathcal{T}}^{(k,l)} \boldsymbol{\sigma}_{\mathcal{T}}^{(m,n)} (\check{\theta}_{\mathcal{T}}^{(k,l)})^{-1}$$

is an automorphism of $(\check{X}_{\mathcal{T}}^{(k,l)}, \check{\sigma}_{\mathcal{T}}^{(k,l)})$, and $(\check{X}_{\mathcal{T}}^{(k,l)}, \psi_{\mathcal{T},(k,l)}^{(m,n)})$ is topologically conjugate to $(U_{\mathcal{T}}, \boldsymbol{\sigma}_{\mathcal{T}}^{(m,n)})$.

(2) If both $\check{\theta}_{\mathcal{T}}^{(k,l)}$ and $\check{\theta}_{\mathcal{T}}^{(m,n)}$ are 1-1, then $(\check{X}_{\mathcal{T}}^{(k,l)}, \psi_{\mathcal{T},(k,l)}^{(m,n)})$ is topologically conjugate to $(\check{X}_{\mathcal{T}}^{(m,n)}, \check{\sigma}_{\mathcal{T}}^{(m,n)})$ by the topological conjugacy that maps $\check{\theta}_{\mathcal{T}}^{(k,l)}(s)$ to $\check{\theta}_{\mathcal{T}}^{(m,n)}(s)$ for $s \in U_{\mathcal{T}}$.

(3) If $\mathcal{T}$ is 1-1, then $(X_{\mathcal{T}}, \varphi_{\mathcal{T}}^m \sigma_{\mathcal{T}}^n)$ is topologically conjugate to $(U_{\mathcal{T}}, \boldsymbol{\sigma}_{\mathcal{T}}^{(m,n)})$.

Proposition 4.7. Let $\mathcal{T}$ be a sofic textile system Let $k \in \mathbf{Z}$ with $k \neq 0$. Then $(U_{\mathcal{T}}, \boldsymbol{\sigma}_{\mathcal{T}}^{(k,0)})$ is expansive if and only if there is $n \in \mathbf{N}$ such that $(\mathcal{T}^{[n]})^*$ is 1-1. If $(\mathcal{T}^{[n]})^*$ is 1-1, then $(U_{\mathcal{T}}, \boldsymbol{\sigma}_{\mathcal{T}}^{(k,0)})$ is topologically conjugate to $(X_{(\mathcal{T}^{[n]})^*}^{(k)}, \sigma_{(\mathcal{T}^{[n]})^*}^{(k)})$.

Proof. This proposition follows from the proof of Theorem 4.1 and Fact 4.5. □

We can define the product and the composition for sofic textile systems as natural generalizations of those for textile systems.

Let $\mathcal{T}_1 = (T_1, \lambda_1, \mu)$ and $\mathcal{T}_2 = (T_2, \lambda_2, \mu_2)$ be two sofic textile systems with $T_1 = (p_1, q_1 : \Gamma_1 \to G_1)$ and $T_2 = (p_2, q_2 : \Gamma_2 \to G_2)$. Let $T_1^* = (p_1^{T_1}, q_1^{T_1} : \Gamma_1^{T_1} \to G_1^{T_1})$ and $T_2^* = (p_2^{T_2}, q_2^{T_2} : \Gamma_2^{T_2} \to G_2^{T_2})$. Let $\mathcal{G}_1, \mathcal{G}_2, \mathcal{G}_1^{T_1}$ and $\mathcal{G}_2^{T_2}$ be the λ-graphs defined by $\mathcal{G}_1 = (G_1, \lambda_1), \mathcal{G}_2 = (G_2, \lambda_2), \mathcal{G}_1^{T_1} = (G_1^{T_1}, \mu_1)$ and $\mathcal{G}_2^{T_2} = (G_2^{T_2}, \mu_2)$.

(i) If $\mathcal{G}_1^{T_1} = \mathcal{G}_2^{T_2}$, then we define the *product* $\mathcal{T}_1\mathcal{T}_2$ as the sofic textile system

$$\mathcal{T}_1\mathcal{T}_2 = (T_1T_2, (\lambda_1\lambda_2), \mu),$$

where $(\lambda_1\lambda_2)$ is the labeling of the arcs of G_1G_2 defined by

$$(\lambda_1\lambda_2)(a_1a_2) = \lambda_1(a_1)\lambda_2(a_2), \quad a_1a_2 \in A_{G_1G_2}, a_1 \in A_{G_1}, a_2 \in A_{G_2},$$

and $\mu = \mu_1 = \mu_2$.

(ii) If $\mathcal{G}_1 = \mathcal{G}_2$, then we define the *composition* $\mathcal{T}_1 \circ \mathcal{T}_2$ by

$$\mathcal{T}_1 \circ \mathcal{T}_2 = (\mathcal{T}_1^*\mathcal{T}_2^*)^*.$$

If $\mathcal{G}_1 = \mathcal{G}_2, X_{\mathcal{T}_1} = X_{\mathcal{T}_2}$ and both $\mathcal{T}_1$ and $\mathcal{T}_2$ are one-sided 1-1, then $\mathcal{T}_1 \circ \mathcal{T}_2$ is also one-sided 1-1 and we have

$$\varphi_{\mathcal{T}_1 \circ \mathcal{T}_2} = \varphi_{\mathcal{T}_2} \circ \varphi_{\mathcal{T}_1}.$$

The *product-power* and the *composition-power* of a sofic textile system are defined in the same way as those of a textile system were defined.

5. TOPOLOGICAL CONJUGACY BETWEEN SOFIC SYSTEMS

In this section, we review the structure of topological conjugacy between sofic systems, which is taken from [N2], and give some refinement.

A topological conjugacy between subshifts is called a *symbolic conjugacy* if it is a 1-block map given by a bijection between the underlying alphabets of the subshifts. A subshift (Y, σ) over an alphabet A is said to be *bipartite* if A is partitioned into two disjoint subsets such that if $(a_i)_{i \in \mathbf{Z}} \in Y$ with $a_i \in A$, then a_i and a_{i+1} belong to the different subsets for all $i \in \mathbf{Z}$. For alphabets C and D, let CD denotes the set $\{cd \mid c \in C, d \in D\}$, where cd is the word obtained from c concatenated by d. Let (Y, σ) be a bipartite subshift over an alphabet A. Then there are disjoint subsets C and D of A with $C \cup D = A$ such that if $(a_i)_{i \in \mathbf{Z}} \in Y, a_i \in A$, then for all $i \in \mathbf{Z}, a_i a_{i+1}$ is contained in either CD or DC. Let us consider the second power system $(Y^{(2)}, \sigma^{(2)})$ of (Y, σ). Then $Y^{(2)}$ is partitioned into two disjoint closed sets $\acute{Y}$ and $\grave{Y}$ which are defined by

$$\begin{aligned} \acute{Y} &= \{(c_i d_i)_{i \in \mathbf{Z}} \mid \exists (a_i)_{i \in \mathbf{Z}} \in Y, \forall i \in \mathbf{Z}, c_i = a_{2i} \in C \text{ and } d_i = a_{2i+1} \in D\} \\ \grave{Y} &= \{(d_i c_{i+1})_{i \in \mathbf{Z}} \mid \exists (a_i)_{i \in \mathbf{Z}} \in Y, \forall i \in \mathbf{Z}, c_i = a_{2i} \in C \text{ and } d_i = a_{2i+1} \in D\} \end{aligned}$$

Therefore we have two subshifts $(\acute{Y}, \acute{\sigma})$ and $(\grave{Y}, \grave{\sigma})$, which will be called the *induced pair of subshifts* of the bipartite subshift (Y, σ). The conjugacy of $(\acute{Y}, \acute{\sigma})$ onto $(\grave{Y}, \grave{\sigma})$ that maps $(c_i d_i)_{i \in \mathbf{Z}}$ to $(d_i c_{i+1})_{i \in \mathbf{Z}}$ for $(c_i d_i)_{i \in \mathbf{Z}} \in \acute{Y}$, is called the *forward bipartite conjugacy induced* by (Y, σ), or a *forward bipartite conjugacy* . The conjugacy of $(\acute{Y}, \acute{\sigma})$ onto $(\grave{Y}, \grave{\sigma})$ that maps $(c_i d_i)_{i \in \mathbf{Z}}$ to $(d_{i-1} c_i)_{i \in \mathbf{Z}}$, is called the *backward bipartite conjugacy induced* by (Y, σ), or a *backward bipartite conjugacy* .

For an alphabet A, a 1-1 mapping $f : A \to CD$ is called a *bipartite expression* of A, where C and D are alphabets. For a subshift (X, σ) over an alphabet A and a subshift (X', σ') over an alphabet A', a mapping $\psi : X \to X'$ is called a *forward bipartite code* if there are bipartite expressions $f : A \to CD$ and $g : A' \to DC$ such that for $(a_i)_{i \in \mathbf{Z}} \in X, a_i \in A$, if

$$\psi((a_i)_{i \in \mathbf{Z}}) = (a'_i)_{i \in \mathbf{Z}}, \quad a'_i \in A'$$

and

$$f(a_i) = c_i d_i, \quad \text{all } i \in \mathbf{Z},$$

then

$$g(a'_i) = d_i c_{i+1}, \quad \text{all } i \in \mathbf{Z}.$$

In the above, f determines ψ up to a symbolic conjugacy so that ψ is called the *forward bipartite code induced* by f. A *backward bipartite code* is defined similarly, or is defined to be the inverse of a forward bipartite code.

In [N2], it was proved that any topological conjugacy is factorized into the composition of bipartite codes (each of which is either forward or backward). We state this result in the following form.

Theorem 5.1. [N2] Any topological conjugacy ϕ between subshifts is factorized into a composition of the form

$$\phi = \kappa_n \zeta_n \kappa_{n-1} \zeta_{n-1} \cdots \kappa_1 \zeta_1 \kappa_0,$$

where $n \in \mathbf{N}$, $\kappa_0, \cdots, \kappa_n$ are symbolic conjugacies, and $\zeta_1, \cdots, \zeta_n$ are (either forward or backward) bipartite conjugacies.

Let $n \geq 0$. A topological conjugacy ϕ between subshifts is called a *forward conjugacy* of *lag* n if ϕ has a factorization of the form

$$\phi = \kappa_n \zeta_n \kappa_{n-1} \zeta_{n-1} \cdots \kappa_1 \zeta_1 \kappa_0$$

where $\kappa_0, \cdots, \kappa_n$ are symbolic conjugacies and $\zeta_1, \cdots, \zeta_n$ are all forward bipartite conjugacies. This factorization is called a *κ-ζ factorization* of *lag* n of the forward conjugacy ϕ. A *backward conjugacy* is defined to be the inverse of a forward conjugacy. Any topological conjugacy $\phi : (X, \sigma) \to (X', \sigma')$ between subshifts is written in the form

$$\phi = \phi_0 \sigma^l$$

where ϕ_0 is a forward conjugacy and $l \in \mathbf{Z}$. If ϕ_0 is of lag n with $n \geq 0$, then for all $l \geq 0$, $\phi_0 \sigma^l$ is a forward conjugacy and $\phi_0 \sigma^{-l-n}$ is a backward conjugacy.

For a subshift (X, σ) over an alphabet A and for $n \in \mathbf{N}$, we define a conjugacy $\rho_{X,n} : (X, \sigma) \to (X^{[n]}, \sigma^{[n]})$ by

$$\rho_{X,n}((a_i)_{i \in \mathbf{Z}}) = (a_i a_{i+1} \cdots a_{i+n-1})_{i \in \mathbf{Z}}, \quad (a_i)_{i \in \mathbf{Z}} \in X, \ a_i \in A.$$

The proof of Theorem 4.1 in [N2] gives the following refinement, which we shall use later.

Theorem 5.1′ [N2]. Let $\phi : (X, \sigma) \to (X', \sigma')$ be a topological conjugacy between subshifts. Then there are $m, n \in \mathbf{N}$, alphabets $A_i, i = 0, \cdots, n$, subshifts (X_i, σ_i) over $A_i, i = 0, \cdots, n$, with $(X_0, \sigma_0) = (X, \sigma)$ and $(X_n, \sigma_n) = (X'^{[n]}, \sigma'^{[n]})$, and bipartite codes $\psi_i : X_{i-1} \to X_i, i = 1, \cdots, n$, such that

$$\rho_{X',m} \phi = \psi_n \cdots \psi_1$$

and each ψ_i is induced by a bipartite expression $f_i : A_i \to C_iD_i$ either defined by

$$f_i(a) = ae_i(a), \quad a \in A_i,$$

for some mapping $e_i : A_i \to D_i$, with $C_i = A_i$, or defined by

$$f_i(a) = e_i(a)a, \quad a \in A_i,$$

for some mapping $e_i : A_i \to C_i$, with $D_i = A_i$.

We consider representing λ-graphs by matrices. For a λ-graph $\mathcal{G} = (G, \lambda), \mathcal{M}_\mathcal{G}$ denotes the matrix with indexing set $V_G \times V_G$ such that for $u, v \in V_G$, the (u, v)-component of $\mathcal{M}_\mathcal{G}$ is equal to the formal sum

$$\sum \lambda(\alpha),$$

where the summation is taken over all $\alpha \in A_G$ with $i_G(\alpha) = u$ and $t_G(\alpha) = v$. For example, the λ-graph $\mathcal{G}$ described as

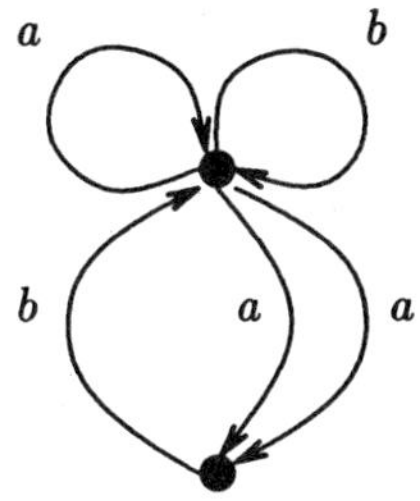

is represented by the matrix

$$\mathcal{M}_\mathcal{G} = \begin{pmatrix} a+b & 2a \\ b & 0 \end{pmatrix}.$$

Note that $2a$ simply means $a + a$. For a λ-graph $\mathcal{G}$, $\mathcal{M}_\mathcal{G}$ is called the *representation matrix* of $\mathcal{G}$. More generally a rectangular matrix whose components are formal sums of symbols, are called a *representation matrix.* For a representation matrix $\mathcal{M}$, the nonnegative integral matrix obtained by substituting 1 to all symbols that appear in $\mathcal{M}$, is called the *support* of $\mathcal{M}$. A square representation matrix represents a λ-graph and its support is the adjacency matrix of the support of the λ-graph. For two λ-graphs $\mathcal{G} = (G, \lambda)$ and $\mathcal{H} = (H, \lambda)$ with $V_G = V_H$, the product $\mathcal{M}_\mathcal{G}\mathcal{M}_\mathcal{H}$ of $\mathcal{M}_\mathcal{G}$ and $\mathcal{M}_\mathcal{H}$ is defined to be $\mathcal{M}_{\mathcal{G}\mathcal{H}}$. The way of calculating $\mathcal{M}_\mathcal{G}\mathcal{M}_\mathcal{H}$ is naturally generalized to obtain the product of two rectangular representation matrices in such a form that one is $m \times n$ and the other is $n \times p$. For example,

$$\begin{pmatrix} a & b \\ 0 & a \\ b & 0 \end{pmatrix} \begin{pmatrix} a & 0 & 0 \\ 0 & a & b \end{pmatrix} = \begin{pmatrix} aa & ba & bb \\ 0 & aa & ab \\ ba & 0 & 0 \end{pmatrix}.$$

For two λ-graphs $\mathcal{G}$ and $\mathcal{H}$, we define the *disjoint union* of $\mathcal{G}$ and $\mathcal{H}$, denoted by $\mathcal{G} \sqcup \mathcal{H}$, as the λ-graph whose representation matrix is given by

$$\begin{pmatrix} \mathcal{M}_{\mathcal{G}} & 0 \\ 0 & \mathcal{M}_{\mathcal{H}} \end{pmatrix}.$$

A λ-graph $\mathcal{G} = (G, \lambda)$ is said to be *bipartite* if G is *bipartite*, that is, A_G and V_G are respectively partitioned into disjoint subsets $\acute{A}_G$ and $\grave{A}_G$ and into disjoint subsets $\acute{V}_G$ and $\grave{V}_G$ such that

$$\acute{V}_G = i_G(\acute{A}_G) = t_G(\grave{A}_G)$$

and

$$\grave{V}_G = t_G(\acute{A}_G) = i_G(\grave{A}_G).$$

These bipartite partitions are not always unique for a bipartite graph, but when we mention a bipartite graph or a bipartite λ-graph, we assume that the bipartite partitions are specified. Let $\mathcal{G} = (G, \lambda)$ be a bipartite λ-graph with $\acute{A}_G, \grave{A}_G, \acute{V}_G$ and $\grave{V}_G$ as above. Then $\mathcal{M}_{\mathcal{G}}$ is written in the form

$$\mathcal{M}_{\mathcal{G}} = \begin{pmatrix} 0 & \mathcal{P} \\ \mathcal{Q} & 0 \end{pmatrix},$$

where $\mathcal{P}$ and $\mathcal{Q}$ are rectangular representation matrices with indexing sets $\acute{V}_G \times \grave{V}_G$ and $\grave{V}_G \times \acute{V}_G$, respectively. Since

$$\mathcal{M}_{\mathcal{G}^2} = (\mathcal{M}_{\mathcal{G}})^2 = \begin{pmatrix} \mathcal{P}\mathcal{Q} & 0 \\ 0 & \mathcal{Q}\mathcal{P} \end{pmatrix},$$

$\mathcal{G}^2$ is the disjoint union of the λ-graphs $\acute{\mathcal{G}}$ and $\grave{\mathcal{G}}$ that are defined by

$$\mathcal{M}_{\acute{\mathcal{G}}} = \mathcal{P}\mathcal{Q} \quad \text{and} \quad \mathcal{M}_{\grave{\mathcal{G}}} = \mathcal{Q}\mathcal{P}.$$

Alternatively, $\acute{\mathcal{G}}$ is defined as the λ-graph $\acute{\mathcal{G}} = (\acute{G}, \acute{\lambda})$ such that

$$\begin{aligned} V_{\acute{G}} &= \acute{V}_G \\ A_{\acute{G}} &= \{\alpha\beta \mid \alpha \in \acute{A}_G, \beta \in \grave{A}_G, t_G(\alpha) = i_G(\beta)\} \end{aligned}$$

and for $\alpha\beta \in A_{\acute{G}}$

$$\begin{aligned} i_{\acute{G}}(\alpha\beta) &= i_G(\alpha),\ t_{\acute{G}}(\alpha\beta) = t_G(\beta) \text{ and} \\ \acute{\lambda}(\alpha\beta) &= \lambda(\alpha)\lambda(\beta); \end{aligned}$$

$\grave{\mathcal{G}}$ is defined as the λ-graph $\grave{\mathcal{G}} = (\grave{G}, \grave{\lambda})$ such that

$$\begin{aligned} V_{\grave{G}} &= \grave{V}_G \\ A_{\grave{G}} &= \{\beta\alpha \mid \beta \in \grave{A}_G, \alpha \in \acute{A}_G, t_G(\beta) = i_G(\alpha)\} \end{aligned}$$

and $i_{\grave{G}}, t_{\grave{G}}$ and $\grave{\lambda}$ are given similarly. We call the pair of $\acute{\mathcal{G}}$ and $\grave{\mathcal{G}}$ the *induced pair of λ-graphs* of $\mathcal{G}$. Define $\zeta_G : X_{\acute{G}} \to X_{\grave{G}}$ by

$$\zeta_G((\alpha_i\beta_i)_{i\in\mathbf{Z}}) = (\beta_i\alpha_{i+1})_{i\in\mathbf{Z}}, \quad (\alpha_i\beta_i)_{i\in\mathbf{Z}} \in X_{\acute{G}}, \ \alpha_i \in \acute{A}_G, \ \beta_i \in \grave{A}_G.$$

Then ζ_G is the forward bipartite conjugacy of $(X_{\acute{G}}, \sigma_{\acute{G}})$ onto $(X_{\grave{G}}, \sigma_{\grave{G}})$ induced by the bipartite topological Markov shift (X_G, σ_G). Define $\zeta_{\mathcal{G}} : X_{\acute{\mathcal{G}}} \to X_{\grave{\mathcal{G}}}$ by

$$\zeta_{\mathcal{G}}((\lambda(\alpha_i)\lambda(\beta_i))_{i\in\mathbf{Z}}) = (\lambda(\beta_i)\lambda(\alpha_{i+1}))_{i\in\mathbf{Z}}, \quad (\alpha_i\beta_i)_{i\in\mathbf{Z}} \in X_{\acute{G}}, \ \alpha_i \in \acute{A}_G, \ \beta_i \in \grave{A}_G.$$

Then $\zeta_{\mathcal{G}}$ is a forward bipartite code of $(X_{\acute{\mathcal{G}}}, \sigma_{\acute{\mathcal{G}}})$ onto $(X_{\grave{\mathcal{G}}}, \sigma_{\grave{\mathcal{G}}})$. If we assume without loss of generality that $\lambda(\acute{A}_G) \cap \lambda(\grave{A}_G) = \emptyset$, then $\zeta_{\mathcal{G}}$ is the forward bipartite conjugacy induced by the bipartite sofic system $(X_{\mathcal{G}}, \sigma_{\mathcal{G}})$. It is clear that the following diagram commutes:

$$\begin{array}{ccc} X_{\acute{G}} & \xrightarrow{\ \zeta_G\ } & X_{\grave{G}} \\ {\scriptstyle \pi_{\acute{\mathcal{G}}}}\big\downarrow & & \big\downarrow{\scriptstyle \pi_{\grave{\mathcal{G}}}} \\ X_{\acute{\mathcal{G}}} & \xrightarrow[\ \zeta_{\mathcal{G}}\ ]{} & X_{\grave{\mathcal{G}}} \end{array}$$

Note that $\pi_{\acute{\mathcal{G}}}$ and $\pi_{\grave{\mathcal{G}}}$ denote the sofic covers defined by $\acute{\mathcal{G}}$ and $\grave{\mathcal{G}}$(see the beginning of Section 4).

If $\mathcal{G}$ is the canonical λ-graph of a sofic system (Y, σ) and $n \in \mathbf{N}$, then $\mathcal{G}^n$ is the canonical λ-graph of $(Y^{(n)}, \sigma^{(n)})$. (The same fact was pointed out in [BK2] for $\mathcal{K}_Y^+$ and $\mathcal{K}_Y^-$ and other "canonical λ-graphs" introduced in [Kr2].) If $\mathcal{G}_i$ is the canonical λ-graph of a sofic system (Y_i, σ_i) for $i = 1, 2$ with $Y_1 \cap Y_2 = \emptyset$, then $\mathcal{G}_1 \sqcup \mathcal{G}_2$ is the canonical λ-graph for the sofic system $(Y_1 \cup Y_2, \sigma)$.

Let ζ be a forward bipartite conjugacy between sofic systems. Let (Y, σ) be the bipartite subshift that induces ζ. Let C and D be the disjoint alphabets such that (Y, σ) is over $C \cup D$, for the induced pair of subshifts $(\acute{Y}, \acute{\sigma})$ and $(\grave{Y}, \grave{\sigma})$, $(\acute{Y}, \acute{\sigma})$ is over CD and $(\grave{Y}, \grave{\sigma})$ is over DC, and ζ is the conjugacy of $(\acute{Y}, \acute{\sigma})$ onto $(\grave{Y}, \grave{\sigma})$ defined by

$$\zeta((c_id_i)_{i\in\mathbf{Z}}) = (d_ic_{i+1})_{i\in\mathbf{Z}}, \quad (c_id_i)_{i\in\mathbf{Z}} \in \acute{Y}, \quad c_i \in C, \ \ d_i \in D.$$

Since $(\acute{Y},\acute{\sigma})$ is a sofic system, it follows that (Y,σ) is a sofic system. Let $\mathcal{G}=(G,\lambda)$ be a λ-graph that defines (Y,σ). Then $\mathcal{G}$ is a bipartite λ-graph with A_G partitioned into

$$\acute{A}_G=\{\alpha\in A_G \mid \lambda(\alpha)\in C\} \quad\text{and}\quad \grave{A}_G=\{\alpha\in A_G \mid \lambda(\alpha)\in D\},$$

and with V_G partitioned into $\acute{V}_G$ and $\grave{V}_G$ where

$$\acute{V}_G=i_G(\acute{A}_G)=t_G(\grave{A}_G) \quad\text{and}\quad \grave{V}_G=t_G(\acute{A}_G)=i_G(\grave{A}_G).$$

Clearly $(X_{\acute{\mathcal{G}}},\sigma_{\acute{\mathcal{G}}})=(\acute{Y},\acute{\sigma}),(X_{\grave{\mathcal{G}}},\sigma_{\grave{\mathcal{G}}})=(\grave{Y},\grave{\sigma})$ and $\zeta_{\mathcal{G}}=\zeta$. Since the canonical λ-graph of (Y,σ) is unique, a canonical bipartite λ-graph $\mathcal{G}$ such that $\zeta=\zeta_{\mathcal{G}}$ is unique. If $\mathcal{G}$ is the canonical λ-graph of (Y,σ), then $\mathcal{G}^2$ is the canonical λ-graph of $(Y^{(2)},\sigma^{(2)})$. Since $Y^{(2)}=\acute{Y}\cup\grave{Y}$ with $L_1(\acute{Y})\cap L_1(\grave{Y})=\emptyset$ and $\mathcal{G}^2=\acute{\mathcal{G}}\sqcup\grave{\mathcal{G}}$, it follows that if $\mathcal{G}$ is the canonical λ-graph of (Y,σ), then $\acute{\mathcal{G}}$ and $\grave{\mathcal{G}}$ are the canonical λ-graphs of $(\acute{Y},\acute{\sigma})$ and $(\grave{Y},\grave{\sigma})$, respectively. Thus we have proved the following facts.

(1) If ζ is a forward bipartite conjugacy between sofic systems, then there is a unique canonical bipartite λ-graph $\mathcal{G}$ such that $\zeta=\zeta_{\mathcal{G}}$.

(2) If $\mathcal{G}$ is a canonical bipartite λ-graph, then both $\acute{\mathcal{G}}$ and $\grave{\mathcal{G}}$ are canonical.

Let $\kappa:(Y,\sigma)\to(Y',\sigma')$ be a symbolic conjugacy between sofic systems. Let A and A' be the underlying alphabets of (Y,σ) and (Y',σ'), respectively. Then there is a bijection $k:A\to A'$ that gives κ. Let $\mathcal{G}=(G,\lambda)$ and $\mathcal{G}'=(G',\lambda')$ be the canonical λ-graph that defines (Y,σ) and (Y',σ'), respectively. Then by the construction of the canonical λ-graph, there is a graph-isomorphism $h:G\to G'$ such that the following diagram commutes:

$$\begin{array}{ccc} A_G & \xrightarrow{\;h_A\;} & A_{G'} \\ {\scriptstyle\lambda}\downarrow & & \downarrow{\scriptstyle\lambda'} \\ A & \xrightarrow[\;k\;]{} & A' \end{array}$$

Recall that $h=(h_A,h_V)$, where h_A is the arc-map and h_V is the vertex-map.

Generally, for two λ-graphs $\mathcal{G}=(G,\lambda)$ over an alphabet A and $\mathcal{G}'=(G',\lambda')$ over an alphabet A', a pair (h,k) of a graph-isomorphism $h:G\to G'$ and a bijection $k:A\to A'$ such that the diagram above commutes, is called a *λ-graph-isomorphism* $\mathcal{G}$ onto $\mathcal{G}'$ and denoted by

$$(h,k):\mathcal{G}\simeq\mathcal{G}'.$$

If there is a λ-graph-isomorphism between λ-graphs $\mathcal{G}$ and $\mathcal{G}'$, then $\mathcal{G}$ and $\mathcal{G}'$ are said to be *isomorphic*. We have shown the following fact:

(3) If κ is a symbolic conjugacy between sofic systems, then there is a unique λ-graph-isomorphism $(h,k) : \mathcal{G} \simeq \mathcal{G}'$ such that $\mathcal{G}$ and $\mathcal{G}'$ are canonical λ-graphs and κ is a symbolic conjugacy of $(X_{\mathcal{G}}, \sigma_{\mathcal{G}})$ onto $(X_{\mathcal{G}'}, \sigma_{\mathcal{G}'})$ given by k.

If two λ-graphs $\mathcal{G}$ and $\mathcal{G}'$ are isomorphic through a λ-graph-isomorphism (h,k), then $\mathcal{M}_{\mathcal{G}}$ is the same as $\mathcal{M}_{\mathcal{G}'}$ if every symbol a appearing in the entries of $\mathcal{M}_{\mathcal{G}}$ is identified with $k(a)$ and the indexing for $\mathcal{M}_{\mathcal{G}'}$ is adjusted according to h_V.

In general, for two representation matrices over alphabets A and A', respectively, and a bijection $k : A \to A'$, we say that $\mathcal{M}$ and $\mathcal{N}$ are *equivalent under specification* k, write

$$\mathcal{M} \overset{k}{\simeq} \mathcal{N},$$

and call this a *specified equivalence* from $\mathcal{M}$ to $\mathcal{N}$ if $\mathcal{N}$ can be obtained from $\mathcal{M}$ by replacing every symbol a appearing $\mathcal{M}$ by $k(a)$ (without changing the indexing of $\mathcal{M}$ or $\mathcal{N}$). If $\mathcal{M} \overset{k}{\simeq} \mathcal{N}$ for some specification k, we say that $\mathcal{M}$ and $\mathcal{N}$ are *equivalent* and write

$$\mathcal{M} \simeq \mathcal{N}.$$

If $\mathcal{M}$ and $\mathcal{N}$ are square representation matrices and $\mathcal{M} \overset{k}{\simeq} \mathcal{N}$ is a specified equivalence, then for the λ-graphs $\mathcal{G}$ and $\mathcal{G}'$ with $\mathcal{M}_{\mathcal{G}} = \mathcal{M}$ and $\mathcal{M}_{\mathcal{G}'} = \mathcal{N}$, there is a λ-graph-isomorphism $(h,k) : \mathcal{G} \simeq \mathcal{G}'$. Generally, h is not uniquely determined by $\mathcal{M} \overset{k}{\simeq} \mathcal{N}$, because it is possible that h_A is not uniquely determined (because $\mathcal{G}$ and $\mathcal{G}'$ may have multiple arcs having the same initial vertex, the same terminal vertex, and the same label). But we note that h_V is uniquely determined by the natural correspondence between the indices of the matrices. Boyle and Krieger [BK2] introduced a certain ring such that the representation matrices can be presented as elements of the ring, with this presentation the equivalence relation $\simeq$ becomes equality, and the equations of "strong shift equivalence" and "shift equivalence" (see below and Section 9) hold with respect to multiplication in the ring. But, in this memoir we use representation matrices and $\simeq$ because we consider specified equivalences.

A sequence of specified equivalences of representation matrices of the form

$$\mathcal{M} \overset{k_0}{\simeq} \mathcal{P}_1\mathcal{Q}_1, \quad \mathcal{Q}_1\mathcal{P}_1 \overset{k_2}{\simeq} \mathcal{P}_2\mathcal{Q}_2, \cdots, \quad \mathcal{Q}_{n-1}\mathcal{P}_{n-1} \overset{k_{n-1}}{\simeq} \mathcal{P}_n\mathcal{Q}_n, \quad \mathcal{Q}_n\mathcal{P}_n \overset{k_n}{\simeq} \mathcal{N}$$

with $n \in \mathbf{N}$, $\mathcal{M}$ and $\mathcal{N}$ square and $\mathcal{P}_1, \cdots, \mathcal{P}_n$ and $\mathcal{Q}_1, \cdots, \mathcal{Q}_n$ rectangular, is called a *specified strong shift equivalence* of *lag* n from $\mathcal{M}$ to $\mathcal{N}$. If we eliminate the specifications k_i's in the sequence above, then we have a sequence of equivalences

$$\mathcal{M} \simeq \mathcal{P}_1\mathcal{Q}_1, \mathcal{Q}_1\mathcal{P}_1 \simeq \mathcal{P}_2\mathcal{Q}_2, \cdots, \mathcal{Q}_{n-1}\mathcal{P}_{n-1} \simeq \mathcal{P}_n\mathcal{Q}_n, \mathcal{Q}_n\mathcal{P}_n \simeq \mathcal{N}$$

which is called a *strong shift equivalence* of *lag* n from $\mathcal{M}$ to $\mathcal{N}$ after Williams [Wi]. A sequence of equalities of nonnegative integral matrices of the form

$$M = P_1Q_1, Q_1P_1 = P_2Q_2, \cdots, Q_{n-1}P_{n-1} = P_nQ_n, Q_nP_n = N$$

with $n \in \mathbf{N}$, M and N square and $P_1, \cdots P_n$ and $Q_1, \cdots, Q_n$ rectangular, is called a *strong shift equivalence* of *lag n* from M to N, which was defined by Williams [Wi].

A λ-graph $\mathcal{G} = (G, \lambda)$ is said to *represent a graph* if $\lambda(a) \neq \lambda(b)$ for any distinct arcs a, b of G. For a graph G, let $\tilde{G}$ denote the λ-graph (G, id_{A_G}) and let $\tilde{M}_G$ mean $\mathcal{M}_{\tilde{G}}$. In general, for a nonnegative integral rectangular matrix M, let $\tilde{M}$ denote the representation matrix such that all the symbols appearing in $\tilde{M}$ are distinct and M can be obtained from $\tilde{M}$ by substituting 1 for all the symbols in $\tilde{M}$. For example if

$$M = \begin{pmatrix} 2 & 2 \\ 1 & 0 \end{pmatrix},$$

then

$$\tilde{M} = \begin{pmatrix} a+b & c+d \\ e & 0 \end{pmatrix}.$$

Theorem 5.2. Let $\phi : (Y, \sigma) \to (Y', \sigma')$ be a forward conjugacy between sofic systems. Let n be a nonnegative integer. Let

$$\phi = \kappa_n \zeta_n \cdots \kappa_1 \zeta_1 \kappa_0$$

be a κ-ζ factorization of ϕ (i.e, $\kappa_0, \cdots, \kappa_n$ are symbolic conjugacies and $\zeta_1, \cdots, \zeta_n$ are forward bipartite conjugacies). Let $\mathcal{G}$ and $\mathcal{G}'$ be the canonical λ-graphs of (Y, σ) and (Y', σ'), respectively. Then there are unique bipartite canonical λ-graphs

$$\mathcal{G}_i, \quad i = 1, \cdots, n$$

and unique λ-graph-isomorphisms

$$(h_i, k_i) : \grave{\mathcal{G}}_i \simeq \acute{\mathcal{G}}_{i+1}, \quad i = 0, \cdots, n$$

with $\grave{\mathcal{G}}_0 = \mathcal{G}$ and $\acute{\mathcal{G}}_{n+1} = \mathcal{G}'$ (for convenience sake) such that κ_i is the symbolic conjugacy given by k_i for $i = 0, \cdots, n$, and

$$\zeta_{\mathcal{G}_i} = \zeta_i, \quad i = 1, \cdots, n.$$

Proof. This follows from (1),(2) and (3) above and the uniqueness of the canonical λ-graph of a sofic system. □

Let $\mathcal{G} = (G, \lambda)$ and $\mathcal{G}' = (G', \lambda')$ be (not necessarily canonical) λ-graphs. Let n be a nonnegative integer. Suppose that there are (not necessarily canonical) bipartite λ-graphs.

$$\mathcal{G}_i = (G_i, \lambda_i), \quad i = 1, \cdots, n \tag{5.1}$$

and λ-graph-isomorphisms

$$(h_i, k_i) : \grave{\mathcal{G}}_i \simeq \acute{\mathcal{G}}_{i+1} \quad i = 0, \cdots, n \tag{5.2}$$

with $\grave{\mathcal{G}}_0 = (\grave{G}_0, \grave{\lambda}_0) = \mathcal{G}$ and $\acute{\mathcal{G}}_{n+1} = (\acute{G}_{n+1}, \acute{\lambda}_{n+1}) = \mathcal{G}'$ (for convenience sake). Put, for $i = 0, \cdots, n+1$,

$$\begin{array}{cc} (X_{\acute{\mathcal{G}}_i}, \sigma_{\acute{\mathcal{G}}_i}) = (\acute{Y}_i, \acute{\sigma}_i), & (X_{\grave{\mathcal{G}}_i}, \sigma_{\grave{\mathcal{G}}_i}) = (\grave{Y}_i, \grave{\sigma}_i) \\ (X_{\acute{G}_i}, \sigma_{\acute{G}_i}) = (\acute{X}_i, \acute{\varsigma}_i), & (X_{\grave{G}_i}, \sigma_{\grave{G}_i}) = (\grave{X}_i, \grave{\varsigma}_i) \\ \pi_{\acute{\mathcal{G}}_i} = \acute{\pi}_i, & \pi_{\grave{\mathcal{G}}_i} = \grave{\pi}_i. \end{array}$$

Let $\kappa_i : \grave{Y}_i \to \acute{Y}_{i+1}$ be the symbolic conjugacy given by k_i for $i = 0, \cdots, n$. Putting

$$h_{A_{G_i}} = \bar{k}_i, \quad i = 0, \cdots, n,$$

let $\bar{\kappa}_i : \grave{X}_i \to \acute{X}_{i+1}$ be the symbolic conjugacy given by $\bar{k}_i$ for $i = 0, \cdots, n$. Put, for $i = 1, \cdots, n$,

$$\zeta_{\mathcal{G}_i} = \zeta_i \quad \text{and} \quad \zeta_{G_i} = \bar{\zeta}_i.$$

Then we have the following commutative diagram:

(5.3)

$$\begin{array}{ccccccccccccc} \grave{X}_0 & \xrightarrow{\bar{\kappa}_0} & \acute{X}_1 & \xrightarrow{\bar{\zeta}_1} & \grave{X}_1 & \xrightarrow{\bar{\kappa}_1} & \acute{X}_2 & \cdots & \acute{X}_n & \xrightarrow{\bar{\zeta}_n} & \grave{X}_n & \xrightarrow{\bar{\kappa}_n} & \acute{X}_{n+1} \\ \downarrow{\scriptstyle \grave{\pi}_0} & & \downarrow{\scriptstyle \acute{\pi}_1} & & \downarrow{\scriptstyle \grave{\pi}_1} & & \downarrow{\scriptstyle \acute{\pi}_2} & \cdots & \downarrow{\scriptstyle \acute{\pi}_n} & & \downarrow{\scriptstyle \grave{\pi}_n} & & \downarrow{\scriptstyle \acute{\pi}_{n+1}} \\ \grave{Y}_0 & \xrightarrow[\kappa_0]{} & \acute{Y}_1 & \xrightarrow[\zeta_1]{} & \grave{Y}_1 & \xrightarrow[\kappa_1]{} & \acute{Y}_2 & \cdots & \acute{Y}_n & \xrightarrow[\zeta_n]{} & \grave{Y}_n & \xrightarrow[\kappa_n]{} & \acute{Y}_{n+1}. \end{array}$$

Put $X_{\mathcal{G}} = Y$, put $X_{\mathcal{G}'} = Y'$, let

$$\phi = \kappa_n \zeta_n \cdots \kappa_1 \zeta_1 \kappa_0,$$

and let

$$\bar{\phi} = \bar{\kappa}_n \bar{\zeta}_n \cdots \bar{\kappa}_1 \bar{\zeta}_1 \bar{\kappa}_0.$$

If $\mathcal{G} = \mathcal{K}_Y^-$ and $\mathcal{G}' = \mathcal{K}_{Y'}^-$, then

$$\bar{\phi} = \phi_0,$$

where ϕ_0 is the unique topological conjugacy with $\pi_{\mathcal{K}_{Y'}^-} \phi_0 = \phi \pi_{\mathcal{K}_Y^-}$ given by Krieger (Theorem 4.2). Therefore it follows from the proof of Theorem 3.1 of [N3] and Theorem 4.3 that if $\mathcal{G}$ and $\mathcal{G}'$ are canonical, that is, $\mathcal{G} = (\mathcal{K}_Y^-)_+$ and $\mathcal{G}' = (\mathcal{K}_{Y'}^-)_+$, then

$$\bar{\phi} = (\phi_0)_+,$$

where $(\phi_0)_+$ is the unique topological conjugacy such that $\pi_{\mathcal{G}'}(\phi_0)_+ = \phi \pi_{\mathcal{G}}$ and $\gamma_+^{\mathcal{G}'}((\phi_0)_+(x)) = \phi_0(\gamma_+^{\mathcal{G}}(x))$ for all $x \in X_G$ (cf. Theorem 4.3).

Assume that $n \geq 1$ in the above. Then since $\mathcal{G}_i = (G_i, \lambda_i)$ is a bipartite λ-graph, we can write

$$\begin{aligned} \mathcal{M}_{\mathcal{G}_i} &= \begin{pmatrix} 0 & \mathcal{P}_i \\ \mathcal{Q}_i & 0 \end{pmatrix}, \quad i = 1, \cdots, n \\ M_{G_i} &= \begin{pmatrix} 0 & P_i \\ Q_i & 0 \end{pmatrix}, \quad i = 1, \cdots, n \\ \tilde{M}_{G_i} &= \begin{pmatrix} 0 & \tilde{P}_i \\ \tilde{Q}_i & 0 \end{pmatrix}, \quad i = 1, \cdots, n. \end{aligned}$$

Then the existence of the λ-graph-isomorphisms (5.2) implies the existence of the following specified strong shift equivalences and strong shift equivalences.

$$(5.4) \quad \mathcal{M}_{\mathcal{G}} \overset{k_0}{\simeq} \mathcal{P}_1\mathcal{Q}_1,\ \mathcal{Q}_1\mathcal{P}_1 \overset{k_1}{\simeq} \mathcal{Q}_2\mathcal{P}_2,\ \cdots,\ \mathcal{Q}_{n-1}\mathcal{P}_{n-1} \overset{k_{n-1}}{\simeq} \mathcal{P}_n\mathcal{Q}_n,\ \mathcal{Q}_n\mathcal{P}_n \overset{k_n}{\simeq} \mathcal{M}_{\mathcal{G}'}$$

$$(5.5) \quad \mathcal{M}_{\mathcal{G}} \simeq \mathcal{P}_1\mathcal{Q}_1,\ \mathcal{Q}_1\mathcal{P}_1 \simeq \mathcal{Q}_2\mathcal{P}_2,\ \cdots,\ \mathcal{Q}_{n-1}\mathcal{P}_{n-1} \simeq \mathcal{P}_n\mathcal{Q}_n,\ \mathcal{Q}_n\mathcal{P}_n \simeq \mathcal{M}_{\mathcal{G}'}$$

$$(5.6) \quad \tilde{M}_G \overset{\bar{k}_0}{\simeq} \tilde{P}_1\tilde{Q}_1,\ \tilde{Q}_1\tilde{P}_1 \overset{\bar{k}_1}{\simeq} \tilde{P}_2\tilde{Q}_2,\ \cdots,\ \tilde{Q}_{n-1}\tilde{P}_{n-1} \overset{\bar{k}_{n-1}}{\simeq} \tilde{P}_n\tilde{Q}_n,\ \tilde{Q}_n\tilde{P}_n \overset{\bar{k}_n}{\simeq} \tilde{M}_{G'}$$

$$(5.7) \quad M_G = P_1Q_1,\ Q_1P_1 = P_2Q_2,\ \cdots,\ Q_{n-1}P_{n-1} = P_nQ_n,\ Q_nP_n = M_{G'}$$

We call (5.6) the *support* of (5.4) (though it is possible that the specifications in (5.6) are not unique for (5.4)). We also call (5.7) the *support* of (5.5). The κ-ζ factorization

$$\phi = \kappa_n\zeta_n \ \cdots \ \kappa_1\zeta_1\kappa_0$$

is uniquely determined by (5.4) and hence is said to be *given by* that specified strong shift equivalence. Clearly the κ-ζ factorization

$$\bar{\phi} = \bar{\kappa}_n\bar{\zeta}_n \cdots \bar{\kappa}_1\bar{\zeta}_1\kappa_0$$

is given by (5.6).

Recall that for a square nonnegative integral matrix M, (X_M, σ_M) means (X_G, σ_G) where G is the graph such that $M_G = M$. Similarly, for a square representation matrix $\mathcal{M}$, let $(X_{\mathcal{M}}, \sigma_{\mathcal{M}})$ mean $(X_{\mathcal{G}}, \sigma_{\mathcal{G}})$, where $\mathcal{G}$ is the λ-graph such that $\mathcal{M}_{\mathcal{G}} = \mathcal{M}$.

A square representation matrix is said to be *canonical* if it represents a canonical λ-graph. A specified strong shift equivalence between representation matrices

$$\mathcal{M} \overset{k_0}{\simeq} \mathcal{P}_1\mathcal{Q}_1,\ \mathcal{Q}_1\mathcal{P}_1 \overset{k_1}{\simeq} \mathcal{P}_2\mathcal{Q}_2,\ \cdots,\ \mathcal{Q}_{n-1}\mathcal{P}_{n-1} \overset{k_{n-1}}{\simeq} \mathcal{P}_n\mathcal{Q}_n,\ \mathcal{Q}_n\mathcal{P}_n \overset{k_n}{\simeq} \mathcal{N}$$

with $n \in \mathbf{N}$, is said to be *canonical* if $\begin{pmatrix} 0 & \mathcal{P}_i \\ \mathcal{Q}_i & 0 \end{pmatrix}$ is canonical for all $i = 1, \cdots, n$. The definition of a *canonical strong shift equivalence* is given in the same way.

Theorem 5.3. (1) Let $\mathcal{M}$ and $\mathcal{N}$ be two canonical representation matrices and let $\phi : (X_{\mathcal{M}}, \sigma_{\mathcal{M}}) \to (X_{\mathcal{N}}, \sigma_{\mathcal{N}})$ be a forward conjugacy. Then there are $n \in \mathbf{N}$ and a canonical specified strong shift equivalence of lag n from $\mathcal{M}$ to $\mathcal{N}$

$$\mathcal{M} \overset{k_0}{\simeq} \mathcal{P}_1\mathcal{Q}_1,\ \mathcal{Q}_1\mathcal{P}_1 \overset{k_1}{\simeq} \mathcal{P}_2\mathcal{Q}_2,\ \cdots,\ \mathcal{Q}_{n-1}\mathcal{P}_{n-1} \overset{k_{n-1}}{\simeq} \mathcal{P}_n\mathcal{Q}_n,\ \mathcal{Q}_n\mathcal{P}_n \overset{k_n}{\simeq} \mathcal{N}$$

which gives a κ-ζ factorization of lag n of ϕ. For each κ-ζ factorization of lag greater than 0 of ϕ, the canonical specified strong shift equivalence giving it is unique.

(2) Let M and N be two square nonnegative integral matrices and let $\phi : (X_M, \sigma_M) \to (X_N, \sigma_N)$ be a forward conjugacy, then there are $n \in \mathbf{N}$ and a specified strong shift equivalence of lag n from $\tilde{M}$ to $\tilde{N}$

$$\tilde{M} \overset{k_0}{\simeq} \tilde{P}_1\tilde{Q}_1,\ \tilde{Q}_1\tilde{P}_1 \overset{k_1}{\simeq} \tilde{P}_2\tilde{Q}_2,\ \cdots,\ \tilde{Q}_{n-1}\tilde{P}_{n-1} \overset{k_{n-1}}{\simeq} \tilde{P}_n\tilde{Q}_n,\ \tilde{Q}_n\tilde{P}_n \overset{k_n}{\simeq} \tilde{N}$$

which gives a κ-ζ factorization of ϕ, where

$$M = P_1Q_1,\ Q_1P_1 = P_2Q_2,\ \cdots,\ Q_{n-1}P_{n-1} = P_nQ_n,\ Q_nP_n = N$$

is a strong shift equivalence from M to N.

Proof. We note that if ϕ is of lag 0, then it is of lag 1. For if ϕ is a symbolic conjugacy given by a bijection $k_0 : A \to B$, where $A = L_1(X_{\mathcal{M}})$ and $B = L_1(X_{\mathcal{N}})$, then ϕ is the bipartite code induced by the bipartite expression f of A defined by $f(a) = sk_0(a), a \in A$, where s is a symbol. Therefore (1) is proved by Theorem 5.2 and the discussion above. We note that neither the indexing of $\mathcal{M}$ nor that of $\mathcal{N}$ needs to be changed.

To show (2), let $\mathcal{M} = \tilde{M}$, let $\mathcal{N} = \tilde{N}$ and apply (1). Then by the discussion above, we know that the support of the canonical specified strong shift equivalence from $\mathcal{M}$ to $\mathcal{N}$ giving a κ-ζ factorization of ϕ, say $\phi = \kappa_n\zeta_n \cdots \kappa_1\zeta_1\kappa_0$, gives the κ-ζ factorization

$$\bar{\phi} = \bar{\kappa}_n\bar{\zeta}_n \cdots \bar{\kappa}_1\bar{\zeta}_1\bar{\kappa}_0.$$

Therefore since $\phi = \bar{\phi}$, (2) is proved. □

Here we give a proof for a fact stated in Section 3.

Proposition 5.4. Let $T = (p, q : \Gamma \to G)$ and $T' = (p', q' : \Gamma' \to G')$ be nondegenerate textile systems and assume that M_Γ and M_G are irreducible and have the same maximal eigenvalue. If T and T' are topologically conjugate, then $L(T) = L(T')$ and $R(T) = R(T')$.

Proof. Since T and T' are topologically conjugate and they are nondegenerate, there are topological conjugacies $\psi : (X_G, \sigma_G) \to (X_{G'}, \sigma_{G'})$ and $\Psi : (X_\Gamma, \sigma_\Gamma) \to (X_{\Gamma'}, \sigma_{\Gamma'})$

such that the following diagram commutes:

$$\begin{array}{ccc}
X_G & \xrightarrow{\psi} & X_{G'} \\
{\scriptstyle \xi_T}\uparrow & & \uparrow{\scriptstyle \xi_{T'}} \\
X_\Gamma & \xrightarrow{\Psi} & X_{\Gamma'} \\
{\scriptstyle \eta_T}\downarrow & & \downarrow{\scriptstyle \eta_{T'}} \\
X_G & \xrightarrow[\psi]{} & X_{G'}
\end{array} .$$

Observing the proof of Theorem 1 of [HN], we see that there are factorizations

$$\Psi = \Psi_n \cdots \Psi_1$$

and

$$\psi = \psi_n \cdots \psi_1$$

into bipartite codes $\Psi_i : Z_{i-1} \to Z_i, i = 1, \cdots, n$, and $\psi_i : X_{i-1} \to X_i, i = 1, \cdots, n$, and onto 1-block maps $\xi_i : Z_i \to X_i$ and $\eta_i : Z_i \to X_i$, $i = 0, \cdots, n$, such that $Z_0 = X_\Gamma, Z_n = X_{\Gamma'}, X_0 = X_G, X_n = X_{G'}, \xi_0 = \xi_T, \xi_n = \xi_{T'}, \eta_0 = \eta_T, \eta_n = \eta_{T'}$, and the diagrams

$$\begin{array}{ccc}
X_i & \xrightarrow{\psi_{i+1}} & X_{i+1} \\
{\scriptstyle \xi_i}\uparrow & & \uparrow{\scriptstyle \xi_{i+1}} \\
Z_i & \xrightarrow{\Psi_{i+1}} & Z_{i+1} \\
{\scriptstyle \eta_i}\downarrow & & \downarrow{\scriptstyle \eta_{i+1}} \\
X_i & \xrightarrow[\psi_{i+1}]{} & X_{i+1}
\end{array} \qquad i = 0, \cdots, n-1$$

commute. We may assume that all Ψ_i and ψ_i are forward bipartite codes. Clearly the subshifts (X_i, σ_i) and (Z_i, ς_i) are subshifts of finite type for all $i = 0, \cdots, n$. Therefore if it is valid that $(*)$ $R(T^{[m]}) = R(T)$ and $L(T^{[m]}) = L(T)$ for any $m \in \mathbf{N}$ and for any nondegenerate textile system T having graphs whose adjacency matrices are irreducible, then passing through higher block systems of the same sufficiently large order for all i, we may assume that (X_i, σ_i) and (Z_i, ς_i) are topological Markov shifts and ξ_i and η_i are given by graph-homomorphisms for all $i = 0, \cdots, n$. Furthermore, if under this assumption, it is proved that $L(T) = L(T')$ and $R(T) = R(T')$, then $(*)$ is valid. Thus it suffices to prove the proposition for the case where ψ and Ψ are forward bipartite codes.

Observing the proof of Theorem 2 of [HN], we see that there are bipartite graphs Γ_B, G_B and graph-homomorphisms $p_B, q_B : \Gamma_B \to G_B$ such that $p_B(\acute{A}_{\Gamma_B}) \subset \acute{A}_{G_B}, p_B(\grave{A}_{\Gamma_B}) \subset \grave{A}_{G_B}, q_B(\acute{A}_{\Gamma_B}) \subset \acute{A}_{G_B}$ and $q_B(\grave{A}_{\Gamma_B}) \subset \grave{A}_{G_B}; \acute{\Gamma}_B = \Gamma, \grave{\Gamma}_B = \Gamma', \acute{G}_B = G$ and $\grave{G}_B = G'$; for $\alpha\beta \in A_\Gamma$ with $\alpha \in \acute{A}_B$ and $\beta \in \grave{A}_B, p(\alpha\beta) = p_B(\alpha)p_B(\beta)$ and $q(\alpha\beta) = q_B(\alpha)q_B(\beta)$, and for $\beta\alpha \in A_{\Gamma'}$ with $\alpha \in \acute{A}_{\Gamma_B}$ and $\beta \in \grave{A}_{\Gamma_B}, p'(\beta\alpha) = p_B(\beta)p_B(\alpha)$ and $q'(\beta\alpha) = q_B(\beta)q_B(\alpha)$. Let λ_B be the maximal eigenvalue of M_{Γ_B} and let l_B be a left eigenvector of M_{Γ_B} corresponding to λ_B. Corresponding to the bipartite partition $\{\acute{V}_{\Gamma_B}, \grave{V}_{\Gamma_B}\}$ of V_{Γ_B}, l_B is written in the form $l_B = (\acute{l}, \grave{l})$. It is easy to see that λ_B^2 is the maximal eigenvalue of both M_Γ and $M_{\Gamma'}$ and $\acute{l}$ is a left eigenvector of M_Γ corresponding to λ_B^2 and $\grave{l}$ is a left eigenvector of $M_{\Gamma'}$ corresponding to λ_B^2. Let $(l_0)_B = (\acute{l}_0, \grave{l}_0)$ be defined similarly for G_B. Then similar facts to the above also hold. For $v \in V_{\Gamma_B}$, let $l_B(v)$ denote the component of l_B corresponding to v, and for $U \subset V_{\Gamma_B}$, let $l_B(U) = \sum_{v \in U} l_B(v)$. Let similar notation be used for $\acute{l}$, $\grave{l}$, $(l_0)_B, \acute{l}_0$, and $\grave{l}_0$. Let U be a maximal left compatible set for p and U' a maximal left compatible set for p'. Then $U \subset \acute{V}_{\Gamma_B}$ and $U' \subset \grave{V}_{\Gamma_B}$ and U and U' are maximal left compatible sets for p_B (as shown in the proof of Theorem 3.1 of [N3]). Therefore, using the property of maximal compatible sets for graph-homomorphisms stated in Section 3, we have

$$\frac{\acute{l}(U)}{\acute{l}_0(p_V(U))} = \frac{l_B(U)}{(l_0)_B((p_B)_V(U))} = \frac{l_B(U')}{(l_0)_B((p_B)_V(U'))} = \frac{\grave{l}(U')}{\grave{l}_0(p'_V(U'))}.$$

Similarly we have

$$\frac{\acute{l}(V)}{\acute{l}_0(q_V(V))} = \frac{\grave{l}(V')}{\grave{l}_0(q'_V(V'))}$$

for a maximal left compatible set V for q and a maximal left compatible set V' for q'. Thus

$$L(T) = \frac{\dfrac{\acute{l}(V)}{\acute{l}_0(q_V(V))}}{\dfrac{\acute{l}(U)}{\acute{l}_0(p_V(U))}} = \frac{\dfrac{\grave{l}(V')}{\grave{l}_0(q'_V(V'))}}{\dfrac{\grave{l}(U')}{\grave{l}_0(p'_V(U'))}} = L(T').$$

We also have $R(T) = R(T')$, because $\deg(T) = \deg(T')$ and $L(T)R(T)\deg(T) = 1 = L(T')R(T')\deg(T')$. □

6. LR TEXTILE SYSTEMS

Let $T = (p, q : \Gamma \to G)$ be an LR textile system with $T^* = (p^T, q^T : \Gamma^T \to G^T)$. Put $G^T = H$. The property that p is left resolving, is interpreted as follows: for any $ab' \in A_{GH}$ with $a \in A_G$ and $b' \in A_H$, there is a unique $\alpha \in A_\Gamma = A_{\Gamma^T}$ such that $p(\alpha) = a$ and $q^T(\alpha) = b'$. The property that q is right resolving, is also interpreted as follows: for any $ba' \in A_{HG}$ with $b \in A_H$ and $a' \in A_G$, there is a unique $\alpha \in A_\Gamma = A_{\Gamma^T}$ such that $p^T(\alpha) = b$ and $q(\alpha) = a'$. Thus we have a bijection $k : A_{GH} \to A_{HG}$ by

$$k(p(\alpha)q^T(\alpha)) = p^T(\alpha)q(\alpha), \quad \alpha \in A_\Gamma.$$

It is clear that

$$i_{GH}(ab') = i_{HG}(k(ab')) \quad \text{and} \quad t_{GH}(ab') = t_{HG}(k(ab'))$$

for $ab' \in A_{GH}$ with $a \in A_G$ and $b' \in A_H$. Thus we have the specified equivalence

(6.1) $$\tilde{M}_G\tilde{M}_H \overset{k}{\simeq} \tilde{M}_H\tilde{M}_G,$$

which is called the *specified equivalence associated* with T. Note that this is also considered as a graph-isomorphism $k : GH \to HG$ which fixes the vertices. By (6.1) we have

$$M_GM_H = M_HM_G.$$

Conversely assume that we are given graphs G and H with $V_G = V_H$ and a specified equivalence

(6.2) $$\tilde{M}_G\tilde{M}_H \overset{k'}{\simeq} \tilde{M}_H\tilde{M}_G.$$

Then we have an LR textile system $T' = (p', q' : \Gamma' \to G)$ defined as follows:

$$A_{\Gamma'} = \{(b, b', a, a') \in A_H \times A_H \times A_G \times A_G \mid ab' \in A_{GH}, k'(ab') = ba'\}$$

and for $\alpha = (b, b', a, a') \in A_{\Gamma'}, i_{\Gamma'}(\alpha) = b, t_{\Gamma'}(\alpha) = b', p'(\alpha) = a$ and $q'(\alpha) = a'$. We call T' the *LR textile system associated* with the specified equivalence (6.2). It is clear that if we take k above as k', then $T' = T$ and if we take T' as T in the above, then $k = k'$. Therefore we have the following proposition:

Proposition 6.1. Let G and H be two graphs with the same vertex-set. Then there is an LR textile system T over G with $G^T = H$ if and only if

$$M_GM_H = M_HM_G.$$

Moreover the relation of being associated with each other gives a 1-1 correspondence between the LR textile systems T over G with $G^T = H$ and the specified equivalences of the form

$$\tilde{M}_G\tilde{M}_H \overset{k}{\simeq} \tilde{M}_H\tilde{M}_G.$$

In this proposition and in what follows, for a textile system T over a graph G, G^T denotes the graph over which T^* is defined.

Recall the notation defined in Section 2. Then we have the following lemma.

Lemma 6.2. Let T be an LR textile system over a graph G. Let $k, l \in \mathbf{N}$. Then

$$(6.3)\qquad (\check{X}_T^{(k,0)}, \check{\sigma}_T^{(k,0)}) = (X_{(G^T)^k}, \sigma_{(G^T)^k})$$

$$(6.4)\qquad (\check{X}_T^{(0,l)}, \check{\sigma}_T^{(0,l)}) = (X_{G^l}, \sigma_{G^l})$$

$$(6.5)\qquad (\check{X}_T^{(k,l)}, \check{\sigma}_T^{(k,l)}) = (X_{(G^T)^k G^l}, \sigma_{(G^T)^k G^l})$$

and $\check{\theta}_T^{(k,l)}$ is 1-1.

Proof. The equations (6.3) and (6.4) directly follow from the fact that an LR textile system is nondegenerate. The proof of the remainder is clear by observing the figure of a textile like Fig. 2.2, though its description given below is rather complicated. Let

$$x = ((b_0^{(i)} \cdots b_{k-1}^{(i)})(a_0^{(i)} \cdots a_{l-1}^{(i)}))_{i\in\mathbf{Z}} \in X_{H^kG^l},$$

where $H = G^T$ and for all $i \in \mathbf{Z}$

$$b_0^{(i)} \cdots b_{k-1}^{(i)} \in L_k(H) \qquad \text{with } b_j^{(i)} \in A_H$$

and

$$a_0^{(i)} \cdots a_{l-1}^{(i)} \in L_l(G) \qquad \text{with } a_j^{(i)} \in A_G.$$

Then there is a unique $t \in U_T$ such that

$$\check{\theta}_T^{(k,l)}(t) = x.$$

In fact, this is proved as follows. Let $T = (p, q : \Gamma \to G)$ with $T^* = (p^T, q^T : \Gamma^T \to G^T)$. For $(i,j) \in \mathbf{Z}^2$, let

$$F(i,j) = \{(i',j') \in \mathbf{Z}^2 \mid i \le i' \le i+k-1, j \le j' \le j+l-1\}.$$

Since q is right resolving, for each $m \in \mathbf{Z}$ there is a unique well-weaved subconfiguration

$$(\alpha_{ij})_{(i,j)\in F(mk,ml)}$$

such that

$$p^T(\alpha_{mk,ml}) \cdots p^T(\alpha_{(m+1)k-1,ml}) \quad = b_0^{(m)} \cdots b_{k-1}^{(m)}$$

and

$$q(\alpha_{(m+1)k-1,ml}) \cdots q(\alpha_{(m+1)k-1,(m+1)l-1}) \quad = a_0^{(m)} \cdots a_{l-1}^{(m)}.$$

Since q is right resolving, it follows that for any $n \geq 0$, if we have a subconfiguration

$$(\alpha_{ij})_{(i,j)\in F(mk,(m+n)l)}$$

well weaved by T for all $m \in \mathbf{Z}$, then we have a unique subconfiguration

$$(\alpha_{ij})_{(i,j)\in F(mk,(m+n+1)l)}$$

well weaved by T such that

$$\begin{aligned} &p^T(\alpha_{mk,(m+n+1)l}) \cdots p^T(\alpha_{(m+1)k-1,(m+n+1)l}) \\ &= q^T(\alpha_{mk,(m+n+1)l-1}) \cdots q^T(\alpha_{(m+1)k-1,(m+n+1)l-1}) \end{aligned}$$

and

$$\begin{aligned} &q(\alpha_{(m+1)k-1,(m+n+1)l}) \cdots q(\alpha_{(m+1)k-1,(m+n+2)l-1}) \\ &= p(\alpha_{(m+1)k,(m+n+1)l}) \cdots p(\alpha_{(m+1)k,(m+n+2)l-1}), \end{aligned}$$

for all $m \in \mathbf{Z}$. Therefore letting

$$\check{E}^{(k,l)} = \bigcup F(mk, nl),$$

where the union is taken over all $m, n \in \mathbf{Z}$ with $m \leq n$, we have a well-weaved subconfiguration

$$(\alpha_{ij})_{(i,j)\in\check{E}^{(k,l)}}$$

such that for all $m \in \mathbf{Z}$

$$p^T(\alpha_{mk,ml}) \cdots p^T(\alpha_{(m+1)k-1,ml}) \quad = b_0^{(m)} \cdots b_{k-1}^{(m)}$$

and

$$q(\alpha_{(m+1)k-1,ml}) \cdots q(\alpha_{(m+1)k-1,(m+1)l-1}) \quad = a_0^{(m)} \cdots a_{l-1}^{(m)}.$$

Similarly since p is left resolving, we have a unique subconfiguration

$$(\alpha_{ij})_{(i,j)\in\hat{E}^{(k,l)}} \quad \text{with } \hat{E}^{(k,l)} = \mathbf{Z}^2 - \check{E}^{(k,l)}$$

well weaved by T such that for all $m \in \mathbf{Z}$

$$p(\alpha_{mk,(m-1)l}) \cdots p(\alpha_{mk,ml-1}) = a_0^{(m-1)} \cdots a_{l-1}^{(m-1)}$$

and

$$q^T(\alpha_{mk,ml-1}) \cdots q^T(\alpha_{(m+1)k-1,ml-1}) = b_0^{(m)} \cdots b_{k-1}^{(m)}.$$

Therefore letting $t = (\alpha_{ij})_{i,j \in \mathbf{Z}}$, we have a unique $t \in U_T$ such that $\check{\theta}_T^{(k,l)}(t) = x$. Thus we have proved (6.5) together with the fact that $\check{\theta}_T^{(k,l)}$ is 1-1. □

By the preceding lemma and Proposition 2.3 we have the following theorem.

Theorem 6.3. Let T be an LR textile system over a graph G. Let $k, l \in \mathbf{Z}$ with $kl > 0$. Then the following statements are valid.

(1) If T is 1-1, then $(U_T, \boldsymbol{\sigma}_T^{(0,l)})$ is topologically conjugate to

$$(\check{X}_T^{(0,l)}, \check{\sigma}_T^{(0,l)}) = (X_{G^l}, \sigma_{G^l})$$

and otherwise $\boldsymbol{\sigma}_T^{(0,l)}$ is not expansive.

(2) If T^* is 1-1, then $(U_T, \boldsymbol{\sigma}_T^{(k,0)})$ is topologically conjugate to

$$(\check{X}_T^{(k,0)}, \check{\sigma}_T^{(k,0)}) = (X_{(G^T)^k}, \sigma_{(G^T)^k})$$

and otherwise $\boldsymbol{\sigma}_T^{(k,0)}$ is not expansive.

(3) $(U_T, \boldsymbol{\sigma}_T^{(k,l)})$ is topologically conjugate to

$$(\check{X}_T^{(k,l)}, \check{\sigma}_T^{(k,l)}) = (X_{(G^T)^k G^l}, \sigma_{(G^T)^k G^l}).$$

Let T be an LR textile system over a graph G. We shall consider the *diagonal bias shift* $(\check{X}_T^{(1,1)}, \check{\sigma}_T^{(1,1)})$ defined by T. By Lemma 6.2, $\check{\theta}_T^{(1,1)}$ is 1-1 and

$$(\check{X}_T^{(1,1)}, \check{\sigma}_T^{(1,1)}) = (X_{G^T G}, \sigma_{G^T G}).$$

For $k, l \in \mathbf{Z}$, let $\psi_T^{(k,l)} : \check{X}_T^{(1,1)} \to \check{X}_T^{(1,1)}$ be defined by

$$\psi_T^{(k,l)} = \check{\theta}_T^{(1,1)} \boldsymbol{\sigma}_T^{(k,l)} (\check{\theta}_T^{(1,1)})^{-1}.$$

Then $\psi_T^{(k,l)}$ is an automorphism of $(\check{X}_T^{(1,1)}, \check{\sigma}_T^{(1,1)})$ by Proposition 2.4 (1). Clearly we have

$$\psi_T^{(k,l)} = (\psi_T^{(1,0)})^k (\psi_T^{(0,1)})^l$$

for all $k, l \in \mathbf{Z}$ and

$$\psi_T^{(1,0)} \psi_T^{(0,1)} = \psi_T^{(0,1)} \psi_T^{(1,0)} = \check{\sigma}_T^{(1,1)}.$$

The following two corollaries are direct consequences of Theorem 6.3 and Proposition 2.4.

Corollary 6.4. Let T be an LR textile system over a graph G with $M_G = M$ and $M_{G^T} = N$. Then the following statements are valid.

(1) If T is 1-1, then $(\check{X}_T^{(1,1)}, \psi_T^{(0,1)})$ is topologically conjugate to (X_M, σ_M) and otherwise $\psi_T^{(1,0)}$ is not expansive.

(2) If T^* is 1-1, then $(\check{X}_T^{(1,1)}, \psi_T^{(1,0)})$ is topologically conjugate to (X_N, σ_N) and otherwise $\psi_T^{(1,0)}$ is not expansive.

(3) For all $k, l \in \mathbf{N}$, $(\check{X}_T^{(1,1)}, \psi_T^{(k,l)})$ is topologically conjugate to $(X_{M^lN^k}, \sigma_{M^lN^k})$.

Corollary 6.5. Let T be a 1-1 LR textile system over a graph G with $M_G = M$ and $M_{G^T} = N$. Let $\varphi = \varphi_T$. Then the following statements are valid.

(1) If T^* is 1-1, then (X_M, φ) is topologically conjugate to (X_N, σ_N) and otherwise φ is not expansive.

(2) For all integers $k \geq 0$ and $l \geq 1$, $(X_M, \varphi^k \sigma_M^l)$ is topologically conjugate to $(X_{M^lN^k}, \sigma_{M^lN^k})$.

Recall the quarter-textile shifts $(\hat{U}_T, \hat{\sigma}_T^{(k,l)})$, $k, l \in \mathbf{N}$, defined by a textile system T (see the paragraph preceding Theorem 2.12). It should be clear that the trivial modifications of Propositions 2.3 and 2.4 and Lemma 6.2 give the following theorem and its corollary. Note that for an LR textile system T, $\tilde{\xi}_T$ is 1-1 if and only if T is one-sided 1-1.

Theorem 6.6. Let T be an LR textile system over a graph G. Let $k, l \in \mathbf{N}$. Then the following statements are valid.

(1) If T is one-sided 1-1, $(\hat{U}_T, \hat{\sigma}_T^{(0,l)})$ is topologically conjugate to $(\tilde{X}_{G^l}, \tilde{\sigma}_{G^l})$, and otherwise $\hat{\sigma}_T^{(0,l)}$ is not expansive.

(2) If T^* is one-sided 1-1, then $(\hat{U}_T, \hat{\sigma}_T^{(k,0)})$ is topologically conjugate to $(\tilde{X}_{(G^T)^k}, \tilde{\sigma}_{(G^T)^k})$, and otherwise $\hat{\sigma}_T^{(k,0)}$ is not expansive.

(3) $(\hat{U}_T, \hat{\sigma}_T^{(k,l)})$ is topologically conjugate to $(\tilde{X}_{(G^T)^kG^l}, \tilde{\sigma}_{(G^T)^kG^l})$.

Corollary 6.7. Let T be a one-sided 1-1 LR textile system over a graph G with $M_G = M$ and $M_{G^T} = N$. Let $\tilde{\varphi} = \tilde{\varphi}_T$. Then the following statements are valid.

(1) If T^* is one-sided 1-1, then $(X_M, \tilde{\varphi})$ is topologically conjugate to $(\tilde{X}_N, \tilde{\sigma}_N)$ and otherwise $\tilde{\varphi}$ is not expansive.

(2) For all integers $k \geq 0$ and $l \geq 1$, $(\tilde{X}_M, \tilde{\varphi}^k \tilde{\sigma}_M^l)$ is topologically conjugate to $(\tilde{X}_{M^lN^k}, \tilde{\sigma}_{M^lN^k})$.

Observation 6.8. Let T be an LR textile system over a graph G with $M_G = M$ and $M_{G^T} = N$. Let

$$\tilde{M}\tilde{N} \overset{k}{\simeq} \tilde{N}\tilde{M} \tag{6.6}$$

be the specified equivalence associated with T. Then this specified equivalence can be viewed as a specified strong shift equivalence of lag 1 from $\tilde{N}\tilde{M}$ to itself, which gives the factorization of $\psi_T^{(1,0)}$:

$$\psi_T^{(1,0)} = \kappa\zeta,$$

where $\zeta : (X_{\tilde{N}\tilde{M}}, \sigma_{\tilde{N}\tilde{M}}) \to (X_{\tilde{M}\tilde{N}}, \sigma_{\tilde{M}\tilde{N}})$ is the forward bipartite conjugacy induced by $(X_{\begin{pmatrix} 0 & \tilde{N} \\ \tilde{M} & 0 \end{pmatrix}}, \sigma_{\begin{pmatrix} 0 & \tilde{N} \\ \tilde{M} & 0 \end{pmatrix}})$, and $\kappa : (X_{\tilde{M}\tilde{N}}, \sigma_{\tilde{M}\tilde{N}}) \to (X_{\tilde{N}\tilde{M}}, \sigma_{\tilde{N}\tilde{M}})$ is the symbolic conjugacy given by k. Similarly $\psi_T^{(0,1)}$ is factorized as

$$\psi_T^{(0,1)} = \zeta'\kappa^{-1}$$

where $\zeta' : (X_{\tilde{M}\tilde{N}}, \sigma_{\tilde{M}\tilde{N}}) \to (X_{\tilde{N}\tilde{M}}, \sigma_{\tilde{N}\tilde{M}})$ is the forward bipartite conjugacy induced by $(X_{\begin{pmatrix} 0 & \tilde{M} \\ \tilde{N} & 0 \end{pmatrix}}, \sigma_{\begin{pmatrix} 0 & \tilde{M} \\ \tilde{N} & 0 \end{pmatrix}})$, which is equal to $\sigma_{\tilde{N}\tilde{M}}\zeta^{-1}$. Let

$$(\tilde{N}\tilde{M})\tilde{N} \overset{i}{\simeq} \tilde{N}(\tilde{M}\tilde{N}) \quad \text{and} \quad (\tilde{M}\tilde{N})\tilde{M} \overset{j}{\simeq} \tilde{M}(\tilde{N}\tilde{M})$$

be the natural specified equivalences (corresponding to associativity). The specified equivalence (6.6) naturally induces specified equivalences

$$\tilde{N}(\tilde{M}\tilde{N}) \overset{k'}{\simeq} \tilde{N}(\tilde{N}\tilde{M}) \quad \text{and} \quad (\tilde{N}\tilde{M})\tilde{M} \overset{(k^{-1})'}{\simeq} (\tilde{M}\tilde{N})\tilde{M}.$$

By composition, we have specified equivalences

$$(\tilde{N}\tilde{M})\tilde{N} \overset{l}{\simeq} \tilde{N}(\tilde{N}\tilde{M}) \tag{6.7}$$

and

$$(\tilde{N}\tilde{M})\tilde{M} \overset{m}{\simeq} \tilde{M}(\tilde{N}\tilde{M}) \tag{6.8}$$

where $l = k'i$ and $m = j(k^{-1})'$. Let T_l and T_m be the LR textile systems associated with (6.7) and (6.8), respectively. Then both T_l and T_m are 1-1 and we have

$$\varphi_{T_l} = \psi_T^{(1,0)} \quad \text{and} \quad \varphi_{T_m} = \psi_T^{(0,1)}.$$

□

Observation 6.9. Let $T_i, i = 1, \cdots, n$, be LR textile systems over a graph G. Let $M = M_G$ and let $N_i = M_{GT_i}$ for $i = 1, \cdots, n$. Let

$$\tilde{M}\tilde{N}_i \overset{l_i}{\simeq} \tilde{N}_i\tilde{M}, \quad i = 1, \cdots, n,$$

be the specified equivalences associated with $T_i, i = 1, \cdots, n$, respectively. These specified equivalences naturally define a sequence of specified equivalences

$$\tilde{M}\tilde{N}_1 \cdots \tilde{N}_n \overset{l_1'}{\simeq} \tilde{N}_1\tilde{M}\tilde{N}_2 \cdots \tilde{N}_n \overset{l_2''}{\simeq} \tilde{N}_1\tilde{N}_2\tilde{M}\tilde{N}_3 \cdots \tilde{N}_n \overset{l_3^{(3)}}{\simeq} \quad \cdots \quad \overset{l_n^{(n)}}{\simeq} \tilde{N}_1 \cdots \tilde{N}_n\tilde{M},$$

and by composing these specified equivalences we have a specified equivalence

$$\tilde{M}(\tilde{N}_1 \cdots \tilde{N}_n) \overset{l}{\simeq} (\tilde{N}_1 \cdots \tilde{N}_n)\tilde{M},$$

where $l = l_n^{(n)} l_{n-1}^{(n-1)} \cdots l_1'$. This specified equivalence is associated with the composition of $T_i, i = 1, \cdots, n$,

$$T_1 \circ T_2 \circ \cdots \circ T_n.$$

□

Let φ be an automorphism of a subshift (X, σ) and n a nonnegative integer. If φ is factorized as

$$\varphi = \kappa_n \zeta_n \cdots \kappa_1 \zeta_1 \kappa_0$$

with $\kappa_0, \cdots, \kappa_n$ symbolic conjugacies and $\zeta_1, \cdots, \zeta_n$ forward bipartite conjugacies, then φ is a forward automorphism of lag n. If φ is a forward automorphism of lag n, then

$$\varphi' = \sigma^n \varphi^{-1}$$

is also a forward automorphism of lag n. A forward automorphism of lag 0 is a symbolic automorphism. If φ is a symbolic automorphism of a topological Markov shift (X_G, σ_G), then there is a graph automorphism $h : G \to G$ that gives φ. If φ is a forward automorphism of lag n of the topological Markov shift (X_G, σ_G) with $n \geq 1$, then by Theorem 5.3, there is a specified strong shift equivalence

$$\tilde{M}_G \overset{k_0}{\simeq} \tilde{P}_1\tilde{Q}_1,\ \tilde{Q}_1\tilde{P}_1 \overset{k_1}{\simeq} \tilde{P}_2\tilde{Q}_2,\ \cdots,\ \tilde{Q}_{n-1}\tilde{P}_{n-1} \overset{k_{n-1}}{\simeq} \tilde{P}_n\tilde{Q}_n,\ \tilde{Q}_n\tilde{P}_n \overset{k_n}{\simeq} \tilde{M}_G$$

that gives a κ-ζ factorization of φ.

Theorem 6.10. Let (X_G, σ_G) be a topological Markov shift with $M_G = M$. Let n be a nonnegative integer. Let φ be a forward automorphism of lag n of (X_G, σ_G). Let $\varphi' = \sigma_G^n \varphi^{-1}$. Assume the following (i) and (ii):

(i) if $n = 0$, then φ is given by a graph-automorphism $h = (h_A, h_V)$ of G;

(ii) if $n \geq 1$, then a specified strong shift equivalence

$$\tilde{M} \overset{k_0}{\simeq} \tilde{P}_1\tilde{Q}_1,\ \tilde{Q}_1\tilde{P}_1 \overset{k_1}{\simeq} \tilde{P}_2\tilde{Q}_2,\ \cdots,\ \tilde{Q}_{n-1}\tilde{P}_{n-1} \overset{k_{n-1}}{\simeq} \tilde{P}_n\tilde{Q}_n,\ \tilde{Q}_n\tilde{P}_n \overset{k_n}{\simeq} \tilde{M} \tag{6.9}$$

gives a κ-ζ factorization of φ.

Let P and Q be the matrices with the indexing set $V_G \times V_G$ defined as follows: if $n = 0$, then P is the permutation matrix corresponding to h_V and $Q = P^{-1}$, and if $n \geq 1$, then

$$P = P_1 \cdots P_n \quad \text{and} \quad Q = Q_n \cdots Q_1.$$

Then we can determine whether φ is expansive or not, and if φ is expansive, then (X_G, φ) is topologically conjugate to (X_P, σ_P). We can also determine whether φ' is expansive or not. If φ' is expansive, then (X_G, φ') is topologically conjugate to (X_Q, σ_Q). Moreover for all integers $k \geq 0$ and $l \geq 1$, $(X_G, \varphi^k \sigma_G^l)$ is topologically conjugate to $(X_{P^k M^l}, \sigma_{P^k M^l})$ and $(X_G, (\varphi')^k \sigma^l)$ is topologically conjugate to $(X_{Q^k M^l}, \sigma_{Q^k M^l})$.

Proof. Case (i). Let T_0 be the textile system defined by $T_0 = (i, h : G \to G)$, where i is the identity graph-automorphism of G. Then T_0 is 1-1 and LR (and RL). Clearly $\varphi_{T_0} = \varphi$ and $M_{G^{T_0}} = P$. Let $T_0' = T_0^{-1}$. Then $\varphi_{T_0'} = \varphi'$ and $M_{G^{T_0'}} = Q$. Since the dual of an LR textile system is LR, we can determine whether it is 1-1 or not, because an LR textile system is nondegenerate. Therefore the conclusions for the case (i) are obtained by Corollary 6.5.

Case (ii). We may assume that k_0 is the identity mapping for (6.9). (For if $\varphi = \kappa_n \zeta_n \cdots \kappa_1 \zeta_1 \kappa_0$ is the κ-ζ factorization given by (6.9), then we may consider $\kappa_0 \varphi \kappa_0^{-1}$ instead of φ. Note that $\tilde{M}$ and $\tilde{P}_1 \tilde{Q}_1$ are the same, up to renaming of the arcs.) Therefore we assume that

$$\varphi = \kappa_n \zeta_n \cdots \kappa_1 \zeta_1$$

and this κ-ζ factorization is given by

$$\text{(6.10)} \quad \tilde{M} = \tilde{P}_1 \tilde{Q}_1, \; \tilde{Q}_1 \tilde{P}_1 \overset{k_1}{\simeq} \tilde{P}_2 \tilde{Q}_2, \; \cdots, \; \tilde{Q}_{n-1} \tilde{P}_{n-1} \overset{k_{n-1}}{\simeq} \tilde{P}_n \tilde{Q}_n, \; \tilde{Q}_n \tilde{P}_n \overset{k_n}{\simeq} \tilde{P}_1 \tilde{Q}_1.$$

Let R and S be the matrices defined by

$$\tilde{R} = \begin{pmatrix} 0 & 0 & \cdots & 0 & \tilde{Q}_n \\ \tilde{Q}_1 & 0 & \cdots & 0 & 0 \\ 0 & \tilde{Q}_2 & \cdots & 0 & 0 \\ \vdots & \vdots & \ddots & \vdots & \vdots \\ 0 & 0 & \cdots & \tilde{Q}_{n-1} & 0 \end{pmatrix}$$

and

$$\tilde{S} = \begin{pmatrix} 0 & \tilde{P}_1 & 0 & \cdots & 0 \\ 0 & 0 & \tilde{P}_2 & \cdots & 0 \\ \vdots & \vdots & \vdots & \ddots & \vdots \\ 0 & 0 & 0 & \cdots & \tilde{P}_{n-1} \\ \tilde{P}_n & 0 & 0 & \cdots & 0 \end{pmatrix}.$$

Then

$$\tilde{R}\tilde{S} = \begin{pmatrix} \tilde{Q}_n \tilde{P}_n & 0 & \cdots & 0 \\ 0 & \tilde{Q}_1 \tilde{P}_1 & \cdots & 0 \\ \vdots & \vdots & \ddots & \vdots \\ 0 & 0 & \cdots & \tilde{P}_{n-1} \tilde{Q}_{n-1} \end{pmatrix}$$

and

$$\tilde{S}\tilde{R} = \begin{pmatrix} \tilde{P}_1\tilde{Q}_1 & 0 & \cdots & 0 \\ 0 & \tilde{P}_2\tilde{Q}_2 & \cdots & 0 \\ \vdots & \vdots & \ddots & \vdots \\ 0 & 0 & \cdots & \tilde{P}_n\tilde{Q}_n \end{pmatrix}.$$

Let

$$\tilde{R}\tilde{S} \overset{k}{\simeq} \tilde{S}\tilde{R} \tag{6.11}$$

be the specified equivalence determined by (6.10). Let T be the textile system associated with (6.11). Then

$$X_{\tilde{R}\tilde{S}} = X_{\tilde{Q}_n\tilde{P}_n} \sqcup X_{\tilde{Q}_1\tilde{P}_1} \sqcup \cdots \sqcup X_{\tilde{Q}_{n-1}\tilde{P}_{n-1}}$$

and

$$X_{\tilde{S}\tilde{R}} = X_{\tilde{P}_1\tilde{Q}_1} \sqcup X_{\tilde{P}_2\tilde{Q}_2} \sqcup \cdots \sqcup X_{\tilde{P}_n\tilde{Q}_n},$$

where $\sqcup$ denotes disjoint union. By Observation 6.8,

$$\psi_T^{(1,0)} = \kappa\zeta,$$

where $\zeta : (X_{\tilde{S}\tilde{R}}, \sigma_{\tilde{S}\tilde{R}}) \to (X_{\tilde{R}\tilde{S}}, \sigma_{\tilde{R}\tilde{S}})$ is the forward bipartite conjugacy induced by $(X_{\begin{pmatrix} 0 & \tilde{S} \\ \tilde{R} & 0 \end{pmatrix}}, \sigma_{\begin{pmatrix} 0 & \tilde{S} \\ \tilde{R} & 0 \end{pmatrix}})$ and $\kappa : (X_{\tilde{R}\tilde{S}}, \sigma_{\tilde{R}\tilde{S}}) \to (X_{\tilde{S}\tilde{R}}, \sigma_{\tilde{S}\tilde{R}})$ is the symbolic conjugacy given by k. Clearly we have

$$\zeta | X_{\tilde{P}_i\tilde{Q}_i} = \zeta_i, \quad i = 1, \cdots, n$$

and

$$\kappa | X_{\tilde{Q}_i\tilde{P}_i} = \kappa_i, \quad i = 1, \cdots n.$$

Since

$$\varphi = \kappa_n\zeta_n \cdots \kappa_1\zeta_1,$$

we have

$$\varphi = (\psi_T^{(1,0)})^n | X_{\tilde{P}_1\tilde{Q}_1}.$$

Let

$$(\tilde{S}\tilde{R})\tilde{S} \overset{l}{\simeq} \tilde{S}(\tilde{S}\tilde{R}) \tag{6.12}$$

be the specified equivalence defined for T in the same way as described in Observation 6.8. Let T_l be the textile system associated with it. Then

$$\varphi_{T_l} = \psi_T^{(1,0)}.$$

Consider the nth composition-power T_l^n of T_l. We have

$$\varphi_{T_l^n} = (\varphi_{T_l})^n$$

so that we have

$$\varphi = (\psi_T^{(1,0)})^n | X_{\tilde{P}_1\tilde{Q}_1} = \varphi_{T_l^n} | X_{\tilde{P}_1\tilde{Q}_1}$$

Applying Observation 6.9 to (6.12), we have the specified equivalence

$$(\tilde{S}\tilde{R})\tilde{S}^n \overset{I}{\simeq} \tilde{S}^n(\tilde{S}\tilde{R}) \tag{6.13}$$

with which T_l^n is associated. Since

$$\tilde{S}^n = \begin{pmatrix} \tilde{P}_1 \cdots \tilde{P}_n & & & \\ & \tilde{P}_2 \cdots \tilde{P}_n \tilde{P}_1 & & 0 \\ 0 & & \ddots & \\ & & & \tilde{P}_n \tilde{P}_1 \cdots \tilde{P}_{n-1} \end{pmatrix}$$

and $\tilde{S}\tilde{R}$ is given as above, taking the restriction of (6.13), we have a specified equivalence

$$(\tilde{P}_1\tilde{Q}_1)(\tilde{P}_1 \cdots \tilde{P}_n) \overset{f}{\simeq} (\tilde{P}_1 \cdots \tilde{P}_n)(\tilde{P}_1\tilde{Q}_1). \tag{6.14}$$

Let T_1 be the LR textile system associated with (6.14). Then

$$M_{GT_1} = P.$$

Since T_l is 1-1, so is T_1. Since $\varphi = \varphi_{T_l^n} | X_{\tilde{P}_1\tilde{Q}_1}$, we have

$$\varphi_{T_1} = \varphi.$$

Since $\psi_T^{(1,0)}\psi_T^{(0,1)} = \sigma_{\tilde{S}\tilde{R}}$, it follows that

$$(\psi_T^{(0,1)})^n = (\sigma_{\tilde{S}\tilde{R}})^n (\psi_T^{(1,0)})^{-n}.$$

Therefore we have

$$(\psi_T^{(0,1)})^n | X_{\tilde{P}_1\tilde{Q}_1} = (\sigma_{\tilde{P}_1\tilde{Q}_1})^n \varphi^{-n} = \varphi'.$$

In the same way as in Observation 6.8, we have the specified equivalence

$$(\tilde{S}\tilde{R})\tilde{R} \overset{m}{\simeq} \tilde{R}(\tilde{S}\tilde{R})$$

such that the LR textile system T_m associated with it is 1-1 and

$$\varphi_{T_m} = \psi_T^{(0,1)}.$$

Therefore, by an analogous argument to the above, we also have the specified equivalence

$$(\tilde{P}_1\tilde{Q}_1)(\tilde{Q}_n \cdots \tilde{Q}_1) \overset{f'}{\simeq} (\tilde{Q}_n \cdots \tilde{Q}_1)(\tilde{P}_1\tilde{Q}_1) \tag{6.15}$$

and the 1-1 LR textile system ${T_1}'$ associated with it, and we also have

$$\varphi_{{T_1}'} = \varphi'$$

and

$$M_{G{T_1}'} = Q.$$

Thus the conclusions for the case (ii) are obtained by applying Corollary 6.5 to T_1 and ${T_1}'$. □

Let the hypothesis and the notation be the same as in Theorem 6.10 and its proof. Let us consider the LR textile system T_0, T_0', T_1 and T_1' in the proof.

Case (i). Let

$$\tilde{M}\tilde{P} \overset{f}{\simeq} \tilde{P}\tilde{M} \quad \text{and} \quad \tilde{M}\tilde{Q} \overset{f'}{\simeq} \tilde{P}\tilde{M}$$

be the specified equivalences associated with T_0 and T_0', respectively. We have the trivial specified equivalences

$$\tilde{I} \overset{g}{\simeq} \tilde{P}\tilde{Q}, \quad \tilde{Q}\tilde{P} \overset{g'}{\simeq} \tilde{I},$$

where I is the identity matrix.

Case (ii). As seen in Observation 6.9, the specified equivalence (6.13) is obtained by composing the specified equivalences in the sequence

$$(\tilde{S}\tilde{R})\tilde{S}^n \overset{l'}{\simeq} \tilde{S}(\tilde{S}\tilde{R})\tilde{S}^{n-1} \overset{l''}{\simeq} \tilde{S}^2(\tilde{S}\tilde{R})\tilde{S}^{n-2} \overset{l^{(3)}}{\simeq} \cdots \overset{l^{(n)}}{\simeq} \tilde{S}^n(\tilde{S}\tilde{R})$$

that is naturally induced by (6.12). Taking the restrictions of those specified equivalences, we have

$$(\tilde{P}_1\tilde{Q}_1)\tilde{P}_1\cdots\tilde{P}_n \overset{k_1'}{\simeq} \tilde{P}_1(\tilde{P}_2\tilde{Q}_2)\tilde{P}_2\cdots\tilde{P}_n \overset{k_2''}{\simeq} \cdots \overset{k_{n-1}^{(n-1)}}{\simeq} \tilde{P}_1\cdots\tilde{P}_{n-1}(\tilde{P}_n\tilde{Q}_n)\tilde{P}_n \overset{k_n^{(n)}}{\simeq} \tilde{P}_1\cdots\tilde{P}_n(\tilde{P}_1\tilde{Q}_1),$$

which is the sequence of specified equivalences naturally induced by the specified strong shift equivalence (6.10). Composing these specified equivalences, we have (6.14) with

$$f = k_n^{(n)}\cdots k_2''k_1'.$$

Therefore, the specified equivalence

$$\tilde{M}\tilde{P} \overset{f}{\simeq} \tilde{P}\tilde{M}$$

associated with T_1 is naturally induced by (6.9).

Similarly the specified equivalence

$$\tilde{M}\tilde{Q} \overset{f'}{\simeq} \tilde{Q}\tilde{M}$$

associated with T_1' is naturally induced by (6.9). The specified equivalence (6.9) also naturally induces specified equivalences

$$\tilde{M}^n \overset{g}{\simeq} \tilde{P}\tilde{Q}, \quad \tilde{Q}\tilde{P} \overset{g'}{\simeq} \tilde{M}^n.$$

As was shown by Boyle ([B4] Lemma 1), these constitute a one-step specified strong shift equivalence which gives $\varphi^{(n)}$. Recall that $\varphi^{(n)}$ denotes the automorphism of $(X_M^{(n)}, \sigma_M^{(n)})$ induced by φ. After Williams [Wi], we should call the set of specified equivalences

$$\tilde{M}\tilde{P} \overset{f}{\simeq} \tilde{P}\tilde{M}, \quad \tilde{M}\tilde{Q} \overset{f'}{\simeq} \tilde{Q}\tilde{M}, \quad \tilde{M}^n \overset{g}{\simeq} \tilde{P}\tilde{Q}, \quad \tilde{Q}\tilde{P} \overset{g'}{\simeq} \tilde{M}^n$$

the *specified shift equivalence of lag n from* $\tilde{M}$ *to itself induced by the specified strong shift equivalence* (6.9).

Let φ be a forward automorphism of a topological Markov shift (X_M, σ_M). Let $n \geq 0$. A set of specified equivalences

$$\tilde{M}\tilde{P} \overset{f}{\simeq} \tilde{P}\tilde{M}, \quad \tilde{M}\tilde{Q} \overset{f'}{\simeq} \tilde{M}\tilde{Q}, \quad \tilde{M}^n \overset{g}{\simeq} \tilde{P}\tilde{Q}, \quad \tilde{Q}\tilde{P} \overset{g'}{\simeq} \tilde{M}^n$$

is called a *specified shift equivalence of lag n associated with* φ if the LR textile system T associated with $\tilde{M}\tilde{P} \overset{f}{\simeq} \tilde{P}\tilde{M}$ is 1-1 and $\varphi_T = \varphi$, the LR textile system T' associated with $\tilde{M}\tilde{Q} \overset{f'}{\simeq} \tilde{Q}\tilde{M}$ is 1-1 and $\varphi_{T'} = \varphi^{-1}\sigma_M^n$, and the one-step specified strong shift equivalence

$$\tilde{M}^n \overset{g}{\simeq} \tilde{P}\tilde{Q}, \quad \tilde{Q}\tilde{P} \overset{g'}{\simeq} \tilde{M}^n$$

gives $\varphi^{(n)}$ when $n \geq 1$. Especially, the specified equivalence

$$\tilde{M}\tilde{P} \overset{f}{\simeq} \tilde{P}\tilde{M}$$

such that the textile system T associated with it is 1-1 and $\varphi_T = \varphi$, is called the *specified equivalence associated with* φ , which is unique for φ, as will be shown in Corollary 7.27. Moreover, in Corollary 7.27, we shall show that the specified shift equivalence of lag n associated with φ is unique for φ and n.

Combining the proof of Theorem 6.10 and the considerations in cases (i) and (ii) above, we have

Corollary 6.11. Let φ be a forward automorphism of a topological Markov shift. Let n be a nonnegative integer. If φ has a κ-ζ factorization of lag n, then there is a specified shift equivalence of lag n associated with φ. If $n \geq 1$, then the specified shift equivalence induced by the specified strong shift equivalence of lag n giving the κ-ζ factorization of φ, is a specified shift equivalence of lag n associated with φ.

Boyle and Krieger have found that for any automorphism φ of any subshift, $\varphi\sigma^n$ is expansive for all n greater than a coding bound for φ and φ^{-1} (Lemma 2.11 of [BK1]). We remark that for any automorphism φ of any subshift, $\varphi\sigma^n$ is a forward automorphism for all sufficiently large n, a forward automorphism can be nonexpansive but we have the following result.

Corollary 6.12. Let $\varphi : (X, \sigma) \to (X, \sigma)$ be a forward automorphism of a subshift. Then for all $k \in \mathbf{N}$, $\varphi\sigma^k$ is expansive.

Proof. By Theorem 6.10, it suffices to show that there is a forward automorphism of a topological Markov shift which is an extension of φ. If $\varphi : (X, \sigma) \to (X, \sigma)$ is a symbolic automorphism and (X, σ) is over an alphabet A, then φ is extended to a symbolic automorphism of the full shift (X_A, σ_A). Therefore we may assume that there are bipartite subshifts (Y_i, σ_i), $i = 1, \cdots, n$, with $n \in \mathbf{N}$, symbolic conjugacies $\kappa_0 : (X, \sigma) \to (\acute{Y}_1, \acute{\sigma}_1)$, $\kappa_1 : (\grave{Y}_1, \grave{\sigma}_1) \to (\acute{Y}_2, \acute{\sigma}_2)$, $\cdots$, $\kappa_{n-1} : (\grave{Y}_{n-1}, \grave{\sigma}_{n-1}) \to (\acute{Y}_n, \acute{\sigma}_n)$, and $\kappa_n : (\grave{Y}_n, \grave{\sigma}_n) \to (X, \sigma)$, and a factorization of φ such that

$$\varphi = \kappa_n \zeta_n \cdots \kappa_1 \zeta_1 \kappa_0,$$

where $\zeta_i : (\acute{Y}_i, \acute{\sigma}_i) \to (\grave{Y}_i, \grave{\sigma}_i)$ is the forward bipartite conjugacy induced by (Y_i, σ_i) for $i = 1, \cdots n$. Let A_i be the underlying alphabet of (Y_i, σ_i) for $i = 1, \cdots, n$. Then for $i = 1, \cdots, n$, there are alphabets C_i and D_i such that $A_i = C_i \cup D_i, C_i \cap D_i = \emptyset, (\acute{Y}_i, \acute{\sigma}_i)$ is over $C_i D_i$ and $(\grave{Y}_i, \grave{\sigma}_i)$ is over $D_i C_i$. We choose alphabets $\bar{C}_i$ and $\bar{D}_i, i = 1, \cdots, n$, all having the same cardinality, say N, such that

$$\bar{C}_i \supset C_i, \quad \bar{D}_i \supset D_i \quad \text{and} \quad \bar{C}_i \cap \bar{D}_i = \emptyset.$$

For $i = 1, \cdots, n$, let G_i be the bipartite graph with two vertices $\acute{v}_i$ and $\grave{v}_i$ such that $A_{G_i} = \bar{C}_i \cup \bar{D}_i$, the arcs in $\bar{C}_i$ go from $\acute{v}_i$ to $\grave{v}_i$ and the arcs in $\bar{D}_i$ go to $\grave{v}_i$ to $\acute{v}_i$. Let A_0 be an alphabet of cardinality N^2 such that $A_0 \supset A$. Then there are extensions $\bar{\kappa}_0 : (X_{A_0}, \sigma_{A_0}) \to (X_{\acute{G}_1}, \sigma_{\acute{G}_1})$, $\bar{\kappa}_1 : (X_{\grave{G}_1}, \sigma_{\grave{G}_1}) \to (X_{\acute{G}_2}, \sigma_{\acute{G}_2})$, $\cdots$, $\bar{\kappa}_{n-1} : (X_{\grave{G}_{n-1}}, \sigma_{\grave{G}_{n-1}}) \to (X_{\acute{G}_n}, \sigma_{\acute{G}_n})$, and $\bar{\kappa}_n : (X_{\grave{G}_n}, \sigma_{\grave{G}_n}) \to (X_{A_0}, \sigma_{A_0})$ of $\kappa_0, \kappa_1, \cdots, \kappa_{n-1}$, and κ_n, respectively, and there are extensions $\bar{\zeta}_i : (X_{\acute{G}_i}, \sigma_{\acute{G}_i}) \to (X_{\grave{G}_i}, \sigma_{\grave{G}_i})$ of ζ_i, $i = 1, \cdots, n$, where $\acute{G}_i$ and $\grave{G}_i$ are the induced pair of graphs of the bipartite graph G_i for $i = 1, \cdots, n$ (see Section 5). Let

$$\bar{\varphi} = \bar{\kappa}_n \bar{\zeta}_n \cdots \bar{\kappa}_1 \bar{\zeta}_1 \bar{\kappa}_0.$$

Then it is clear that $\bar{\varphi}$ is a forward automorphism of the full shift (X_{A_0}, σ_{A_0}) and an extension of φ. □

We can also define *LR sofic textile systems* and give results on them which are generalizations of those on LR textile systems. The proofs of them can be given by straightforward generalization of the proofs of the results on LR textile systems given above and so will be omitted except for some.

Let

$$\mathcal{M}\mathcal{N} \overset{k}{\simeq} \mathcal{N}\mathcal{M} \tag{6.16}$$

be the specified equivalence, where $\mathcal{M}$ and $\mathcal{N}$ are square representation matrices. We define a sofic textile system associated with (6.16) as follows. Let $\mathcal{G} = (G, \lambda)$ and

$\mathcal{H} = (H, \mu)$ be the λ-graphs represented by $\mathcal{M}$ and $\mathcal{N}$, respectively, with $V_G = V_H$. There is a graph-isomorphism $h : GH \to HG$ such that

$$(h, k) : \mathcal{GH} \simeq \mathcal{HG}$$

is a λ-graph-isomorphism. Let $T = (p, q : \Gamma \to G)$ be the LR textile system associated with the specified equivalence

$$\tilde{M}_G \tilde{M}_H \overset{h_A}{\simeq} \tilde{M}_H \tilde{M}_G$$

that is,

$$A_\Gamma = \{(b, b', a, a') \in A_H \times A_H \times A_G \times A_G \mid ab' \in A_{GH}, h_A(ab') = ba'\}$$

and for $\alpha = (b, b', a, a') \in A_\Gamma, i_\Gamma(\alpha) = b, t_\Gamma(\alpha) = b', p_A(\alpha) = a$ and $q_A(\alpha) = a'$. We define

$$\mathcal{T} = (T, \lambda, \mu).$$

We call $\mathcal{T}$ the *LR sofic textile system associated with the specified equivalence* (6.16). A sofic textile system $\mathcal{T}$ is said to be *LR* if there is a specified equivalence with which $\mathcal{T}$ is associated. It is easy to see that the following facts hold:

Fact 6.13. A sofic textile system $\mathcal{T} = (T, \lambda, \mu)$ with $T = (p, q : \Gamma \to G)$ is LR if and only if T is LR and the correspondence

$$\lambda(p_A(\alpha))\mu(t_\Gamma(\alpha)) \leftrightarrow \mu(i_\Gamma(\alpha))\lambda(q_A(\alpha)), \quad \alpha \in A_\Gamma$$

is 1-1.

Fact 6.14. An LR sofic textile system is nondegenerate.

Fact 6.15. If $\mathcal{T}$ is an LR sofic textile system, then $\mathcal{T}^*$ is also LR.

Fact 6.16. If $\mathcal{T}$ is an LR sofic textile system, then $\mathcal{T}^{[n]}$ is also LR for all $n \in \mathbf{N}$.

Lemma 6.2 is generalized to the following lemma:

Lemma 6.17. Let $\mathcal{T} = (T, \lambda, \mu)$ be an LR sofic textile system with a textile system T over a graph G. Let $\mathcal{G}$ and $\mathcal{H}$ be the λ-graphs defined by $\mathcal{G} = (G, \lambda)$ and $\mathcal{H} = (G^T, \mu)$. Let $k, l \in \mathbf{N}$. Then

$$\begin{aligned}
(\check{X}_{\mathcal{T}}^{(k,0)}, \check{\sigma}_{\mathcal{T}}^{(k,0)}) &= (X_{\mathcal{H}^k}, \sigma_{\mathcal{H}^k}) \\
(\check{X}_{\mathcal{T}}^{(0,l)}, \check{\sigma}_{\mathcal{T}}^{(0,l)}) &= (X_{\mathcal{G}^l}, \sigma_{\mathcal{G}^l}) \\
(\check{X}_{\mathcal{T}}^{(k,l)}, \check{\sigma}_{\mathcal{T}}^{(k,l)}) &= (X_{\mathcal{H}^k\mathcal{G}^l}, \sigma_{\mathcal{H}^k\mathcal{G}^l})
\end{aligned}$$

and $\check{\theta}_{\mathcal{T}}^{(k,l)}$ is 1-1.

We have the following theorem.

Theorem 6.18. Let $\mathcal{T} = (T, \lambda, \mu)$ be an LR sofic textile system with T over a graph G. Let $\mathcal{G}$ and $\mathcal{H}$ be the λ-graphs defined by $\mathcal{G} = (G, \lambda)$ and $\mathcal{H} = (G^T, \mu)$. Let $k, l \in \mathbf{Z}$ with $kl > 0$. Then the following statements are valid.

(1) If $\boldsymbol{\sigma}_{\mathcal{T}}^{(k,0)}$ is expansive, then $(U_{\mathcal{T}}, \boldsymbol{\sigma}_{\mathcal{T}}^{(k,0)})$ is topologically conjugate to a sofic system. If $\mathcal{T}^*$ is 1-1, then $(U_{\mathcal{T}}, \boldsymbol{\sigma}_{\mathcal{T}}^{(k,0)})$ is topologically conjugate to $(X_{\mathcal{T}^*}^{(k)}, \sigma_{\mathcal{T}^*}^{(k)}) = (X_{\mathcal{H}^k}, \sigma_{\mathcal{H}^k})$.

(2) If $\boldsymbol{\sigma}_{\mathcal{T}}^{(0,l)}$ is expansive, then $(U_{\mathcal{T}}, \boldsymbol{\sigma}_{\mathcal{T}}^{(0,l)})$ is topologically conjugate to a sofic system. If $\mathcal{T}$ is 1-1, then $(U_{\mathcal{T}}, \boldsymbol{\sigma}_{\mathcal{T}}^{(0,l)})$ is topologically conjugate to $(X_{\mathcal{T}}^{(l)}, \sigma_{\mathcal{T}}^{(l)}) = (X_{\mathcal{G}^l}, \sigma_{\mathcal{G}^l})$.

(3) $(U_{\mathcal{T}}, \boldsymbol{\sigma}_{\mathcal{T}}^{(k,l)})$ is topologically conjugate to $(\check{X}_{\mathcal{T}}^{(k,l)}, \check{\sigma}_{\mathcal{T}}^{(k,l)}) = (X_{\mathcal{H}^k\mathcal{G}^l}, \sigma_{\mathcal{H}^k\mathcal{G}^l})$.

Proof. If $\boldsymbol{\sigma}_{\mathcal{T}}^{(k,0)}$ is expansive, then there is an $n \in \mathbf{N}$ such that $(\mathcal{T}^{[n]})^*$ is 1-1 and $(U_{\mathcal{T}}, \boldsymbol{\sigma}_{\mathcal{T}}^{(k,0)})$ is topologically conjugate to $(X_{(\mathcal{T}^{[n]})^*}^{(k)}, \sigma_{(\mathcal{T}^{[n]})^*}^{(k)})$, by Proposition 4.7. Since $\mathcal{T}$ is LR, it follows from Facts 6.15 and 6.16 that $(\mathcal{T}^{[n]})^*$ is LR. Therefore by Lemma 6.17, $(X_{(\mathcal{T}^{[n]})^*}^{(k)}, \sigma_{(\mathcal{T}^{[n]})^*}^{(k)})$ is topologically conjugate to a sofic system. It also follows from Proposition 4.7 that if $\mathcal{T}^*$ is 1-1, then $(U_{\mathcal{T}}, \boldsymbol{\sigma}_{\mathcal{T}}^{(k,0)})$ is topologically conjugate to $(\check{X}_{\mathcal{T}}^{(k,0)}, \check{\sigma}_{\mathcal{T}}^{(k,0)})$, which is equal to $(X_{\mathcal{H}^k}, \sigma_{\mathcal{H}^k})$ by Lemma 6.17.

By duality, (2) follows from (1).

From Fact 4.5 and Lemma 6.17, (3) follows. □

Let $\mathcal{T} = (T, \lambda, \mu)$ be an LR sofic textile system with T over a graph G. Let $\mathcal{G}$ and $\mathcal{H}$ be the λ-graphs defined by $\mathcal{G} = (G, \lambda)$ and $\mathcal{H} = (G^T, \mu)$. By Lemma 6.17, $\check{\theta}_{\mathcal{T}}^{(1,1)}$ is 1-1 and $(\check{X}_{\mathcal{T}}^{(1,1)}, \check{\sigma}_{\mathcal{T}}^{(1,1)}) = (X_{\mathcal{HG}}, \sigma_{\mathcal{HG}})$. For $k, l \in \mathbf{Z}$, let $\psi_{\mathcal{T}}^{(k,l)} : \check{X}_{\mathcal{T}}^{(1,1)} \to \check{X}_{\mathcal{T}}^{(1,1)}$ be defined by

$$\psi_{\mathcal{T}}^{(k,l)} = \check{\theta}_{\mathcal{T}}^{(1,1)} \boldsymbol{\sigma}_{\mathcal{T}}^{(k,l)} (\check{\theta}_{\mathcal{T}}^{(1,1)})^{-1}.$$

Then $\psi_{\mathcal{T}}^{(k,l)}$ is an automorphisms of $(\check{X}_{\mathcal{T}}^{(1,1)}, \check{\sigma}_{\mathcal{T}}^{(1,1)})$.

The following two corollaries are direct consequences of Theorem 6.18 and Fact 4.6.

Corollary 6.19. Let $\mathcal{T} = (T, \lambda, \mu)$ be an LR sofic textile system with T over a graph G. Let $\mathcal{M}$ and $\mathcal{N}$ be the representation matrices of the λ-graphs (G, λ) and (G^T, μ), respectively. Then the following statements are valid.

(1) If $\psi_{\mathcal{T}}^{(1,0)}$ is expansive, then $(\check{X}_{\mathcal{T}}^{(1,1)}, \psi_{\mathcal{T}}^{(1,0)})$ is topologically conjugate to a sofic system.

(2) If $\psi_{\mathcal{T}}^{(0,1)}$ is expansive, then $(\check{X}_{\mathcal{T}}^{(1,1)}, \psi_{\mathcal{T}}^{(0,1)})$ is topologically conjugate to a sofic system.

(3) For all $k, l \in \mathbf{N}$, $(\check{X}_{\mathcal{T}}^{(1,1)}, \psi_{\mathcal{T}}^{(k,l)})$ is topologically conjugate to $(X_{\mathcal{N}^k\mathcal{M}^l}, \sigma_{\mathcal{N}^k\mathcal{M}^l})$.

Corollary 6.20. Let $\mathcal{T} = (T, \lambda, \mu)$ be a 1-1 LR sofic textile system with T over a graph G. Let $\mathcal{M}$ and $\mathcal{N}$ be the representation matrices of the λ-graphs (G, λ) and (G^T, μ), respectively. Then the following statements are valid.

(1) If φ_T is expansive, then $(X_{\mathcal{M}}, \varphi_T)$ is topologically conjugate to a sofic system.

(2) For all integers $k \geq 0$ and $l \geq 1$, $(X_{\mathcal{M}}, \varphi_T^k \sigma_{\mathcal{M}}^l)$ is topologically conjugate to $(X_{\mathcal{N}^k\mathcal{M}^l}, \sigma_{\mathcal{N}^k\mathcal{M}^l})$.

The following two observation are very analogous to Observations 6.8 and 6.9.

Observation 6.21. Let $\mathcal{T} = (T, \lambda, \mu)$ be an LR sofic textile system with T over a graph G. Let $\mathcal{M}$ and $\mathcal{N}$ be the representation matrices of the λ-graphs (G, λ) and (G^T, μ), respectively. Let

$$\mathcal{M}\mathcal{N} \overset{k}{\simeq} \mathcal{N}\mathcal{M} \tag{6.17}$$

be the specified equivalence with which $\mathcal{T}$ is associated. This can be viewed as a specified strong shift equivalence of lag 1 from $\mathcal{N}\mathcal{M}$ to itself, which gives the following factorization of $\psi_T^{(1,0)}$:

$$\psi_T^{(1,0)} = \kappa\zeta,$$

where $\zeta : (X_{\mathcal{N}\mathcal{M}}, \sigma_{\mathcal{N}\mathcal{M}}) \to (X_{\mathcal{M}\mathcal{N}}, \sigma_{\mathcal{M}\mathcal{N}})$ is the forward bipartite conjugacy induced by $(X_{\begin{pmatrix} 0 & \mathcal{N} \\ \mathcal{M} & 0 \end{pmatrix}}, \sigma_{\begin{pmatrix} 0 & \mathcal{N} \\ \mathcal{M} & 0 \end{pmatrix}})$, and $\kappa : (X_{\mathcal{M}\mathcal{N}}, \sigma_{\mathcal{M}\mathcal{N}}) \to (X_{\mathcal{N}\mathcal{M}}, \sigma_{\mathcal{N}\mathcal{M}})$ is the symbolic conjugacy given by k. Similarly $\psi_T^{(0,1)}$ is factorized as

$$\psi_T^{(0,1)} = \zeta'\kappa^{-1},$$

where $\zeta' : (X_{\mathcal{M}\mathcal{N}}, \sigma_{\mathcal{M}\mathcal{N}}) \to (X_{\mathcal{N}\mathcal{M}}, \sigma_{\mathcal{N}\mathcal{M}})$ is the forward bipartite conjugacy induced by $(X_{\begin{pmatrix} 0 & \mathcal{M} \\ \mathcal{N} & 0 \end{pmatrix}}, \sigma_{\begin{pmatrix} 0 & \mathcal{M} \\ \mathcal{N} & 0 \end{pmatrix}})$, which is equal to $\sigma_{\mathcal{N}\mathcal{M}}\zeta^{-1}$. Let

$$(\mathcal{N}\mathcal{M})\mathcal{N} \overset{i}{\simeq} \mathcal{N}(\mathcal{M}\mathcal{N}) \quad \text{and} \quad (\mathcal{M}\mathcal{N})\mathcal{M} \overset{j}{\simeq} \mathcal{M}(\mathcal{N}\mathcal{M})$$

be the natural specified equivalences. The specified equivalence (6.17) naturally induces specified equivalences

$$\mathcal{N}(\mathcal{M}\mathcal{N}) \overset{k'}{\simeq} \mathcal{N}(\mathcal{N}\mathcal{M}) \quad \text{and} \quad (\mathcal{N}\mathcal{M})\mathcal{M} \overset{(k^{-1})'}{\simeq} (\mathcal{M}\mathcal{N})\mathcal{M}.$$

By composition, we have specified equivalences

$$(\mathcal{N}\mathcal{M})\mathcal{N} \overset{l}{\simeq} \mathcal{N}(\mathcal{N}\mathcal{M}) \tag{6.18}$$

and

$$(\mathcal{N}\mathcal{M})\mathcal{M} \overset{m}{\simeq} \mathcal{M}(\mathcal{N}\mathcal{M}), \tag{6.19}$$

where $l = k'i$ and $m = j(k^{-1})'$. If $\mathcal{T}_l$ and $\mathcal{T}_m$ be the LR sofic textile systems associated with (6.18) and (6.19), respectively, then both $\mathcal{T}_l$ and $\mathcal{T}_m$ are 1-1 and we have

$$\varphi_{\mathcal{T}_l} = \psi_{\mathcal{T}}^{(1,0)} \quad \text{and} \quad \varphi_{\mathcal{T}_m} = \psi_{\mathcal{T}}^{(0,1)}.$$

□

Observation 6.22. Let $\mathcal{T}_i, i = 1, \cdots, n$, be LR sofic textile systems associated with specified equivalences

$$\mathcal{M}\mathcal{N}_i \overset{l_i}{\simeq} \mathcal{N}_i\mathcal{M}, \ i = 1, \cdots, n,$$

respectively. These define a sequence of specified equivalences

$$\mathcal{M}\mathcal{N}_1 \cdots \mathcal{N}_n \overset{l_1'}{\simeq} \mathcal{N}_1\mathcal{M}\mathcal{N}_2 \cdots \mathcal{N}_n \overset{l_2''}{\simeq} \mathcal{N}_1\mathcal{N}_2\mathcal{M}\mathcal{N}_3 \cdots \mathcal{N}_n \overset{l_3^{(3)}}{\simeq} \cdots \overset{l_n^{(n)}}{\simeq} \mathcal{N}_1 \cdots \mathcal{N}_n\mathcal{M}$$

and by composing these specified equivalences, we have a specified equivalence

$$\mathcal{M}(\mathcal{N}_1 \cdots \mathcal{N}_n) \overset{l}{\simeq} (\mathcal{N}_1 \cdots \mathcal{N}_n)\mathcal{M}$$

with $l = l_n^{(n)} l_{n-1}^{(n-1)} \cdots l_1'$. The LR textile system associated with this specified equivalence is

$$\mathcal{T}_1 \circ \mathcal{T}_2 \circ \cdots \circ \mathcal{T}_n.$$

□

Let $\varphi : (Y, \sigma) \to (Y, \sigma)$ be a forward automorphism of lag n of a sofic system, where n is a nonnegative integer. Let $\mathcal{G}$ be the right canonical λ-graph of the sofic system (Y, σ). If $n = 0$, i.e., φ is a symbolic automorphism, then there is a λ-graph-automorphism $(h, k) : \mathcal{G} \simeq \mathcal{G}$ such that k gives φ, by Theorem 5.2. If $n > 0$, then by Theorem 5.3, there is a specified strong shift equivalence

$$\mathcal{M}_{\mathcal{G}} \overset{k_0}{\simeq} \mathcal{P}_1\mathcal{Q}_1, \ \mathcal{Q}_1\mathcal{P}_1 \overset{k_1}{\simeq} \mathcal{P}_2\mathcal{Q}_2, \ \cdots, \ \mathcal{Q}_{n-1}\mathcal{P}_{n-1} \overset{k_{n-1}}{\simeq} \mathcal{P}_n\mathcal{Q}_n, \ \mathcal{Q}_n\mathcal{P}_n \overset{k_n}{\simeq} \mathcal{M}_{\mathcal{G}}$$

which gives a κ-ζ factorization of φ.

Corresponding to Theorem 6.10, we have:

Theorem 6.23. Let $\varphi : (Y, \sigma) \to (Y, \sigma)$ be a forward automorphism of lag n of a sofic system with $n \geq 0$. Let $\mathcal{G} = (G, \lambda)$ be a λ-graph that defines (Y, σ). Let $\mathcal{M} = \mathcal{M}_{\mathcal{G}}$. Let $\varphi' = \varphi^{-1}\sigma^n$. We assume that if $n = 0$, then $(h, k) : \mathcal{G} \simeq \mathcal{G}$ is a λ-graph-automorphism such that k gives φ, and we assume that if $n > 0$, then a specified strong shift equivalence

$$\text{(6.20)} \quad \mathcal{M} \overset{k_0}{\simeq} \mathcal{P}_1\mathcal{Q}_1, \ \mathcal{Q}_1\mathcal{P}_1 \overset{k_1}{\simeq} \mathcal{P}_2\mathcal{Q}_2, \ \cdots, \ \mathcal{Q}_{n-1}\mathcal{P}_{n-1} \overset{k_{n-1}}{\simeq} \mathcal{P}_n\mathcal{Q}_n, \ \mathcal{Q}_n\mathcal{P}_n \overset{k_n}{\simeq} \mathcal{M}$$

gives a κ-ζ factorization of φ. Let $\mathcal{P}$ and $\mathcal{Q}$ be the representation matrices with the indexing set $V_G \times V_G$ defined as follows: (i) if $n = 0$, then $\mathcal{P}$ is the representation matrix obtained from the permutation matrix corresponding to the permutation h_V on V_G by replacing all 1's in the permutation matrix by the same symbol, and $\mathcal{Q}$ is the transpose of $\mathcal{P}$ (i.e., the representation matrix that represents the transpose of the λ-graph represented by $\mathcal{P}$); (ii) if $n > 0$, then

$$\mathcal{P} = \mathcal{P}_1 \cdots \mathcal{P}_n \quad \text{and} \quad \mathcal{Q} = \mathcal{Q}_n \cdots \mathcal{Q}_1.$$

Then if φ is expansive, then (Y, φ) is topologically conjugate to a sofic system and if φ' is expansive, then (Y, φ') is topologically conjugate to a sofic system. Moreover, for all integers $l \geq 0$ and $m \geq 1$, $(Y, \varphi^l \sigma^m)$ is topologically conjugate to $(X_{\mathcal{P}^l \mathcal{M}^m}, \sigma_{\mathcal{P}^l \mathcal{M}^m})$ and $(Y, (\varphi')^l \sigma^m)$ is topologically conjugate to $(X_{\mathcal{Q}^l \mathcal{M}^m}, \sigma_{\mathcal{Q}^l \mathcal{M}^m})$.

Let the hypothesis and the notation be the same as in Theorem 6.23.

Case (i). Let $\mathcal{T}_0$ be the sofic textile system defined by $\mathcal{T}_0 = (T_0, \lambda, \mu)$, where $T_0 = (i, h : G \to G)$ with i being the identity graph-isomorphism and μ is the mapping of V_G onto the set consisting of one symbol, say b. Then $\mathcal{T}_0$ is 1-1 and LR, $\varphi_{\mathcal{T}_0} = \varphi$ and $\mathcal{T}_0$ is associated with the specified equivalence

$$\mathcal{M}\mathcal{P} \overset{l}{\simeq} \mathcal{P}\mathcal{M},$$

where l is the bijection that sends ab to $bk(a)$ for every label $a \in \lambda(A_G)$. Let $\mathcal{T}_0' = (T_0^{-1}, \lambda, \mu)$. Then $\varphi_{\mathcal{T}_0'} = \varphi'$. The specified equivalence associated with $\mathcal{T}_0'$ is of the form

$$\mathcal{M}\mathcal{Q} \overset{l'}{\simeq} \mathcal{Q}\mathcal{M}$$

and we have

$$\mathcal{I} \overset{m}{\simeq} \mathcal{P}\mathcal{Q}, \quad \mathcal{Q}\mathcal{P} \overset{m'}{\simeq} \mathcal{I},$$

where $\mathcal{I}$ is the representation matrix obtained from the identity matrix by replacing all 1's by the same symbol, and m and m' are the trivial specifications.

Case (ii). The specified strong shift equivalence (6.20)(from $\mathcal{M}$ to itself) naturally induces a sequence of specified equivalences

$$\begin{aligned}\mathcal{M}\mathcal{P}_1 \cdots \mathcal{P}_n \overset{k_0'}{\simeq} (\mathcal{P}_1\mathcal{Q}_1)\mathcal{P}_1 \cdots \mathcal{P}_n \overset{k_1'}{\simeq} \mathcal{P}_1(\mathcal{P}_2\mathcal{Q}_2)\mathcal{P}_2 \cdots \mathcal{P}_n \overset{k_2''}{\simeq} \cdots \\ \overset{k_{n-1}^{(n-1)}}{\simeq} \mathcal{P}_1 \cdots \mathcal{P}_{n-1}(\mathcal{P}_n\mathcal{Q}_n)\mathcal{P}_n \overset{k_n^{(n)}}{\simeq} \mathcal{P}_1 \cdots \mathcal{P}_n\mathcal{M}.\end{aligned}$$

Let

$$l = k_n^{(n)} k_{n-1}^{(n-1)} \cdots k_2'' k_1'.$$

Then we have a specified equivalence

$$\mathcal{M}\mathcal{P} \overset{l}{\simeq} \mathcal{P}\mathcal{M}. \tag{6.21}$$

Similarly (6.20) naturally induces specified equivalences of the form

$$(6.22) \qquad \mathcal{M}\mathcal{Q} \overset{l'}{\simeq} \mathcal{Q}\mathcal{M}$$

and of the form

$$(6.23) \qquad \mathcal{M}^n \overset{m}{\simeq} \mathcal{P}\mathcal{Q}, \quad \mathcal{Q}\mathcal{P} \overset{m'}{\simeq} \mathcal{M}^n.$$

Recalling the notion of shift equivalence for representation matrices introduced by Boyle and Krieger [BK2], we call the set of specified equivalences given by (6.21),(6.22) and (6.23) the *specified shift equivalence of lag n from $\mathcal{M}$ to itself induced by* (6.20). Let $\mathcal{T}_1$ be the sofic textile system associated with (6.21) and $\mathcal{T}_1'$ the sofic textile system associated with (6.22). Then we have

$$\varphi_{\mathcal{T}_1} = \varphi \quad \text{and} \quad \varphi_{\mathcal{T}_1'} = \varphi',$$

and we see that (6.23) is a one-step strong shift equivalence which gives $\varphi^{(n)}$ (see Lemma 1 of [B4]).

Let φ be forward automorphism of a sofic system $(X_{\mathcal{M}}, \sigma_{\mathcal{M}})$. Let $n \geq 0$. A set of specified equivalences

$$\mathcal{M}\mathcal{P} \overset{l}{\simeq} \mathcal{P}\mathcal{M}, \quad \mathcal{M}\mathcal{Q} \overset{l'}{\simeq} \mathcal{Q}\mathcal{M}, \quad \mathcal{M}^n \overset{m}{\simeq} \mathcal{P}\mathcal{Q}, \quad \mathcal{Q}\mathcal{P} \overset{m'}{\simeq} \mathcal{M}^n$$

is called a *specified shift equivalence of lag n associated with* φ if the LR sofic textile system $\mathcal{T}$ associated with $\mathcal{M}\mathcal{P} \overset{l}{\simeq} \mathcal{P}\mathcal{M}$ is 1-1 and $\varphi_{\mathcal{T}} = \varphi$ and the LR sofic textile system $\mathcal{T}'$ associated with $\mathcal{M}\mathcal{Q} \overset{l'}{\simeq} \mathcal{Q}\mathcal{M}$ is 1-1 and $\varphi_{\mathcal{T}'} = \varphi^{-1}\sigma_{\mathcal{M}}^n$, and the one-step specified strong shift equivalence

$$\mathcal{M}^n \overset{m}{\simeq} \mathcal{P}\mathcal{Q}, \quad \mathcal{Q}\mathcal{P} \overset{m'}{\simeq} \mathcal{M}^n$$

gives $\varphi^{(n)}$ when $n \geq 1$. Summarizing the above, we have:

Corollary 6.24. Let φ be a forward automorphism of a sofic system. Let n be a nonnegative integer. If φ has a κ-ζ factorization of lag n, then there is a specified shift equivalence of lag n associated with φ. If $n \geq 1$, then the specified shift equivalence induced by the specified strong shift equivalence of lag n giving the κ-ζ factorization of φ, is a specified shift equivalence associated with φ.

By Corollary 6.11 a forward automorphism of a topological Markov shift is an LR automorphism. Also, by Corollary 6.24, a forward automorphism of a sofic system is a "sofic LR automorphism". We can prove that an LR automorphism of a topological Markov shift is forward. The sofic version of this, however, has not yet been proved.

Let $h : \Gamma \to G$ be a right resolving or left resolving graph-homomorphism. For $m \in \mathbf{N}, h$ is said to be m *definite* either if h is right resolving and for any paths $\alpha_1 \cdots \alpha_m, \beta_1 \cdots \beta_m \in L_m(\Gamma), \alpha_i, \beta_i \in A_\Gamma$,

$$h_A(\alpha_1) \cdots h_A(\alpha_m) = h_A(\beta_1) \cdots h_A(\beta_m) \quad \Rightarrow \quad t_\Gamma(\alpha_m) = t_\Gamma(\beta_m)$$

or if h is left resolving and for any paths $\alpha_1 \cdots \alpha_m, \beta_1 \cdots \beta_m \in L_m(\Gamma), \alpha_i, \beta_i \in A_\Gamma$,

$$h_A(\alpha_1) \cdots h_A(\alpha_m) = h_A(\beta_1) \cdots h_A(\beta_m) \quad \Rightarrow \quad i_\Gamma(\alpha_1) = i_\Gamma(\beta_1).$$

We say that h is 0 *definite* if h is a graph-isomorphism. It is clear that if h is m definite with $m \geq 0$, then h is $m+n$ definite for all $n \geq 0$. For $m \in \mathbf{N}$, we say that h is *strictly* m *definite* if h is m definite but not $m-1$ definite. We say that h is *strictly* 0 *definite* if h is 0 definite. We say that h is *definite* if h is m definite for some $m \geq 0$.

A definite right resolving or left resolving graph-homomorphism is an extension of the definite transition diagram of a finite automaton, which was introduced by Perles, Rabin and Shamir [PRS]. The properties and a practical decision procedure for the definiteness of transition diagrams presented in [PRS], can be straightforwardly extended to right resolving or left resolving graph-homomorphisms.

Lemma 6.25. Let $h : \Gamma \to G$ be a right resolving or left resolving, onto graph-homomorphism. Let $\phi : (X_\Gamma, \sigma_\Gamma) \to (X_G, \sigma_G)$ be the 1-block factor map given by h. Then h is a conjugacy if and only if h is definite.

Proof. Assume that ϕ is a conjugacy. Let $m = (\#V_\Gamma)^2$. Let $\alpha_1 \cdots \alpha_m, \beta_1 \cdots \beta_m \in L_m(\Gamma)$ with $\alpha_i, \beta_i \in A_\Gamma$ and assume that $h(\alpha_1) \cdots h(\alpha_m) = h(\beta_1) \cdots h(\beta_m)$. Since $m = (\#V_\Gamma)^2$, there are integers $i, j, 1 \leq i \leq j \leq m$, such that

$$(i_\Gamma(\alpha_i), i_\Gamma(\beta_i)) = (t_\Gamma(\alpha_j), t_\Gamma(\beta_j)).$$

Since ϕ is 1-1, we have $\alpha_i \cdots \alpha_j = \beta_i \cdots \beta_j$. (For otherwise, distinct periodic points would be sent to the same point of X_G by ϕ.) If h is right resolving, then it follows that $t_\Gamma(\alpha_m) = t_\Gamma(\beta_m)$ since $t_\Gamma(\alpha_j) = t_\Gamma(\beta_j)$ and $h(\alpha_{j+1}) \cdots h(\alpha_m) = h(\beta_{j+1}) \cdots h(\beta_m)$ if $m > j$. If h is left resolving, then it similarly follows that $i_\Gamma(\alpha_1) = i_\Gamma(\beta_1)$. Therefore, h is m-definite.

If h is definite, then it is easily seen that ϕ is a conjugacy. □

Lemma 6.26. If $h:\Gamma \to G$ is a strictly m-definite right-resolving graph-homomorphism with $m \geq 1$, then there are strictly 1-definite right-resolving graph-homomorphisms $h_i : \Gamma_i \to \Gamma_{i+1}, i = 0, \cdots, m-1$, such that $\Gamma_0 = \Gamma, \Gamma_m = G$ and

$$h = h_m h_{m-1} \cdots h_1.$$

The statement replaced each "right-resolving" by "left-resolving" also holds.

Proof (Implicit in [PRS]). Assume that h is right resolving. (The proof for the case where h is left resolving is similar.) We may assume that $m \geq 2$. It suffices to show that there are right-resolving graph-homomorphisms $h' : \Gamma \to \Gamma'$ and $h'' : \Gamma' \to G$ such that $h = h''h', h'$ is strictly 1 definite and h'' is strictly $m-1$ definite. Let us say that two vertices $u, v \in V_\Gamma$ are *1-equivalent* if $t_\Gamma(\alpha) = t_\Gamma(\beta)$ for any two arcs $\alpha, \beta \in A_\Gamma$

with $i_\Gamma(\alpha) = u, i_\Gamma(\beta) = v$ and $h_A(\alpha) = h_A(\beta)$. Let us say that two arcs $\alpha, \beta \in A_\Gamma$ are *1-equivalent* if $i_\Gamma(\alpha)$ and $i_\Gamma(\beta)$ are 1-equivalent and $h_A(\alpha) = h_A(\beta)$. For $u \in V_\Gamma$, let $C(u)$ denote the 1-equivalence class containing u, and for $\alpha \in A_\Gamma$ let $D(\alpha)$ denote the 1-equivalence class containing α. Let Γ' be the graph such that

$$V_{\Gamma'} = \{C(u) \mid u \in V_\Gamma\}, \quad A_{\Gamma'} = \{D(\alpha) \mid \alpha \in A_\Gamma\}$$

and

$$i_{\Gamma'}(D(\alpha)) = C(i_\Gamma(\alpha)) \quad \text{and} \quad t_{\Gamma'}(D(\alpha)) = C(t_\Gamma(\alpha)), \quad \alpha \in A_\Gamma.$$

Let $h' : \Gamma \to \Gamma'$ be defined by

$$h'_A(\alpha) = D(\alpha), \quad \alpha \in A_\Gamma$$
$$h'_V(u) = C(u), \quad u \in V_\Gamma,$$

and let $h'' : \Gamma' \to G$ be defined by

$$h''_A(D(\alpha)) = h_A(\alpha), \quad \alpha \in A_\Gamma$$
$$h''_V(C(u)) = h_V(u), \quad u \in V_\Gamma.$$

It is straightforward to check that Γ', h' and h'' are well defined. It is also straightforward to see that h' is right resolving and 1-definite and h'' is right resolving. Since h is m definite with $m \geq 2$, it follows that if $\alpha_1 \cdots \alpha_{m-1}, \beta_1 \cdots \beta_{m-1} \in L_{m-1}(\Gamma)$ with $\alpha_i, \beta_i \in A_\Gamma$ and $h_A(\alpha_1) \cdots h_A(\alpha_{m-1}) = h_A(\beta_1) \cdots h_A(\beta_{m-1})$, then $t_\Gamma(\alpha_{m-1})$ and $t_\Gamma(\beta_{m-1})$ are 1-equivalent. This implies that h'' is $m - 1$ definite. Since h is strictly m definite and $m \geq 2$, there are $\alpha_1 \cdots \alpha_{m-1}, \beta_1 \cdots \beta_{m-1} \in L_{m-1}(\Gamma), \alpha_i, \beta_i \in A_\Gamma$, such that

$$h_A(\alpha_1) \cdots h_A(\alpha_{m-1}) = h_A(\beta_1) \cdots h_A(\beta_{m-1})$$

but

$$t_\Gamma(\alpha_{m-1}) \neq t_\Gamma(\beta_{m-1}).$$

Thus $t_\Gamma(\alpha_{m-1})$ and $t_\Gamma(\beta_{m-1})$ are distinct 1-equivalent vertices. Therefore h' is not 0 definite. Since $t_\Gamma(\alpha_{m-1}) \neq t_\Gamma(\beta_{m-1})$ and $h_A(\alpha_{m-1}) = h_A(\beta_{m-1})$, we have

$$C(i_\Gamma(\alpha_{m-1})) \neq C(i_\Gamma(\beta_{m-1})).$$

Therefore since

$$h''_V(C(i_\Gamma(\alpha_{m-1}))) = h''_V(C(i_\Gamma(\beta_{m-1}))),$$

h'' is not 0 definite so that h'' is not $m - 2$ definite if $m = 2$. If $m > 2$, then

$$h''_A(D(\alpha_1)) \cdots h''_A(D(\alpha_{m-2})) = h''_A(D(\beta_1)) \cdots h''_A(D(\beta_{m-2}))$$

so that h'' is not $m - 2$ definite. □

Lemma 6.27. If $h : \Gamma \to G$ is a 1-definite right-resolving onto graph-homomorphism, then the 1-block conjugacy $\phi : (X_\Gamma, \sigma_\Gamma) \to (X_G, \sigma_G)$ given by h, is a forward bipartite code. If $h : \Gamma \to G$ is a 1-definite left-resolving onto graph-homomorphism, then the 1-block conjugacy $\phi : (X_\Gamma, \sigma_\Gamma) \to (X_G, \sigma_G)$ given by h, is a backward bipartite code.

Proof. Assume that $h:\Gamma \to G$ is a 1-definite right-resolving onto graph-homomorphism. Let $f : A_\Gamma \to V_\Gamma A_G$ and $g : A_G \to A_G V_\Gamma$ be the mappings defined by

$$f(\alpha) = i_\Gamma(\alpha) h_A(\alpha), \quad \alpha \in A_\Gamma$$

and

$$g(h_A(\alpha)) = h_A(\alpha) t_\Gamma(\alpha), \quad \alpha \in A_\Gamma.$$

Since h is right resolving, f is 1-1 so that f is a bipartite expression of A_Γ. Since h is 1 definite, g is well defined. Clearly g is 1-1 so that g is a bipartite expression of A_G. If $\phi((\alpha_i)_{i\in \mathbf{Z}}) = (a_i)_{i\in\mathbf{Z}}$ for $(\alpha_i)_{i\in\mathbf{Z}} \in X_\Gamma$, then $f(\alpha_i) = i_\Gamma(\alpha_i)a_i$ and $g(a_i) = a_i t_\Gamma(\alpha_i)$ for all $i \in \mathbf{Z}$. Therefore ϕ is a forward bipartite code.

Assume that $h : \Gamma \to G$ is a 1-definite left-resolving onto graph-homomorphism. Then let $f' : A_\Gamma \to A_G V_\Gamma$ and $g' : A_G \to V_\Gamma A_G$ be the mappings defined by

$$f'(\alpha) = h_A(\alpha) t_\Gamma(\alpha), \quad \alpha \in A_\Gamma$$

and

$$g'(h_A(\alpha)) = i_\Gamma(\alpha) h_A(\alpha), \quad \alpha \in A_\Gamma.$$

Then f' is 1-1, and g' is well defined and 1-1. If $\phi((\alpha_i)_{i\in\mathbf{Z}}) = (a_i)_{i\in\mathbf{Z}}$ for $(\alpha_i)_{i\in\mathbf{Z}} \in X_\Gamma$, then $f'(\alpha_i) = a_i t_\Gamma(\alpha_i)$ and $g'(a_i) = i_\Gamma(\alpha_i)a_i$ for all $i \in \mathbf{Z}$. Therefore ϕ is a backward bipartite code. □

Proposition 6.28. If $h:\Gamma\to G$ is a definite right-resolving onto graph-homomorphism, then the 1-block conjugacy $\phi : (X_\Gamma, \sigma_\Gamma) \to (X_G, \sigma_G)$ given by h, is forward. If $h : \Gamma \to G$ is a definite left-resolving onto graph-homomorphism, then the 1-block conjugacy ϕ : $(X_\Gamma, \sigma_\Gamma) \to (X_G, \sigma_G)$ given by h, is backward.

Proof. This follows from Lemmas 6.26 and 6.27. □

Theorem 6.29. For an automorphism φ of a topological Markov shift, φ is an LR automorphism if and only if φ is a forward automorphism.

Proof. By Corollary 6.11, the only-if part is valid.

If φ is an LR automorphism, then there is a 1-1 LR textile system $T = (p, q : \Gamma \to G)$ such that $\varphi_T = \varphi$. Let $\xi : (X_\Gamma, \sigma_\Gamma) \to (X_G, \sigma_G)$ and $\eta : (X_\Gamma, \sigma_\Gamma) \to (X_G, \sigma_G)$ be the conjugacies given by p and q, respectively. Then by Lemma 6.25 and Proposition 6.28, ξ is a backward conjugacy and η is a forward conjugacy so that $\varphi = \eta\xi^{-1}$ is a forward automorphism. □

Proposition 6.30. Let $\varphi : (X,\sigma) \to (X,\sigma)$ be an endomorphism of a topological Markov shift. If φ is right closing, then there is an integer $N \geq 0$ such that $\varphi\sigma^n$ is an LR endomorphism for all $n \geq N$. If $\varphi\sigma^n$ is an LR endomorphism for some $n \in \mathbf{Z}$, then φ is right closing. If φ is left closing, then there is an integer $N' \geq 0$ such that $\varphi\sigma^{-n}$ is an RL endomorphism for all $n \geq N'$. If $\varphi\sigma^n$ is an RL endomorphism for some $n \in \mathbf{Z}$, then φ is left closing.

Proof. Assume that φ is right closing. Let G be a graph such that $(X_G, \sigma_G) = (X,\sigma)$. Suppose that φ is a block map of (k,l) type with integers $k,l \geq 0$. Let $f : L_{k+l+1}(G) \to A_G$ be the mapping which gives φ. For $m \in \mathbf{N}$, let $\mathcal{G}^{[m]} = (G^{[k+l+m]}, \lambda_f^{[m]})$ be the λ-graph such that for $a_1 \cdots a_{k+l+m} \in L_{k+l+m}(G)$, $a_i \in A_G$,

$$\lambda_f^{[m]}(a_1 \cdots a_{k+l+m}) = f(a_1 \cdots a_{k+l+1}) \cdots f(a_m \cdots a_{k+l+m}).$$

Clearly $\mathcal{G}^{[m]}$ gives a graph-homomorphism of $G^{[k+l+m]}$ onto $G^{[m]}$. We consider the induced right resolving λ-graph $(\mathcal{G}^{[m]})_+ = (\Gamma_m, \lambda_m)$ of $\mathcal{G}^{[m]}$. Then $(\mathcal{G}^{[m]})_+$ gives a right resolving graph-homomorphism of Γ_m onto $G^{[m]}$. By definition, each arc of Γ_m is of the form (U,w) for a subset U of $L_{k+l+m-1}(G)$ and a path $w \in L_m(G)$. Since φ is right closing, there is $m_0 \in \mathbf{N}$ such that if $m \geq m_0$, then for each arc (U,w) of Γ_m, all paths in U have the same first arc, say a_U (cf. [Ki] or [N1]). Therefore we can define a textile system $T_m = (p_m, q_m : \Gamma_m \to G)$ for $m \geq m_0$ as follows: for each arc (U,w) of Γ_m,

$$p_m((U,w)) = a_U \quad \text{and} \quad q_m((U,w)) = b_w,$$

where b_w is the last arc of the path w. Clearly p_m and q_m are graph-homomorphisms. It is easy to see that p_m is left resolving, q_m is right resolving and the factor map given by p_m is 1-1 (cf.[Ki] or [N1]). Thus T_m is one-sided 1-1 and LR. Clearly $\varphi_{T_m} = \varphi\sigma^{m+k-1}$. Therefore if we put $N = m_0 + k - 1$, then the first claim of the proposition is proved.

The proof of the third claim is similarly given using the induced left resolving λ-graph $(\mathcal{G}^{[m]})_-$ of $\mathcal{G}^{[m]}$.

It is clear that an LR endomorphism is right closing and an RL endomorphism is left closing. Therefore, the second and forth claims are proved. □

Now we shall give entropy results on LR or RL endomorphisms of topological Markov shifts, which include an extension of a result of Boyle and Krieger on the entropy of automorphisms of topological Markov shifts (see Theorem 2.17 of [BK]). The entropy of an LR or RL endomorphism is easily given. We note that an LR or RL endomorphism is not necessarily expansive.

Theorem 6.31. Let $T = (p,q : \Gamma \to G)$ be a one-sided 1-1 textile system. Then the following statements are valid.

(1) If T is LR, then $h(\varphi_T) = h(\tilde{\varphi}_T) = h(\sigma_{G^T})$.

(2) If T is LR and (X_G, σ_G) is irreducible, then $h(\varphi_T) = h(\tilde{\varphi}_T) = -\log R(\varphi_T)$.

(3) If T is RL, then $h(\varphi_T) = h(\sigma_{G^T})$.

(4) If T is RL and (X_G, σ_G) is irreducible, then $h(\varphi_T) = -\log L(\varphi_T)$.

Proof. If T is LR, then $\tilde{\xi}_T$ is 1-1. Hence (1) follows from Theorem 2.13 and Proposition 3.5. Also from these (3) follows.

To prove (2), assume that T is LR and (X_G, σ_G) is irreducible. Since T is LR, T is nondegenerate. Since (X_G, σ_G) is irreducible and ξ_T is a conjugacy, M_Γ and M_G are irreducible and have the same maximal eigenvalue. Therefore $R(T)$ can be defined, and we have $R(\varphi_T) = R(T)$. Let b be an arc of G^T and let $b_1, \cdots, b_k$ be all the arcs that emanate from $t_{G^T}(b)$ in G^T. Note that $b_1, \cdots, b_k$ and b are vertices of Γ and $t_{G^T}(b)$ is a vertex of G. Let $a_1 \cdots a_m$ be in $L_m(G)$ with $m \geq 1$, $a_i \in A_G$, and $t_G(a_m) = t_{G^T}(b)$. Then, since p is left resolving, there are paths $\alpha_1^{(i)} \cdots \alpha_m^{(i)}, i = 1, \cdots, k$, in $L_m(\Gamma)$ such that

$$t_\Gamma(\alpha_m^{(i)}) = b_i \quad \text{and} \quad p(\alpha_1^{(i)}) \cdots p(\alpha_m^{(i)}) = a_1 \cdots a_m, \quad i = 1, \cdots k.$$

Since ξ_T is 1-1, it follows that if m is sufficiently large then

$$i_\Gamma(\alpha_1^{(1)}) = i_\Gamma(\alpha_1^{(2)}) = \cdots = i_\Gamma(\alpha_1^{(k)}).$$

This implies that $\{b_1, \cdots, b_k\}$ is a maximal right-compatible set for p. Since q is right resolving, $\{b\}$ is a maximal right-compatible set for q. Let r be a right eigenvector of M_Γ corresponding to the maximal eigenvalue. For $c \in V_\Gamma$, let $r(c)$ denote the component of r corresponding to c. Then

$$\frac{r(b)}{r(b_1) + \cdots + r(b_k)} = R(T)$$

so that

$$\frac{1}{R(T)} r(b) = r(b_1) + \cdots + r(b_m).$$

This implies that $1/R(T)$ is an eigenvalue of $M_{(G^T)^{[2]}}$ and r is a right eigenvector of $M_{(G^T)^{[2]}}$ corresponding to that eigenvalue, because b is an arbitrary arc of G^T and $b_1, \cdots, b_m$ are all the arcs that emanate from $t_{G^T}(b)$. Since r is strictly positive, $1/R(T)$ is the maximal eigenvalue of $M_{(G^T)^{[2]}}$ (see Corollary 1.12 of [BP]). Therefore $1/R(T)$ is the maximal eigenvalue of M_{G^T}. Therefore, by (1) we have

$$h(\varphi_T) = h(\tilde{\varphi}_T) = -\log R(T)$$

so that

$$h(\varphi_T) = h(\tilde{\varphi}_T) = -\log R(\varphi_T).$$

The proof of (4) is similar. □

In closing this section, we shall add some discussions on "textile relation systems". A *textile relation system* T is an ordered pair of graph-homomorphisms $p : \Gamma \to G_1$

and $q : \Gamma \to G_2$ such that the quadruple $(i_\Gamma(\alpha), t_\Gamma(\alpha), p(\alpha), q(\alpha))$ with $\alpha \in A_\Gamma$ uniquely determines α. We write $T = (p : \Gamma \to G_1, q : \Gamma \to G_2)$. (By analogy, each $\alpha \in A_\Gamma$ is imagined as the square whose upper and lower sides are $p(\alpha)$ and $q(\alpha)$ and whose left and right sides are $i_\Gamma(\alpha)$ and $t_\Gamma(\alpha)$, though T can weave only obis of width 1.) The textile relation system T defines a relation between X_{G_1} and X_{G_2} by $\{\frac{\xi(z)}{\eta(z)} \mid z \in X_\Gamma\}$ ($\frac{\xi(z)}{\eta(z)}$ simply denotes the element $(\xi(z), \eta(z))$ of $X_{G_1} \times X_{G_2}$), where ξ and η are the factor maps given by p and q, respectively. If ξ is a conjugacy, then we define a factor map $\phi_T : (X_{G_1}, \sigma_{G_1}) \to (X_{G_2}, \sigma_{G_2})$ by $\phi_T = \eta\xi^{-1}$. The notions of an *LR textile relation system*, an *LR factor map* and an *LR conjugacy* should be clear by analogy. Proposition 6.1 is straightforwardly extended to LR textile relation systems. (A similar approach is found in [MT]. See also [P].) Every LR textile relation system $T = (p : \Gamma \to G_1, q : \Gamma \to G_2)$ is uniquely associated with a specified equivalence of the form

$$\tilde{M}_{G_1}\tilde{P} \overset{k}{\simeq} \tilde{P}\tilde{M}_{G_2},$$

where P is a nonnegative integral rectangular matrix. Conversely, every specified equivalence of this form is uniquely associated with the LR textile relation system associated with it. Observation 6.9 is also straightforwardly generalized to LR textile relation systems. We easily see that if

$$\tilde{M} \overset{k_0}{\simeq} \tilde{P}\tilde{Q}, \quad \tilde{Q}\tilde{P} \overset{k_1}{\simeq} \tilde{N}$$

is a one-step strong shift equivalence and if $\phi : X_M \to X_N$ is the bipartite code given by it, then the specified equivalence $\tilde{M}\tilde{P} \overset{l}{\simeq} \tilde{P}\tilde{N}$ which is naturally induced by it (i.e., corresponding to it, we naturally have $\tilde{M}\tilde{P} \overset{k_0'}{\simeq} \tilde{P}\tilde{Q}\tilde{P} \overset{k_1'}{\simeq} \tilde{P}\tilde{N}$ and we have $l = k_1'k_0'$), defines an LR textile relation system T with $\phi_T = \phi$. Using this, Theorem 5.3(2), and the generalization of Observation 6.9, we have the generalization of Corollary 6.11 to forward conjugacies. (This approach gives another proof of Theorem 6.10 and its generalization to "sofic LR textile relation systems" gives another proof of Theorem 6.23.) Hence, if $\phi : (X_M, \sigma_M) \to (X_N, \sigma_N)$ is a forward conjugacy a κ-ζ factorization of which is given by a specified strong shift equivalence

$$\tilde{M} \overset{k_0}{\simeq} \tilde{P}_1\tilde{Q}_1, \ \tilde{Q}_1\tilde{P}_1 \overset{k_1}{\simeq} \tilde{P}_2\tilde{Q}_2, \ \cdots, \ \tilde{Q}_{n-1}\tilde{P}_{n-1} \overset{k_{n-1}}{\simeq} \tilde{P}_n\tilde{Q}_n, \ \tilde{Q}_n\tilde{P}_n \overset{k_n}{\simeq} \tilde{N}$$

and if

$$\tilde{M}\tilde{P} \overset{l}{\simeq} \tilde{P}\tilde{N} \quad \text{with } \tilde{P} = \tilde{P}_1 \cdots \tilde{P}_n$$

is the specified equivalence which is naturally induced by that specified strong shift equivalence, then $\phi_T = \phi$ for the textile relation system associated with this specified equivalence. The generalization of Corollary 7.25 which will be remarked at the end of Section 7 implies that an LR textile relation system T with $\phi_T = \phi$ is unique for ϕ and hence the specified equivalence above is unique for ϕ. Furthermore, by the above and Proposition 6.28, we have the generalization of Theorem 6.29 that for a topological conjugacy ϕ between topological Markov shifts, ϕ is forward if and only if ϕ is LR. Proposition 6.30 is also straightforwardly generalized to onto factor maps.

7. RESOLVABLE TEXTILE SYSTEMS

In this section, we do not assume that a textile system is necessarily in the standard form. But we assume that in every textile system $T = (p, q : \Gamma \to G)$, G and Γ are nondegenerate (but p and q are not necessarily onto, that is, in their duals $T^* = (p^T, q^T : \Gamma^T \to G^T)$, G^T and Γ^T do not need to be nondegenerate).

Let $T = (p, q : \Gamma \to G)$ be a textile system. A subconfiguration of the form

$$(\alpha_{ij})_{1 \le i \le m, 1 \le j \le n}, \quad \alpha_{ij} \in A_\Gamma, \ m, n \in \mathbf{N},$$

is called a *cloth* (weaved by T) if it is well weaved by T. For a cloth $C = (\alpha_{ij})_{1 \le i \le m, 1 \le j \le n}$, we call $p(\alpha_{11}) \cdots p(\alpha_{1n})$ the *upper side* of C, $q(\alpha_{m1}) \cdots q(\alpha_{mn})$ the *lower side* of C, $\downarrow i_\Gamma(\alpha_{11}) \cdots i_\Gamma(\alpha_{m1})$ the *left side* of C, and $\downarrow t_\Gamma(\alpha_{1n}) \cdots t_\Gamma(\alpha_{mn})$ the *right side* of C. Two paths x and x' in $L_n(G)$ with $n \in \mathbf{N}$ are said to be *compatible* (with respect to T) if there are two cloths C and C' such that the upper sides of C and C' are the same, the left sides of C and C' are the same, the lower side of C is equal to x, and the lower side of C' is equal to x'. We also say that x and x' in $L_0(G) = V_G$ are *compatible* if $x = x'$. For $n \in \mathbf{N}$, a subconfiguration of the form

$$(\alpha_{ij})_{i \in \mathbf{N}, 1 \le j \le n}, \quad \alpha_{ij} \in A_\Gamma,$$

is called a *curtain* of *length* n (weaved by T) if it is well weaved by T. For $y \in \tilde{X}_{G^T}$

$$\downarrow y$$

is called a *curtain* of *length 0* (weaved by T). For a curtain $(\alpha_{ij})_{i \in \mathbf{N}, 1 \le j \le n}$, $\downarrow (i_\Gamma(\alpha_{i1}))_{i \in \mathbf{N}}$ is called its *left side* or its *initial subcurtain* of *length* 0, $\downarrow (t_\Gamma(\alpha_{i,n}))_{i \in \mathbf{N}}$ is called its *right side* or its *terminal subcurtain of length* 0, $(\alpha_{ij})_{i \in \mathbf{N}, 1 \le j \le n'}$ is called its *initial subcurtain* of *length* n' for $1 \le n' \le n$, $(\alpha_{ij})_{i \in \mathbf{N}, n-n' < j \le n}$ is called its *terminal subcurtain of length* n' for $1 \le n' \le n$, and $p(\alpha_{11}) \cdots p(\alpha_{1n})$ is called its *upper side*. For $(b_i)_{i \in \mathbf{N}} \in \tilde{X}_{G^T}, b_i \in A_{G^T}$, the curtain $\downarrow (b_i)_{i \in \mathbf{N}}$ is the initial and terminal subcurtain of length 0 of itself and its upper side is $i_{G^T}(b_1)$. Two paths x and x' in $L(G)$ are said to be *equivalent* (with respect to T) if for any $y \in \tilde{X}_{G^T}$, there is a curtain whose upper side is x and whose left side is $\downarrow y$ if and only if there is a curtain whose upper side is x' and whose left side is $\downarrow y$.

Note that for a graph G and $x = a_1 \cdots a_n$ in $L(G)$ with $a_i \in A_G$, the initial subpath of length 0 of x means $i_G(a_1)$ and the terminal subpath of length 0 of x means $t_G(a_n)$.

Definition 7.1. Let T be a textile system over a graph G and k a nonnegative integer. We say that T is k *resolvable* if the following three conditions are satisfied:

(i) for any $x \in L_k(G)$ and any $x' \in L(G)$ such that x is an initial subpath of x', x and x' are equivalent;

(ii) if n is an integer greater than k, then for any two curtains of length n with the same upper side and the same left side, their initial subcurtains of length $n-k$ are the same;

(iii) any two compatible paths in $L_k(G)$ are equivalent.

We say that a textile system is *resolvable* if it is k resolvable for some nonnegative integer k.

Remark 7.2. For an integer $k \geq 0$, a k-resolvable textile system is $k+1$ resolvable.

Proof. Assume that T is a k-resolvable textile system. The conditions (i) and (ii) of Definition 7.1 are clearly satisfied for $k+1$ by T. To see that (iii) is satisfied for $k+1$, let x_1 and x_2 be in $L_{k+1}(G)$ and compatible. Let $\bar{x}_1$ and $\bar{x}_2$ be the initial subpaths of length k of x_1 and x_2, respectively. Clearly $\bar{x}_1$ and $\bar{x}_2$ are compatible. Therefore, by (iii) of Definition 7.1 for k, $\bar{x}_1$ and $\bar{x}_2$ are equivalent. By (i) of Definition 7.1 for k, x_i and $\bar{x}_i$ are equivalent for $i = 1, 2$. Thus x_1 and x_2 are equivalent. □

Remark 7.3 Let $T = (p, q : \Gamma \to G)$ be a textile system. Let $\bar{T} = (\bar{p}, \bar{q} : \bar{\Gamma} \to \bar{G})$ be its trimming closure. Then the following statements are valid.

(1) If p is right resolving, then T is 0 resolvable.

(2) If T is 0 resolvable, then $\bar{p}$ is right resolving.

(3) If T is k resolvable with $k \geq 0$, then $\bar{T}$ is k resolvable.

Proof. Since (1) is clear and (2) is easily proved by using (3) (see the proof of Remark 7.5(2)), it suffices to prove (3).

To prove (3), assume that T is k resolvable with $k \geq 0$.

(i) Let C be a curtain of length k weaved by $\bar{T}$ and x the upper side of C. Let $(a_j)_{j\in\mathbf{N}} \in \tilde{X}_{\bar{G}}$ with $a_j \in A_{\bar{G}}$, be any right-infinite sequence such that $a_1 \cdots a_k = x$ if $k \geq 1$, and $i_G(a_1) = x$ if $k = 0$. Since T can weave C and T is k resolvable, it follows from the condition (i) of Definition 7.1 and the compactness of $\hat{U}_T$ that T weaves a well-weaved subconfiguration $D = (\beta_{ij})_{i,j\in\mathbf{N}}$ whose left side i.e., $\downarrow (i_\Gamma(\beta_{i1}))_{i\in\mathbf{N}}$, is the same as that of C and whose upper side i.e., $(p(\beta_{1j}))_{j\in\mathbf{N}}$, is equal to $(a_j)_{j\in\mathbf{N}}$. Since the left side of C is in $\tilde{X}_{\bar{G}^T}$ and $(a_j)_{j\in\mathbf{N}} \in \tilde{X}_{\bar{G}}$, it follows that $\beta_{ij} \in A_{\bar{\Gamma}}$ for all $i, j \in \mathbf{N}$ so that $\bar{T}$ can weave D. This implies that $\bar{T}$ satisfies the condition (i) of Definition 7.1 for k.

(ii) It is clear that $\bar{T}$ satisfies the condition (ii) of Definition 7.1 for k.

(iii) We may assume that $k \geq 1$. Let x_1 and x_2 be compatible paths in $L_k(\bar{G})$ with respect to $\bar{T}$. Let C_1 be a curtain weaved by $\bar{T}$ whose upper side is x_1. Then, since T is k resolvable, x_1 and x_2 are equivalent with respect to T, so that T weaves a curtain

C_2 whose left side is the same as that of C_1. Since T satisfies the condition (i) of Definition 7.1 for k, for any right infinite sequence $(a_j)_{j\in\mathbf{N}} \in \tilde{X}_{\bar{G}}$, $a_j \in A_{\bar{G}}$, such that $a_1 \cdots a_k = x_2$, T weaves a well-weaved subconfiguration $(\beta_{ij})_{i,j\in\mathbf{N}}$ whose upper side is $(a_j)_{j\in\mathbf{N}}$ and whose left side is the same as that of C_1. For the same reason as in (i), $\beta_{ij} \in A_{\bar{\Gamma}}$ for all $i,j \in \mathbf{N}$. Hence, in particular, the curtain $(\beta_{ij})_{i\in\mathbf{N},1\le j\le k}$ is weaved by $\bar{T}$.

Thus, if $\bar{T}$ weaves a curtain whose upper side is x_1, then $\bar{T}$ weaves a curtain whose left side is the same and whose upper side is x_2. By symmetry, we conclude that x_1 and x_2 are equivalent with respect to $\bar{T}$. ☐

It is easily seen that the following remark holds.

Remark 7.4. Let k be a nonnegative integer. If T is a k-resolvable textile system, then $T^{[n]}$ is k resolvable for all $n \in \mathbf{N}$.

Remark 7.5. Let $T = (p, q : \Gamma \to G)$ be a textile system. Let ξ be the factor map given by p. Then the following statements are valid.

(1) If ξ is right closing, then (i) of Definition 7.1 implies (ii) of it.

(2) If T is resolvable and ξ is onto, then ξ is right closing. In particular, if T is 0 resolvable and ξ is onto, then p is right resolving.

(3) If T is k resolvable with $k \ge 1$ and if $\overline{\text{trim}}(T^{[k]}) = (\bar{p}, \bar{q} : \bar{\Gamma} \to \bar{G})$, then the factor map $\bar{\xi}$ given by $\bar{p}$ is onto.

Proof. (1) Assume that (i) of Definition 7.1 holds and ξ is right closing. Let $n > k$. Let C and D be two curtains of length n with the same upper side, say w, and with the same left side. Let $\bar{C}$ and $\bar{D}$ be the initial subcurtains of length $n-k$ of C and D, respectively. Let $(a_j)_{j\in\mathbf{N}} \in \tilde{X}_G$ with $a_j \in A_G$ and $a_1 \cdots a_n = w$. Let y be the right side of $\bar{C}$. The it follows from (i) of Definition 7.1 (and the compactness of $\hat{U}_T$) that the existence of terminal subcurtain of length k of C implies the existence of a well-weaved subconfiguration $(\alpha_{ij})_{i\in\mathbf{N},j>n-k}$, $\alpha_{ij} \in A_\Gamma$, such that

$$\downarrow (i_\Gamma(\alpha_{i,n-k+1}))_{i\in\mathbf{N}} = y \quad \text{and} \quad (p(\alpha_{1,j}))_{j>n-k} = (a_j)_{j>n-k},$$

so that we have a well-weaved subconfiguration $(\alpha_{ij})_{i,j\in\mathbf{N}}$ with $\alpha_{ij} \in A_\Gamma$ such that

$$(\alpha_{ij})_{i\in\mathbf{N},1\le j\le n-k} = \bar{C} \quad \text{and} \quad (p(\alpha_{1,j}))_{j\in\mathbf{N}} = (a_j)_{j\in\mathbf{N}}.$$

Similarly we have a well-weaved subconfiguration $(\beta_{ij})_{i,j\in\mathbf{N}}$ with $\beta_{ij} \in A_\Gamma$ such that

$$(\beta_{ij})_{i\in\mathbf{N},1\le j\le n-k} = \bar{D} \quad \text{and} \quad (p(\beta_{1,j})_{j\in\mathbf{N}} = (a_j)_{j\in\mathbf{N}}.$$

Since $\bar{C}$ and $\bar{D}$ have the same left side, we have

$$i_\Gamma(\alpha_{i1}) = i_\Gamma(\beta_{i1}) \quad \text{for all} \quad i \in \mathbf{N}.$$

Therefore since ξ is right closing and

$$(p(\alpha_{1,j}))_{j\in\mathbf{N}} = (p(\beta_{1,j}))_{j\in\mathbf{N}},$$

it follows by induction on i that

$$(\alpha_{ij})_{j\in\mathbf{N}} = (\beta_{ij})_{j\in\mathbf{N}} \quad \text{for all } \ i \in \mathbf{N}.$$

Therefore we have $\bar{C} = \bar{D}$.

(2) Suppose that T is k resolvable with $k \geq 0$ and ξ is onto. Assume that ξ is not right closing. Then there are two paths $\alpha_1 \cdots \alpha_{k+1}, \beta_1 \cdots \beta_{k+1}$ in $L_{k+1}(\Gamma), \alpha_j, \beta_j \in A_\Gamma$, such that

$$i_\Gamma(\alpha_1) = i_\Gamma(\beta_1), \quad \alpha_1 \neq \beta_1 \quad \text{and} \quad p(\alpha_1)\cdots p(\alpha_{k+1}) = p(\beta_1)\cdots p(\beta_{k+1}).$$

Since ξ is onto, there is a curtain $C = (\alpha_{ij})_{i\in\mathbf{N},1\leq j\leq k+1}$ such that $\alpha_{11}\cdots\alpha_{1,k+1} = \alpha_1\cdots\alpha_{k+1}$. Since $q(\alpha_1)\cdots q(\alpha_{k+1})$ and $q(\beta_1)\cdots q(\beta_{k+1})$ are compatible, it follows from (iii) of Definition 7.1 that there is a curtain $D = (\beta_{ij})_{i\in\mathbf{N},1\leq j\leq k+1}$ such that $\beta_{11}\cdots\beta_{1,k+1} = \beta_1\cdots\beta_{k+1}$ and the left side of D is the same as that of C. Since C and D have the same upper side and the same left side and are of length $k+1$, it follows from (ii) of Definition 7.1 that they have the same initial subcurtain of length 1, which contradicts that $\alpha_{11} = \alpha_1 \neq \beta_1 = \beta_{11}$.

Assume that $k = 0$. Let $u \in V_\Gamma$ and let $a \in A_G$ with $i_G(a) = p_V(u)$. Since ξ is onto, p_V is onto. Therefore there is a curtain $\downarrow (b_i)_{i\in\mathbf{N}}$ with $(b_i)_{i\in\mathbf{N}} \in \tilde{X}_{G^T}$, $b_1 = u$ and $b_i \in A_{G^T} = V_\Gamma$ for $i \geq 2$. By the condition (i) of Definition 7.1 for $k = 0$, there is a curtain $(\alpha_{i1})_{i\in\mathbf{N}}$ weaved by T whose upper side is a and whose left side is $\downarrow (b_i)_{i\in\mathbf{N}}$. Therefore, if we let $\alpha = \alpha_{11}$, then we have $\alpha \in A_\Gamma$ such that $i_\Gamma(\alpha) = u$ and $p_A(\alpha) = a$. Such α is unique, by the above. Thus p is right resolving.

(3) Assume that T is k resolvable with $k \geq 1$. Let $\bar{T} = \overline{\mathrm{trim}}(T^{[k]}) = (\bar{p}, \bar{q} : \bar{\Gamma} \to \bar{G})$. Let $(c_j)_{j\in\mathbf{Z}} \in X_{\bar{G}}$ with $c_j \in A_{\bar{G}}$. There is $(a_j)_{j\in\mathbf{Z}}$ with $a_j \in A_G$ such that $c_j = a_j\cdots a_{j+k-1}$ for $j \in \mathbf{Z}$. To see that $\bar{\xi}$ is onto, it suffices to show that there is a subconfiguration $(\gamma_{ij})_{i\in\mathbf{N},j\in\mathbf{Z}}$ which is well weaved by $\bar{T}$ and whose upper side is $(c_j)_{j\in\mathbf{Z}}$, i.e., $\bar{p}(\gamma_{1j}) = c_j$ for all $j \in \mathbf{Z}$.

Let $j \in \mathbf{Z}$. Since $\bar{T}$ is in the standard form, $\bar{p}$ is an onto graph-homomorphism so that $\bar{T}$ weaves a curtain $(\delta_i^{(j)})_{i\in\mathbf{N}}$ of length 1 whose upper side is c_j. Put

$$\delta_i^{(j)} = \beta_{ij}^{(j)} \cdots \beta_{i,j+k-1}^{(j)} \quad \text{with } \beta_{il}^{(j)} \in A_G,\ l = j, \cdots, j+k-1.$$

Then the subconfiguration $(\beta_{il}^{(j)})_{i\in\mathbf{N},j\leq l\leq j+k-1}$ is a curtain weaved by T. Since T is k resolvable, it follows from the condition (i) of Definition 7.1 and the compactness of $\hat{U}_T$, there is a subconfiguration $(\alpha_{il}^{(j)})_{i\in\mathbf{N},l\geq j}$ which is well weaved by T and whose upper side is $(a_l)_{l\geq j}$.

By the above and the compactness of $\bar{U}_T$, we conclude that there is a subconfiguration $(\alpha_{ij})_{i\in\mathbf{N},j\in\mathbf{Z}}$ which is well weaved by T and whose upper side is $(a_j)_{j\in\mathbf{Z}}$. Let $\gamma_{ij} = \alpha_{ij}\cdots\alpha_{i,j+k-1}$ for $i \in \mathbf{N}$ and $j \in \mathbf{Z}$. Then $T^{[k]}$ weaves the well-weaved subconfiguration $(\gamma_{ij})_{i\in\mathbf{N},j\in\mathbf{Z}}$. Since the upper side of $(\gamma_{ij})_{i\in\mathbf{N},j\in\mathbf{Z}}$ is $(c_j)_{j\in\mathbf{Z}}$ which is in $X_{\bar{G}}$, it follows that $\gamma_{ij} \in A_{\bar{\Gamma}}$ for all $i \in \mathbf{N}$ and $j \in \mathbf{Z}$. Thus $(\gamma_{ij})_{i\in\mathbf{N},j\in\mathbf{Z}}$ is well weaved by $\bar{T}$ and its upper side is $(c_j)_{j\in\mathbf{Z}}$. □

Let $T = (p, q : \Gamma \to G)$ be a textile system. Then T defines the λ-graph $\mathcal{G}_T = (\Gamma, p/q)$ with the labeling $p/q : A_\Gamma \to B$, where $B = \{p(\alpha)/q(\alpha) \mid \alpha \in A_\Gamma\}$ ($p(\alpha)/q(\alpha)$ simply means the pair $(p(\alpha), q(\alpha))$; we shall often write $\frac{p(\alpha)}{q(\alpha)}$ instead of $p(\alpha)/q(\alpha)$). This λ-graph is called the *representation* λ*-graph* of T. We call its representation matrix the *representation matrix* of T (see Section 10). Let γ be the sofic cover defined by $\mathcal{G}_T$. Let $\xi : (X_\Gamma, \sigma_\Gamma) \to (X_G, \sigma_G)$ and $\eta : (X_\Gamma, \sigma_\Gamma) \to (X_G, \sigma_G)$ be the factor maps given by p and q, respectively. Then for $z \in X_\Gamma, \gamma(z)$ is naturally identified with $\xi(z)/\eta(z)$ (this simply means the pair $(\xi(z), \eta(z))$). Define

$$R_T^0 = \gamma(X_\Gamma) \quad \text{and} \quad R_T = \gamma(Z_T),$$

which will be called the *0th relation* and the *relation*, respectively, given by T.

Lemma 7.6. Let $T = (p, q : \Gamma \to G)$ be a textile system. Let T_+ be the textile system represented by the induced right resolving λ-graph $(\mathcal{G}_T)_+$ of $\mathcal{G}_T$. If the factor map ξ given by p is right closing, then T is resolvable if and only if T_+ is resolvable.

Proof. Let $T_+ = (p_+, q_+ : \Gamma_+ \to G_+)$ with $G_+ = G$ and let $T_+^* = (p_+^{T_+}, q_+^{T_+} : \Gamma_+^{T_+} \to G_+^{T_+})$. Since ξ is right closing, there is an integer $l \geq 1$ such that p is l *right-closing* , that is, for any $m \geq l$ and any two paths $\alpha_1\cdots\alpha_m, \alpha'_1\cdots\alpha'_m$ in $L_m(\Gamma), \alpha_i, \alpha'_i \in A_\Gamma$, with $i_\Gamma(\alpha_1) = i_\Gamma(\alpha'_1)$, if $p(\alpha_1)\cdots p(\alpha_m) = p(\alpha'_1)\cdots p(\alpha'_m)$, then the initial subpaths of length $m - l$ of $\alpha_1\cdots\alpha_m$ and $\alpha'_1\cdots\alpha'_m$ are the same. Each arc $B \in A_{\Gamma_+}$ is written in the form

$$B = (U, \frac{a}{b}),$$

where U is a maximal compatible set or a successor of a maximal compatible set for $\mathcal{G}_T = (\Gamma, p/q)$ and $a, b \in A_G$ such that there is $\alpha \in A_\Gamma$ with $i_\Gamma(\alpha) \in U$ and $p(\alpha)/q(\alpha) = a/b$. We can identify B with the set of all $\alpha \in A_\Gamma$ such that $i_\Gamma(\alpha) \in U$ and $p(\alpha)/q(\alpha) = a/b$.

(a) (if part) Assume that T_+ is k resolvable with $k \geq 0$. Let $m = k + l$. To see that T is m resolvable, it suffices to check (i) and (iii) of Definition 7.1, by Remark 7.5.

(i) Let $C = (\alpha_{ij})_{i\in\mathbf{N},1\leq j\leq m}$ be a curtain of length m weaved by T with $\alpha_{ij} \in A_\Gamma$. Let $x \in L_n(G)$ with $n \geq m$ and assume that the initial subpath of length m of x is equal to the upper side of C. Let U_i be a maximal compatible set for $\mathcal{G}_T$ such that $U_i \ni i_\Gamma(\alpha_{i1})$ for $i \in \mathbf{N}$. The existence of C implies that there is a curtain $D = (B_{ij})_{i\in\mathbf{N},1\leq j\leq m}$ weaved by T_+ with $B_{ij} \in A_{\Gamma_+}$ such that the left side of D is $\downarrow (U_i)_{i\in\mathbf{N}}$ and $B_{ij} \ni \alpha_{ij}$ for all $i \in \mathbf{N}$ and $1 \le j \le m$. In fact, we define

$$B_{ij} = (U_{i,j-1}, \frac{p(\alpha_{ij})}{q(\alpha_{ij})}), \quad i \in \mathbf{N},\ 1 \le j \le m,$$

where for all $i \in \mathbf{N}, U_{i,0} = U_i$ and $U_{i,j}$ is the $(p(\alpha_{ij})/q(\alpha_{ij}))$-successor of $U_{i,j-1}$ for $j = 1, \cdots, m$. Clearly the upper side of D is equal to the initial subpath of length m of x. Let $\bar{D}$ be the initial subcurtain of length l of D. Let $(V_i)_{i\in\mathbf{N}}$ be the right side of $\bar{D}$ with $V_i \in V_{\Gamma_+}, i \in \mathbf{N}$. Since T_+ is k resolvable, the existence of the terminal subcurtain of length k of D implies the existence of a curtain D' whose left side is $\downarrow (V_i)_{i\in\mathbf{N}}$ and whose upper side is equal to the terminal subpath of length $n - l$ of x. Therefore T_+ weaves the curtain $\tilde{D} = (\tilde{B}_{ij})_{i\in\mathbf{N},1\le j\le n}, \tilde{B}_{ij} \in A_{\Gamma_+}$, such that the initial subcurtain of length l of $\tilde{D}$ is equal to $\bar{D}$ and the terminal subcurtain of $\tilde{D}$ of length $n - l$ is equal to D'. Clearly the upper side of $\tilde{D}$ is equal to x and $\tilde{B}_{ij} = B_{ij}$ for all $i \in \mathbf{N}$ and $1 \le j \le l$. Since $\tilde{B}_{i1} \cdots \tilde{B}_{in} \in L_n(\Gamma_+)$ for $i \in \mathbf{N}$, there is a path $\tilde{\alpha}_{i1} \cdots \tilde{\alpha}_{in}$ in $L_n(\Gamma)$ such that $\tilde{\alpha}_{ij} \in \tilde{B}_{ij}, j = 1, \cdots, n$, for all $i \in \mathbf{N}$. Let $\tilde{C} = (\tilde{\alpha}_{ij})_{i\in\mathbf{N},1\le j\le n}$. Then $\tilde{C}$ is a curtain weaved by T. For each $i \in \mathbf{N}, \alpha_{i1} \cdots \alpha_{il}, \tilde{\alpha}_{i1} \cdots \tilde{\alpha}_{il} \in L_l(\Gamma), B_{i1} \cdots B_{il} \in L_l(\Gamma_+)$, α_{ij} and $\tilde{\alpha}_{ij}$ are contained in B_{ij} for $1 \le j \le l$, and $i_\Gamma(\alpha_{i1})$ and $i_\Gamma(\alpha'_{i1})$ are in the same compatible set U_i for $\mathcal{G}_T$. Therefore since p is l right-closing we have $i_\Gamma(\alpha_{i1}) = i_\Gamma(\tilde{\alpha}_{i1})$ for all $i \in \mathbf{N}$. Therefore the left side of $\tilde{C}$ is the same as that of C. Since the upper side of $\tilde{D}$ is equal to x, the upper side of $\tilde{C}$ is equal to x. Thus we have proved that (i) of Definition 7.1 is satisfied by T for $k + l$.

(iii) Let $C = (\alpha_{ij})_{i\in\mathbf{N},1\le j\le m}, \alpha_{ij} \in A_\Gamma$, be a curtain weaved by T. Let x be the upper side of C and $\downarrow (b_i)_{i\in\mathbf{N}}$ the left side of C with $b_i \in A_{GT}$. Let $n \in \mathbf{N}$ and let $W = (\alpha_{ij})_{1\le i\le n,1\le j\le m}$ and let w be the lower side of W. Let $W' = (\alpha'_{ij})_{1\le i\le n,1\le j\le m}, \alpha'_{ij} \in A_\Gamma$, be a cloth weaved by T such that its upper side is equal to x and its left side is equal to $\downarrow b_1 \cdots b_n$. Let w' be the lower side of W'. Then w and w' are compatible with respect to T. It suffices to show that W' can be extended to a curtain $(\alpha'_{ij})_{i\in\mathbf{N},1\le j\le m}$ whose left side is $\downarrow (b_i)_{i\in\mathbf{N}}$. Let U_i be a maximal compatible set for $\mathcal{G}_T$ containing b_i (note that $b_i \in V_\Gamma$) for $i \in \mathbf{N}$. Then $\downarrow (U_i)_{i\in\mathbf{N}} \in X_{G_+^{T_+}}$. The existence of C implies that T_+ can weave a curtain $C^+ = (B_{ij})_{i\in\mathbf{N},1\le j\le m}, B_{ij} \in A_{\Gamma_+}$, such that the left side of C^+ is $\downarrow (U_i)_{i\in\mathbf{N}}$ and $B_{ij} \ni \alpha_{ij}$ for $i \in \mathbf{N}$ and $1 \le j \le m$. Let $W^+ = (B_{ij})_{1\le i\le n,1\le j\le m}$. The existence of W' implies that T_+ can weave a cloth $(W')^+ = (B'_{ij})_{1\le i\le n,1\le j\le m}, B'_{ij} \in A_{\Gamma_+}$, such that the left side of $(W')^+$ is $\downarrow U_1 \cdots U_n$ and $B'_{ij} \ni \alpha'_{ij}$ for $1 \le i \le n$ and $1 \le j \le m$.

Then W^+ and $(W')^+$ have the same upper side x and the same left side $\downarrow U_1 \cdots U_n$, w is the lower side of W^+ and w' is the lower side of $(W')^+$ so that w and w' are compatible with respect to T^+. Therefore, since T^+ is k resolvable, there is a curtain $(B'_{ij})_{i>n,1\le j\le m}, B'_{ij} \in A_{\Gamma_+}$, whose upper side is w' and whose left side is $(U_i)_{i>n}$ (by Remark 7.2). Hence we have a curtain $(\alpha'_{ij})_{i>n,1\le j\le m}$ weaved by T such that $\alpha'_{ij} \in B'_{ij}$ for all $i > n$ and $1 \le j \le m$. Thus joining this to W', we have a curtain $C' = (\alpha'_{ij})_{i\in\mathbf{N},1\le j\le m}$

weaved by T. Let $\downarrow (b'_i)_{i\in\mathbf{N}}$ be the left side of C' with $b'_i \in V_\Gamma$. Then $b'_i \in U_i$ for all $i \in \mathbf{N}$. Let $(a_j)_{j\in\mathbf{N}} \in \tilde{X}_G$ with $a_j \in A_G$ and $a_1 \cdots a_m = x$. Then by the fact proved in (i) above and the compactness of $\hat{U}_T$, we see that the existence of C implies the existence of a subconfiguration $(\beta_{ij})_{i,j\in\mathbf{N}}, \beta_{ij} \in A_\Gamma$, well weaved by T such that

$$(p(\beta_{1j}))_{j\in\mathbf{N}} = (a_j)_{j\in\mathbf{N}} \quad \text{and} \quad (i_\Gamma(\beta_{i1}))_{i\in\mathbf{N}} = (b_i)_{i\in\mathbf{N}},$$

and we also see that the existence of C' implies the existence of a subconfiguration $(\beta'_{ij})_{i,j\in\mathbf{N}}, \beta'_{ij} \in A_\Gamma$, well weaved by T such that

$$(p(\beta'_{1j}))_{j\in\mathbf{N}} = (a_j)_{j\in\mathbf{N}} \quad \text{and} \quad (i_\Gamma(\beta'_{i1}))_{i\in\mathbf{N}} = (b'_i)_{i\in\mathbf{N}}.$$

Since ξ is right closing and since $(p(\beta_{1j}))_{j\in\mathbf{N}} = (p(\beta'_{1j}))_{j\in\mathbf{N}}$ and $i_\Gamma(\beta_{11})$ and $i_\Gamma(\beta'_{11})$ are contained in the same compatible set U_1 for $\mathcal{G}_T$, it follows that

$$(\beta_{1j})_{j\in\mathbf{N}} = (\beta'_{1j})_{j\in\mathbf{N}}$$

so that

$$(p(\beta_{2j}))_{j\in\mathbf{N}} = (q(\beta_{1j}))_{j\in\mathbf{N}} = (q(\beta'_{1j}))_{j\in\mathbf{N}} = (p(\beta_{2j}))_{j\in\mathbf{N}}.$$

Therefore since $i_\Gamma(\beta_{21})$ and $i_\Gamma(\beta'_{21})$ are contained in the same compatible set U_2, it follows that

$$(\beta_{2j})_{j\in\mathbf{N}} = (\beta'_{2j})_{j\in\mathbf{N}}.$$

Continuing in this way we have

$$(\beta_{ij})_{j\in\mathbf{N}} = (\beta'_{ij})_{j\in\mathbf{N}} \quad \text{for all } i \in \mathbf{N}$$

so that we have

$$(b_i)_{i\in\mathbf{N}} = (b'_i)_{i\in\mathbf{N}}.$$

Thus C' has $(b_i)_{i\in\mathbf{N}}$ as its left side, so that we have an extension of W' as required.

(b) (only-if part) Assume that T is resolvable. By Remark 7.2, we may assume that T is k resolvable with $k \geq l$.

(i) Let $D = (B_{ij})_{i\in\mathbf{N},1\leq j\leq k}$ be a curtain weaved by T_+ with $B_{ij} \in A_{\Gamma_+}$. Let $x \in L_n(G)$ with $n \geq k$, and assume that the initial subpath of length k of x is equal to the upper side of D. There is a curtain $C = (\alpha_{ij})_{i\in\mathbf{N},1\leq j\leq k}$ such that $\alpha_{ij} \in B_{ij}$ for all $i \in \mathbf{N}$ and $1 \leq j \leq k$. The upper side of C is the same as that of D. Since T is k resolvable, there is a curtain $C' = (\alpha'_{ij})_{i\in\mathbf{N},1\leq j\leq n}$ whose upper side is x and whose left side is the same as that of C. Let $U_i = i_{\Gamma_+}(B_{i1})$ for $i \in \mathbf{N}$. Then U_i is either a maximal compatible set or a successor of a maximal compatible set. Since $i_\Gamma(\alpha'_{i1}) \in U_i$ for $i \in \mathbf{N}$, there is a path $B'_{i1} \cdots B'_{in} \in L_n(\Gamma_+)$ such that $i_{\Gamma_+}(B'_{i1}) = U_i$ and $\alpha'_{ij} \in B'_{ij}, j = 1, \cdots, n$, for all $i \in \mathbf{N}$. Let $D' = (B'_{ij})_{i\in\mathbf{N},1\leq j\leq n}$. Then D' is a curtain weaved by T_+, its upper side is x and its left side is $\downarrow (U_i)_{i\in\mathbf{N}}$, which is the left side of D. Thus we have proved that (i) of Definition 7.1 is satisfied by T_+ for k.

(ii) To see that (ii) of Definition 7.1 is satisfied by T_+ for k, it suffices to show that the factor map ξ_+ given by p_+ is right closing, by the above (i) and Remark 7.5. Assume that there are $(B_j)_{j\in\mathbf{Z}}$ and $(B'_j)_{j\in\mathbf{Z}}$ in $X_{\Gamma_+}, B_j, B'_j \in A_{\Gamma_+}$, such that $B_j = B'_j$ for $j \le 0$ and $p_+(B_j) = p_+(B'_j)$ for $j \in \mathbf{N}$. There are $(\alpha_j)_{j\in\mathbf{Z}}$ and $(\alpha'_j)_{j\in\mathbf{Z}}$ in X_Γ such that $\alpha_j \in B_j$ and $\alpha'_j \in B'_j$ for $j \in \mathbf{Z}$. Let $U = t_{\Gamma_+}(B_0) = t_{\Gamma_+}(B'_0)$. Since U is a compatible set for $\mathcal{G}_T$, there are $u \in V_\Gamma$ and $(a_{-m}/b_{-m})\cdots(a_0/b_0) \in L(\mathcal{G}_T)$ with $m \in \mathbf{N}$ and $a_i, b_i \in A_G$ such that U is the $(a_{-m}/b_{-m})\cdots(a_0/b_0)$-successor of $\{u\}$. Thus there are $(\beta_j)_{j\in\mathbf{Z}}$ and $(\beta'_j)_{j\in\mathbf{Z}}$ in X_Γ such that $\beta_j = \beta'_j$ for $j < -m$, $i_\Gamma(\beta_{-m}) = i_\Gamma(\beta'_{-m}) = u, (p(\beta_{-m})/q(\beta_{-m}))\cdots(p(\beta_0)/q(\beta_0)) = (a_{-m}/b_{-m})\cdots(a_0/b_0)$, and $\beta_j = \alpha_j$ and $\beta'_j = \alpha'_j$ for $j \in \mathbf{N}$. Clearly $(\beta_j)_{j\in\mathbf{Z}}$ and $(\beta'_j)_{j\in\mathbf{Z}}$ are left asymptotic and $\xi((\beta_j)_{j\in\mathbf{Z}}) = \xi((\beta'_j)_{j\in\mathbf{Z}})$. Since ξ is right closing, we have $(\beta_j)_{j\in\mathbf{Z}} = (\beta'_j)_{j\in\mathbf{Z}}$ so that $(\alpha_j)_{j\in\mathbf{N}} = (\alpha'_j)_{j\in\mathbf{N}}$. Since $t_\Gamma(B_j)$ and $t_\Gamma(B'_j)$ are the $(p(\alpha_1)/q(\alpha_1))\cdots(p(\alpha_j)/q(\alpha_j))$-successor of U, we have $B_j = B'_j$ for all $j \in \mathbf{N}$. Thus ξ_+ is right closing.

(iii) Let $D = (B_{ij})_{i\in\mathbf{N},1\le j\le k}$ be a curtain weaved by T_+. Let $m \in \mathbf{N}$ and let $\bar{D} = (B_{ij})_{1\le i\le m,1\le j\le k}$. Let $\bar{D}' = (B'_{ij})_{1\le i\le m,1\le j\le k}$ be a cloth weaved by T_+ and having the same upper and left sides as those of $\bar{D}$. To see that T_+ satisfies the condition (iii) of Definition 7.1 for k, it suffices to show that T_+ can weave a curtain $(B'_{ij})_{i>m,1\le j\le k}$ whose upper side is equal to the lower side of $\bar{D}'$ and whose left side is equal to that of $(B_{ij})_{i>m,1\le j\le k}$.

There is a curtain $C = (\alpha_{ij})_{i\in\mathbf{N},1\le j\le k}$ weaved by T such that $\alpha_{ij} \in B_{ij}$ for all $i \in \mathbf{N}$ and $1 \le j \le k$. There is also a cloth $\bar{C}' = (\alpha'_{ij})_{1\le i\le m,1\le j\le k}$ weaved by T such that $\alpha'_{ij} \in B'_{ij}$ for all $1 \le i \le m$ and $1 \le j \le k$. Clearly C and $\bar{C}'$ have the same upper side. Put $i_{\Gamma_+}(B_{i1}) = U_i$ for $i \in \mathbf{N}$. Then $i_\Gamma(\alpha_{i1}), i_\Gamma(\alpha'_{i1}) \in U_i$ for $i = 1, \cdots, m$. Since U_1 is a compatible set for $\mathcal{G}_T$ with $i_\Gamma(\alpha_{11}), i_\Gamma(\alpha'_{11}) \in U_1$ and

$$p(\alpha_{11})\cdots p(\alpha_{1k}) = p(\alpha'_{11})\cdots p(\alpha'_{1k})$$

and since p is l right closing with $l < k$, if follows that

$$i_\Gamma(\alpha_{11}) = i_\Gamma(\alpha'_{11}).$$

Therefore $q(\alpha_{11})\cdots q(\alpha_{1k})$ and $q(\alpha'_{11})\cdots q(\alpha'_{1k})$ are compatible with respect to T, so that $p(\alpha_{21})\cdots p(\alpha_{2k})$ and $p(\alpha'_{21})\cdots p(\alpha'_{2k})$ are compatible with respect to T. Since T is k resolvable and there is the curtain $(\alpha_{ij})_{i\ge 2,1\le j\le k}$ whose upper side is $p(\alpha_{21})\cdots p(\alpha_{2k})$, there is a curtain $(\alpha^{(2)}_{ij})_{i\ge 2,1\le j\le k}$ whose upper side is $p(\alpha'_{21})\cdots p(\alpha'_{2k})$ and whose left side is the same as that of $(\alpha_{ij})_{i\ge 2,1\le j\le k}$. Hence $i_\Gamma(\alpha^{(2)}_{21}) = i_\Gamma(\alpha_{21})$ and $p(\alpha^{(2)}_{21})\cdots p(\alpha^{(2)}_{2k}) = p(\alpha'_{21})\cdots p(\alpha'_{2k})$. Therefore since $i_\Gamma(\alpha^{(2)}_{21}), i_\Gamma(\alpha'_{21}) \in U_2$ and U_2 is a compatible set for $\mathcal{G}_T$ and since p is l right closing with $l < k$, we have $i_\Gamma(\alpha^{(2)}_{21}) = i_\Gamma(\alpha'_{21})$, so that we have

$$i_\Gamma(\alpha_{21}) = i_\Gamma(\alpha'_{21}).$$

Thus $q(\alpha_{21})\cdots q(\alpha_{2k})$ and $q(\alpha'_{21})\cdots q(\alpha'_{2k})$ are compatible paths with respect to T, so that $p(\alpha_{31})\cdots p(\alpha_{3k})$ and $p(\alpha'_{31})\cdots p(\alpha'_{3k})$ are compatible with respect to T. Continuing in this way, we finally find that $q(\alpha_{m1})\cdots q(\alpha_{mk})$ and $q(\alpha'_{m1})\cdots q(\alpha'_{mk})$ are compatible with respect to T.

Since T is k resolvable and it weaves the curtain $(\alpha_{ij})_{i>m,1\le j\le k}$ whose upper side is $q(\alpha_{m1})\cdots q(\alpha_{mk})$, it weaves a curtain $(\alpha'_{ij})_{i>m,1\le j\le k}$ whose upper side is $q(\alpha'_{m1})\cdots q(\alpha'_{mk})$ and whose left side is the same as that of $(\alpha_{ij})_{i>m,1\le j\le k}$. Therefore there is a curtain $(B'_{ij})_{i>m,1\le j\le k}$ weaved by T_+ such that its left side is $\downarrow (U_i)_{i>m}$ and $B'_{ij} \ni \alpha'_{ij}$ for all $i > m$ and $1 \le j \le k$. Clearly $(B'_{ij})_{i>m,1\le j\le k}$ has the desired property. □

We shall consider very familiar notions in automata theory. Let $\mathcal{G} = (G, \lambda)$ be a right-resolving λ-graph. Let E be an equivalence relation on V_G such that if $v\ E\ v'$ for $v, v' \in V_G$, then the following two conditions are satisfied: (i) for any $a \in A_G$ with $i_G(a) = v$, there is $a' \in A_G$ with $i_G(a') = v'$ and $\lambda(a) = \lambda(a')$; (ii) if $a, a' \in A_G, i_G(a) = v, i_G(a') = v'$, and $\lambda(a) = \lambda(a')$, then $t_G(a)\ E\ t_G(a')$. We call E a *state-equivalence* of $\mathcal{G}$. For $v \in V_G$, let $E(v)$ denote the equivalence class of E containing v. We define a λ-graph $\mathcal{G}_E = (G_E, \lambda_E)$, which is called the *quotient of $\mathcal{G}$ with respect to E*, as follows:

$$V_{G_E} = \{E(v) \mid v \in V_G\},$$
$$A_{G_E} = \{(E(v), \lambda(a)) \mid v \in V_G, a \in A_G, i_G(a) = v\},$$

and for $(E(v), \lambda(a)) \in A_{G_E}$,

$$i_{G_E}((E(v), \lambda(a))) = E(v), \quad t_{G_E}((E(v), \lambda(a))) = E(t_G(a))$$

and

$$\lambda_E((E(v), \lambda(a))) = \lambda(a).$$

As is easily seen, $\mathcal{G}_E$ is well defined and right resolving and $L(\mathcal{G}) = L(\mathcal{G}_E)$. (The definition of $L(\mathcal{G})$ for a λ-graph $\mathcal{G}$ was given in Section 4.)

Lemma 7.7. Let T be a textile system such that $\mathcal{G}_T$ is right resolving. Let E be a state-equivalence of $\mathcal{G}_T$. Let T_E be the textile system such that $\mathcal{G}_{T_E} = (\mathcal{G}_T)_E$. Then for any nonnegative integer k, T is k resolvable if and only if T_E is k resolvable.

Proof. Let $T = (p, q : \Gamma \to G)$ and let $T_E = (p_E, q_E : \Gamma_E \to G)$. For $u \in V_\Gamma$, let $E(u)$ denote the equivalence class of E containing u. If C is a curtain of length 0 weaved by T with $C = \downarrow (u_i)_{i\in\mathbf{N}}, u_i \in A_{GT} = V_\Gamma$, then we let

$$E(C) = \downarrow (E(u_i))_{i\in\mathbf{N}}.$$

If C is a curtain of length greater than 0 weaved by T with $C = (\alpha_{ij})_{i\in\mathbf{N},1\le j\le l}, \alpha_{ij} \in A_\Gamma$, then let $E(C)$ be the subconfiguration over A_{Γ_E} defined by

$$E(C) = ((E(i_\Gamma(\alpha_{ij})), \frac{p(\alpha_{ij})}{q(\alpha_{ij})}))_{i\in\mathbf{N},1\le j\le l}.$$

If C is a cloth weaved by T, we define $E(C)$ similarly. For a curtain C weaved by T, $E(C)$ is a curtain weaved by T_E. For a curtain D weaved by T_E, if $\downarrow (U_i)_{i\in\mathbf{N}}, U_i \in V_{\Gamma_E} = \{E(u) \mid u \in V_\Gamma\}$, is the left side of D, then for any sequence $(u_i)_{i\in\mathbf{N}}$ with $u_i \in U_i$, there is a curtain C weaved by T such that the left side of C is $\downarrow (u_i)_{i\in\mathbf{N}}$ and $E(C) = D$. Similar facts also hold for cloths. By using these properties, the lemma can be straightforwardly proved and hence the detail is left to the reader. □

Lemma 7.8. Let $\mathcal{G} = (\Gamma, \lambda)$ be a λ-graph. Then the following (i) and (ii) are valid. (i) If the sofic cover $\pi_\mathcal{G}$ defined by $\mathcal{G}$ is a topological conjugacy, then there are $n \in \mathbf{N}$, a state-equivalence E of $(\mathcal{G}^{[n]})_+$ and a state-equivalence E' of $((\mathcal{G}^{[n]})_-)^{-1}$ such that $((\mathcal{G}^{[n]})_+)_E$ and $((((\mathcal{G}^{[n]})_-)^{-1})_{E'})^{-1}$ represent the same graph. (ii) If $\pi_\mathcal{G}$ is biclosing, then there are $n \in \mathbf{N}$, a state-equivalence E of $(\mathcal{G}^{[n]})_+$ and a state-equivalence E' of $((\mathcal{G}^{[n]})_-)^{-1}$ such that $((\mathcal{G}^{[n]})_+)_E$ and $((((\mathcal{G}^{[n]})_-)^{-1})_{E'})^{-1}$ are the same biresolving λ-graph.

Proof. Since $\pi_\mathcal{G}$ is a conjugacy, $(X_\mathcal{G}, \sigma_\mathcal{G})$ is a subshift of finite type. Therefore, there is $n \in \mathbf{N}$ such that $(X_\mathcal{G}^{[n]}, \sigma_\mathcal{G}^{[n]})$ is a topological Markov shift. Let G be the graph such that $(X_G, \sigma_G) = (X_\mathcal{G}^{[n]}, \sigma_\mathcal{G}^{[n]})$. Let $\mathcal{G}^{[n]} = (\Gamma^{[n]}, \lambda^{[n]})$ and let $(\mathcal{G}^{[n]})_+ = ((\Gamma^{[n]})_+, (\lambda^{[n]})_+)$. Let $\tilde{G}$ be the λ-graph representing G, i.e., $\tilde{G} = (G, \mathrm{id}_{A_G})$. Then $(\tilde{G})_+ = \tilde{G}$ and the sofic cover $\pi_{\tilde{G}}$ defined by $\tilde{G}$ is the identity $\mathrm{id}_{X_G} : (X_G, \sigma_G) \to (X_G, \sigma_G)$. Let $\phi_0 = \pi_{\mathcal{G}^{[n]}}$ and let $\phi = \mathrm{id}_{X_G}$. Then ϕ and ϕ_0 are topological conjugacies with

$$\phi\pi_{\mathcal{G}^{[n]}} = \pi_{\tilde{G}}\phi_0.$$

Therefore by Theorem 4.3, we have a topological conjugacy $(\phi_0)_+ : (X_{(\Gamma^{[n]})_+}, \sigma_{(\Gamma^{[n]})_+}) \to (X_G, \sigma_G)$ such that

$$\phi\pi_{(\mathcal{G}^{[n]})_+} = \pi_{(\tilde{G})_+}(\phi_0)_+.$$

Therefore $\pi_{(\mathcal{G}^{[n]})_+} = (\phi_0)_+$ so that $\pi_{(\mathcal{G}^{[n]})_+}$ is a topological conjugacy. By definition, $(\mathcal{G}^{[n]})_+$ is a right-resolving λ-graph. Since $\mathcal{G}^{[n]}$ defines a graph-homomorphism of $\Gamma^{[n]}$ onto G, $(\mathcal{G}^{[n]})_+$ defines the graph-homomorphism $h : (\Gamma^{[n]})_+ \to G$ such that $h_A = (\lambda^{[n]})_+$. Clearly $\pi_{(\mathcal{G}^{[n]})_+}$ is the factor map given by h. We want to show that h is a right resolving graph-homomorphism: we shall show that for any $u \in V_{(\Gamma^{[n]})_+}$ and any $a \in A_G$ with $h_V(u) = i_G(a)$, there is an arc $\alpha \in A_{(\Gamma^{[n]})_+}$ such that $i_\Gamma(\alpha) = u$ and $h_A(\alpha) = a$ (the uniqueness of such α follows from the fact that $(\mathcal{G}^{[n]})_+$ is a right resolving λ-graph). Since $\pi_{(\mathcal{G}^{[n]})_+}$ is a topological conjugacy and $(\mathcal{G}^{[n]})_+$ is right resolving, it follows that there is $k \in \mathbf{N}$ such that for any two paths $\alpha_1 \cdots \alpha_k, \beta_1 \cdots \beta_k \in L_k((\Gamma^{[n]})_+)$ with $\alpha_i, \beta_i \in A_{(\Gamma^{[n]})_+}$ if

$$h_A(\alpha_1) \cdots h_A(\alpha_k) = h_A(\beta_1) \cdots h_A(\beta_k),$$

then

$$t_{(\Gamma^{[n]})_+}(\alpha_k) = t_{(\Gamma^{[n]})_+}(\beta_k).$$

Let $u \in V_{(\Gamma^{[n]})_+}$ and let $a \in A_G$ with $h_V(u) = i_G(a)$. Since $(\Gamma^{[n]})_+$ is nondegenerate, there is a path $\alpha_1 \cdots \alpha_k$ in $L_k((\Gamma^{[n]})_+)$ such that $t_{(\Gamma^{[n]})_+}(\alpha_k) = u$. Since $h_V(u) = i_G(a)$,

$$h_A(\alpha_1) \cdots h_A(\alpha_k) a \in L_{k+1}(G).$$

Since $\pi_{(\mathcal{G}^{[n]})_+}$ is onto, there is a path $\beta_1 \cdots \beta_{k+1} \in L_{k+1}((\Gamma^{[n]})_+)$ such that

$$h_A(\beta_1) \cdots h_A(\beta_k) h_A(\beta_{k+1}) = h_A(\alpha_1) \cdots h_A(\alpha_k) a.$$

Therefore by the above, we have

$$t_{(\Gamma^{[n]})_+}(\beta_k) = t_{(\Gamma^{[n]})_+}(\alpha_k) = u.$$

Thus we have

$$i_{(\Gamma^{[n]})_+}(\beta_{k+1}) = u \quad \text{and} \quad h_A(\beta_{k+1}) = a.$$

Let E be the equivalence relation on $V_{(\Gamma^{[n]})_+}$ such that for $v, v' \in V_{(\Gamma^{[n]})_+}, v \mathrel{E} v'$ if and only if $h_V(v) = h_V(v')$. Clearly E is a state-equivalence of $(\mathcal{G}^{[n]})_+$ and

$$((\mathcal{G}^{[n]})_+)_E = \tilde{G}.$$

By the above, there is a state-equivalence E' of $((\mathcal{G}^{-1})^{[n]})_+$ such that

$$(((\mathcal{G}^{-1})^{[n]})_+)_{E'} = \tilde{G}^{-1}.$$

Therefore since

$$((\mathcal{G}^{[n]})_-)^{-1} = ((\mathcal{G}^{[n]})^{-1})_+ = ((\mathcal{G}^{-1})^{[n]})_+,$$

the proof of (i) of the lemma is completed.

(ii) By Proposition 3.11 of [N3], there are $n \in \mathbf{N}$, a biresolving λ-graph $\mathcal{G}' = (G, \lambda')$, and a topological conjugacy $\phi_0 : (X_{\Gamma^{[n]}}, \sigma_{\Gamma^{[n]}}) \to (X_G, \sigma_G)$ such that $\pi_{\mathcal{G}'}\phi_0 = \pi_{\mathcal{G}^{[n]}}$, where $\mathcal{G}^{[n]} = (\Gamma^{[n]}, \lambda^{[n]})$. Observing the proof of Proposition 3.11 of [N3], we know that ϕ_0 is given by a graph-homomorphism $g : \Gamma^{[n]} \to G$ such that $\lambda' g_A = \lambda^{[n]}$. Let $\mathcal{G}_g$ be the λ-graph defined by $\mathcal{G}_g = (\Gamma^{[n]}, g_A)$. Then $(\mathcal{G}_g)_+$ defines the graph-homomorphism $h : \bar{\Gamma} \to G$ such that $(\mathcal{G}_g)_+ = (\bar{\Gamma}, h_A)$. For any two paths $b_1 \cdots b_l, b'_1 \cdots b'_l \in L(\Gamma^{[n]}), l \in \mathbf{N}$, $b_i, b'_i \in A_{\Gamma^{[n]}}$ with $i_{\Gamma^{[n]}}(b_1) = i_{\Gamma^{[n]}}(b'_1)$,

$$\lambda^{[n]}(b_1) \cdots \lambda^{[n]}(b_l) = \lambda^{[n]}(b'_1) \cdots \lambda^{[n]}(b'_l) \quad \Leftrightarrow \quad g_A(b_1) \cdots g_A(b_l) = g_A(b'_1) \cdots g_A(b'_l),$$

because $\lambda' g_A = \lambda^{[n]}$ and $\mathcal{G}' = (G, \lambda')$ is right resolving. Hence if $(\mathcal{G}^{[n]})_+ = ((\Gamma^{[n]})_+, (\lambda^{[n]})_+)$, then $\bar{\Gamma} = (\Gamma^{[n]})_+$ and $\lambda' h_A = (\lambda^{[n]})_+$. By the same argument as in (i) above (substitute $\mathcal{G}_g$ for $\mathcal{G}^{[n]}$ in (i)), we conclude that $h : (\Gamma^{[n]})_+ \to G$ is a right resolving graph-homomorphism, and obtain the state-equivalence E of $(\mathcal{G}_g)_+$ induced by h. Clearly E is a state-equivalence of $(\mathcal{G}^{[n]})_+$ and we have

$$((\mathcal{G}^{[n]})_+)_E = \mathcal{G}'.$$

By the above, there is a state-equivalence E' of $((\mathcal{G}^{-1})^{[n]})_+$ such that

$$(((\mathcal{G}^{-1})^{[n]})_+)_{E'} = (\mathcal{G}')^{-1}.$$

Thus for the same reason as in the proof of (i), the proof of (ii) of the lemma is completed. □

Lemma 7.9. Let $T = (p, q : \Gamma \to G)$ be a textile system such that the factor map given by p is right closing and the factor map given by q is onto. Let k be a nonnegative integer. If T^m is k resolvable for some $m \in \mathbf{N}$, then T is k resolvable.

Proof. It is straightforward to see that the condition (i) of Definition 7.1 is satisfied by T. Thus by Remark 7.5, the condition (ii) of Definition 7.1 is satisfied by T. Since the factor map given by q is onto, any two compatible paths in $L_k(G)$ with respect to T are compatible paths with respect to T^m. (In fact, any two cloths with the same upper side, say w, and the same left side can be extended upward by any length adding a cloth whose lower side is w.) Therefore it is easily seen that T satisfies the condition (iii) of Definition 7.1. □

Lemma 7.10. Let $T = (p, q : \Gamma \to G)$ be a textile system such that the factor map given by p is right closing. If $T^{(m)}$ is resolvable for some $m \in \mathbf{N}$, then T is resolvable.

Proof. Assume that $T^{(m)}$ is k resolvable with $k \geq 0$. Then by using Remark 7.5 it is straightforwardly seen that T is km resolvable. □

In connection with the two lemmas above, we remark that if T is a resolvable textile system, then T^m and $T^{(m)}$ are resolvable for all $m \in \mathbf{N}$.

For a λ-graph $\mathcal{G} = (G, \lambda)$ and $v \in V_G$, a word w of length ≥ 1, say k, in $L(\mathcal{G})$ is called a *homing magic word for* v if for any $a_1 \cdots a_k \in L(G), a_i \in A_G$,

$$\lambda(a_1) \cdots \lambda(a_k) = w \quad \Rightarrow \quad t_G(a_k) = v$$

and w is called a *tracing magic word for* v if for any $a_1 \cdots a_k \in L(G), a_i \in A_G$,

$$\lambda(a_1) \cdots \lambda(a_k) = w \quad \Rightarrow \quad i_G(a_1) = v.$$

A λ-graph $\mathcal{G} = (G, \lambda)$ is called a *graphoid* if $\mathcal{G}$ is biresolving and every vertex in V_G has both a homing magic word and a tracing magic word for itself. As will be proved at the end of this section, a graphoid is canonical (cf. Section 4). It is easy to see that if a λ-graph $\mathcal{G}$ is a graphoid, then $\mathcal{G}^{[n]}$ is a graphoid for all $n \in \mathbf{N}$. Clearly a λ-graph that represents a graph is a graphoid. If (Y, σ) is a one-sided transitive sofic system such that $\mathcal{F}_Y^+ = \mathcal{F}_Y^-$, then $\mathcal{F}_Y^+$ is a graphoid (cf. [F1]).

A λ-graph $\mathcal{G}$ is called an *extension of a graphoid* if there are a graphoid $\mathcal{H}$, $n \in \mathbf{N}$, a state-equivalence E of $(\mathcal{G}^{[n]})_+$ and a state-equivalence E' of $((\mathcal{G}^{[n]})_-)^{-1}$ such that

$$((\mathcal{G}^{[n]})_+)_E = ((((\mathcal{G}^{[n]})_-)^{-1})_{E'})^{-1} = \mathcal{H}.$$

By Lemma 7.8, we know that a λ-graph $\mathcal{G}$ such that the sofic cover $\pi_{\mathcal{G}}$ defined by $\mathcal{G}$ is a topological conjugacy, is an extension of a graphoid, to say more exactly, an extension of a λ-graph that represents a graph. Using Lemma 7.8, it is easily proved that if $\mathcal{G}$ is an irreducible λ-graph such that $\pi_{\mathcal{G}}$ is biclosing, then $\mathcal{G}$ is an extension of a graphoid.

Lemma 7.11 Let $\mathcal{G}$ be a λ-graph and let $n \in \mathbf{N}$. Let E be a state-equivalence of $\mathcal{G}_+$. Then there is a state-equivalence E' of $(\mathcal{G}^{[n]})_+$ such that

$$((\mathcal{G}_+)_E)^{[n]} = ((\mathcal{G}^{[n]})_+)_{E'}.$$

Proof. Since $\mathcal{G}^{[m+1]} = (\mathcal{G}^{[2]})^{[m]}$ for every $m \in \mathbf{N}$ and every λ-graph $\mathcal{G}$, it suffices to show the lemma for $n = 2$.

(1) It is easily seen that

$$(\mathcal{G}^{[2]})_+ = (\mathcal{G}_+)^{[2]}.$$

(2) If $\mathcal{G} = (\Gamma, \lambda)$ is right resolving and E is a state-equivalence of $\mathcal{G}$, then it is straightforward to see that the equivalence relation E' on A_Γ defined by

$$\alpha \; E' \; \beta \quad \Leftrightarrow \quad i_\Gamma(\alpha) \; E \; i_\Gamma(\beta) \text{ and } \lambda(\alpha) = \lambda(\beta), \quad \alpha, \beta \in A_\Gamma,$$

is a state-equivalence of $\mathcal{G}^{[2]}$ and that

$$(\mathcal{G}^{[2]})_{E'} = (\mathcal{G}_E)^{[2]}.$$

Thus by (1) and (2) the lemma follows. □

Lemma 7.12. Let $T = (p, q : \Gamma \to G)$ be a textile system such that the factor map given by p is right closing and the factor map given by q is onto. Let $k \in \mathbf{N}$ and assume that $\mathcal{G}_{T^k}$ is an extension of a graphoid. Let $T_0 = (p_0, q_0 : \Gamma_0 \to G)$ be a textile system such that p_0 is right resolving and $\mathcal{G}_{T_0}$ is an extension of a graphoid. If $R^0_{T_0} = R^0_{T^k}$, then $T^{[n]}$ is resolvable for some $n \in \mathbf{N}$.

Proof. Since p_0 is right resolving, $\mathcal{G}_{T_0}$ is right resolving. Since $\mathcal{G}_{T_0}$ is an extension of a graphoid, there are $n \in \mathbf{N}$ and a state-equivalence E_0 of $(\mathcal{G}^{[n]}_{T_0})_+$ such that $((\mathcal{G}^{[n]}_{T_0})_+)_{E_0}$ is a graphoid. Since $\mathcal{G}_{T_0}$ is right resolving, so is $\mathcal{G}^{[n]}_{T_0}$ so that $(\mathcal{G}^{[n]}_{T_0})_+ = \mathcal{G}^{[n]}_{T_0}$. Since T_0 is 0 resolvable, so is $T^{[n]}_0$, by Remark 7.4. Therefore, since $\mathcal{G}^{[n]}_{T_0} = \mathcal{G}_{T^{[n]}_0}$, the textile system represented by $\mathcal{G}^{[n]}_{T_0}$ is 0 resolvable, so that the textile system represented by $(\mathcal{G}^{[n]}_{T_0})_{E_0}$ is 0 resolvable by Lemma 7.7. Put $T^k = T_1 = (p_1, q_1 : \Gamma_1 \to G)$. Since the factor map given by p is right closing, it follows that the factor map given by p_1 is right closing. Since $\mathcal{G}_{T_1}$ is an extension of a graphoid, there is a state-equivalence E_1 of $(\mathcal{G}_{T_1}^{[n]})_+$ such that $((\mathcal{G}_{T_1}^{[n]})_+)_{E_1}$ is a graphoid. Here we can take the same n as in the above, by Lemma 7.11 and the fact that for a graphoid $\mathcal{G}$, $\mathcal{G}^{[l]}$ is a graphoid for all $l \in \mathbf{N}$. Since $R^0_{T_0^{[n]}} = R^0_{T_1^{[n]}}$, $((\mathcal{G}_{T_0}^{[n]})_+)_{E_0}$ and $((\mathcal{G}_{T_1}^{[n]})_+)_{E_1}$ define the same sofic system. Therefore since these two λ-graphs are graphoids and a graphoid is canonical (by Proposition 7.24 given at the end of this section), we have

$$(\mathcal{G}_{T_0}^{[n]})_{E_0} = ((\mathcal{G}_{T_1}^{[n]})_+)_{E_1} = ((\mathcal{G}_{T_1^{[n]}})_+)_{E_1}.$$

(For if (Y,σ) is a sofic system, then the canonical λ-graph $(\mathcal{K}_{\bar{Y}})_+$ is unique for (Y,σ).) Therefore since the textile system represented by $(\mathcal{G}_{T_0}^{[n]})_{E_0}$ is 0 resolvable, it follows from Lemma 7.7 that the textile system represented by $(\mathcal{G}_{T_1^{[n]}})_+$ is 0 resolvable. If we let $T_1^{[n]} = (p_1^{[n]}, q_1^{[n]} : \Gamma_1^{[n]} \to G^{[n]})$, then the factor map given by $p_1^{[n]}$ is right closing, because the factor map given by p_1 is right closing. Therefore, by Lemma 7.6, $T_1^{[n]}$ is resolvable. Thus $(T^k)^{[n]}$ is resolvable. Every arc of Γ_1 can be considered as a cloth of the form $(\alpha_{i1})_{1\le i\le k}$, $\alpha_{i1} \in A_\Gamma$, weaved by T. (For it follows that there are no such distinct cloths with the same upper, lower, left and right sides, since the factor map given by p is right closing.) Therefore it follows that $(T^k)^{[n]}$ can be identified with $(T^{[n]})^k$. Thus $(T^{[n]})^k$ is resolvable. If $T^{[n]} = (p^{[n]}, q^{[n]} : \Gamma^{[n]} \to G^{[n]})$, then the factor map given by $p^{[n]}$ is right closing and the factor map given by $q^{[n]}$ is onto. Therefore, since $(T^{[n]})^k$ is resolvable, it follows from Lemma 7.9 that $T^{[n]}$ is resolvable. □

Lemma 7.13. Let $T = (p, q : \Gamma \to G)$ be a textile system such that the factor map given by p is right closing. Let $k \in \mathbf{N}$ and assume that $\mathcal{G}_{T^{(k)}}$ is an extension of a graphoid. Let $T_0 = (p_0, q_0 : \Gamma_0 \to G^k)$ be a textile system such that p_0 is right resolving and $\mathcal{G}_{T_0}$ is an extension of a graphoid. If $R^0_{T_0} = R^0_{T^{(k)}}$, then $T^{[m]}$ is resolvable for some $m \in \mathbf{N}$.

Proof. Put $T^{(k)} = T_1 = (p_1, q_1 : \Gamma_1 \to G^k)$. Then, since the factor map given by p is right closing, the factor map given by p_1 is right closing. By the same arguments as in the proof of Lemma 7.12, it is derived that there is $n \in \mathbf{N}$ such that $T_1^{[n]} = (T^{(k)})^{[n]}$ is resolvable. Since the factor map given by p is right closing, it readily follows that $\mathcal{G}_{T^{(k)}}$ is λ-graph-isomorphic to $\mathcal{G}_T^{(k)}$. Therefore $\mathcal{G}_{(T^{(k)})^{[n]}} = (\mathcal{G}_{T^{(k)}})^{[n]}$ is λ-graph-isomorphic to $(\mathcal{G}_T^{(k)})^{[n]}$. We notice that $(\mathcal{G}_T^{(k)})^{[n]}$ is λ-graph-isomorphic to $(\mathcal{G}_T^{[kn-k+1]})^{(k)}$. If we let $T^{[m]} = (p^{[m]}, q^{[m]} : \Gamma^{[m]} \to G^{[m]})$ with $m = kn - k + 1$, then $p^{[m]}$ is right closing because p is right closing. Hence $\mathcal{G}_{(T^{[m]})^{(k)}}$ is λ-graph-isomorphic to $(\mathcal{G}_{T^{[m]}})^{(k)} = (\mathcal{G}_T^{[m]})^{(k)}$. Therefore $\mathcal{G}_{(T^{[m]})^{(k)}}$ is λ-graph-isomorphic to $\mathcal{G}_{(T^{(k)})^{[n]}}$. Since $(T^{(k)})^{[n]}$ is resolvable, we know that $(T^{[m]})^{(k)}$ is resolvable. Therefore, since the factor map given by $p^{[m]}$ is right closing, it follows from Lemma 7.10 that $T^{[m]}$ is resolvable. □

Now we shall use an approach which is found in Section 3 of [N3], for a special case.

Let $\mathcal{G} = (G, \lambda)$ be a biresolving λ-graph over an alphabet A. We consider a bipartite expression f of A which is defined either in the form

$$f(a_0) = e(a_0)a_0, \quad a_0 \in A \tag{7.1}$$

or in the form

$$f(a_0) = a_0 e(a_0), \quad a_0 \in A, \tag{7.2}$$

where e is a mapping of A onto an alphabet. We define a bipartite λ-graph $\mathcal{B}_f(\mathcal{G})$, which will be called the *bipartite λ-graph induced by $\mathcal{G}$ and f*, in the following (1) and (2). (In the following, for convenience sake we use the same elements to define arcs and vertices, which should naturally be distinguished.)

(1) Assume that f is given by (7.1). Then for $a \in A_G$, let $\bar{e}(a)$ denote the pair

$$(i_G(a), e(\lambda(a))).$$

Define

$$\mathcal{B}_f(\mathcal{G}) = (H, \mu),$$

where

$$\begin{aligned} V_H &= V_G \cup \{\bar{e}(a) \mid a \in A_G\} \\ A_H &= \{\bar{e}(a) \mid a \in A_G\} \cup A_G \end{aligned}$$

and for $a \in A_G$,

$$\begin{aligned} i_H(\bar{e}(a)) = i_G(a), \quad t_H(\bar{e}(a)) = \bar{e}(a), \quad \mu(\bar{e}(a)) = e(\lambda(a)) \\ i_H(a) = \bar{e}(a), \quad t_H(a) = t_G(a), \quad \text{and} \quad \mu(a) = \lambda(a). \end{aligned}$$

(2) Assume that f is given by (7.2). Then for $a \in A_G$, let $\bar{e}(a)$ denote the pair

$$(e(\lambda(a)), t_G(a)).$$

Let

$$\mathcal{B}_f(\mathcal{G}) = (H, \mu),$$

where

$$\begin{aligned} V_H &= V_G \cup \{\bar{e}(a) \mid a \in A_G\} \\ A_H &= A_G \cup \{\bar{e}(a) \mid a \in A_G\} \end{aligned}$$

and for $a \in A_G$,

$$\begin{aligned} i_H(a) = i_G(a), \quad t_H(a) = \bar{e}(a), \quad \mu(a) = \lambda(a) \\ i_H(\bar{e}(a)) = \bar{e}(a), \quad t_H(\bar{e}(a)) = t_G(a), \quad \text{and} \quad \mu(\bar{e}(a)) = e(\lambda(a)). \end{aligned}$$

It is easy to see that $\mathcal{B}_f(\mathcal{G})$ is biresolving. We have the induced pair $(\acute{\mathcal{B}}_f(\mathcal{G}), \grave{\mathcal{B}}_f(\mathcal{G}))$ of the bipartite λ-graph $\mathcal{B}_f(\mathcal{G})$. Clearly $\acute{\mathcal{B}}_f(\mathcal{G})$ is isomorphic to $\mathcal{G}$. We shall identify $\acute{\mathcal{B}}_f(\mathcal{G})$ with $\mathcal{G}$ and we define a λ-graph $\mathcal{G}_f$ called the *λ-graph induced by $\mathcal{G}$ and f* by

$$\mathcal{G}_f = \grave{\mathcal{B}}_f(\mathcal{G}).$$

We write

$$\mathcal{G}_f = (G_f, \lambda_f).$$

Since $\mathcal{B}_f(\mathcal{G})$ is biresolving, $\mathcal{G}_f$ is biresolving. If every vertex of $\mathcal{G}$ has a homing magic word and a tracing magic word for itself, then so does every vertex of $\mathcal{B}_f(\mathcal{G})$ so that every vertex of $\mathcal{G}_f$ has a homing magic word and a tracing magic word for itself. Thus if $\mathcal{G}$ is a graphoid, then $\mathcal{G}_f$ is a graphoid. Moreover we can see that if $\mathcal{G}$ represents a graph, then so does $\mathcal{G}_f$.

In fact, if $\mathcal{G} = (G, \lambda)$ represents a graph, then we may assume that $\lambda = \mathrm{id}_{A_G}$. Assume that f is defined by

$$f(a) = e(a)a, \quad a \in A_G. \tag{7.3}$$

Then for $a \in A_G$,

$$\bar{e}(a) = (i_G(a), e(a))$$

For $\mathcal{G}_f = (G_f, \lambda_f)$, each element in the image of λ_f is of the form $ae(b)$ with $a, b \in A_G$ and $t_G(a) = i_G(b)$, which uniquely corresponds to

$$a\bar{e}(b) = a(i_G(b), e(b)) \in A_{G_f}.$$

Therefore $\mathcal{G}_f$ represents G_f. In addition, we know that we may use $e(a)$ instead of $\bar{e}(a)$ assuming without loss of generality that for $a, a' \in A_G$

$$e(a) = e(a') \;\Rightarrow\; i_G(a) = i_G(a'). \tag{7.4}$$

Similarly if $\mathcal{G}$ represents a graph and f is defined by

$$f(a) = ae(a), \quad a \in A_G, \tag{7.5}$$

then $\mathcal{G}_f$ represents a graph and we may use e instead of $\bar{e}$ assuming without loss of generality that for $a, a' \in A_G$

$$e(a) = e(a') \Rightarrow t_G(a) = t_G(a'). \tag{7.6}$$

Put

$$\mathcal{B}_f(\mathcal{G}) = (H, \mu) = \mathcal{B}.$$

Let ζ_H and ζ_B be the forward bipartite conjugacies defined as in Section 5. Then as described in Section 5, we have the commutative diagram

$$\begin{array}{ccc} X_{\acute{H}} & \xrightarrow{\zeta_H} & X_{\grave{H}} \\ {\scriptstyle \pi_{\acute{\mathcal{B}}}}\downarrow & & \downarrow{\scriptstyle \pi_{\grave{\mathcal{B}}}} \\ X_{\acute{\mathcal{B}}} & \xrightarrow[\zeta_B]{} & X_{\grave{\mathcal{B}}} \end{array}$$

with $\acute{H} = G, \acute{\mathcal{B}} = \mathcal{G}, \grave{H} = G_f$ and $\grave{\mathcal{B}} = \mathcal{G}_f$. Since $\zeta_{\mathcal{B}}$ is the bipartite code induced by f, it will be denoted by ψ_f. Let ζ_H be denoted by $\Psi_{\mathcal{G},f}$. Thus we have

$$\psi_f \pi_{\mathcal{G}} = \pi_{\mathcal{G}_f} \Psi_{\mathcal{G},f}. \tag{7.7}$$

If $\mathcal{G}$ represents a graph, we can identify ψ_f with $\Psi_{\mathcal{G},f}$.

Let $T = (p, q : \Gamma \to G)$ be a textile system such that $\mathcal{G}_T$ is biresolving. Let f be a bipartite expression of A_G defined by either (7.3) or (7.5). We assume without loss of generality either (7.4) or (7.6) accordingly to use e instead of $\bar{e}$. Let $\tilde{G}$ denote the λ-graph defined by $\tilde{G} = (G, \mathrm{id}_{A_G})$. Let $(\tilde{G})_f = (G_f, \lambda_f)$. Then $(\tilde{G})_f$ represents G_f. As stated above, the image of the bipartite code ψ_f induced by f is X_{G_f}. Let

$$B = \{\frac{p(\alpha)}{q(\alpha)} \mid \alpha \in A_\Gamma\}$$

For $a/b \in B, a, b \in A_G$, we define

$$F(\frac{a}{b}) = \frac{e(a)}{e(b)}\frac{a}{b}, \quad \text{if } f \text{ is defined by (7.3)}$$

and

$$F(\frac{a}{b}) = \frac{a}{b}\frac{e(a)}{e(b)}, \quad \text{if } f \text{ is defined by (7.5).}$$

Then F is a bipartite expression of B having the form either (7.1) or (7.2). Since $\mathcal{G}_T$ is biresolving, we can define $\mathcal{B}_F(\mathcal{G}_T)$ and $(\mathcal{G}_T)_F$. Let T_f be defined as the textile system represented by $(\mathcal{G}_T)_F$, i.e.,

$$\mathcal{G}_{T_f} = (\mathcal{G}_T)_F.$$

We call T_f the *textile system induced by T and f*. We can write

$$T_f = (p_f, q_f : \Gamma_f \to G_f).$$

We note that G_f has already been defined in the above. Since Γ and G are nondegenerate by our assumption, Γ_f and G_f are nondegenerate. By the above, $\mathcal{G}_{T_f}$ is biresolving. If $\mathcal{G}_T$ is a graphoid, then $\mathcal{G}_{T_f}$ is also a graphoid. In particular, if $\mathcal{G}_T$ represents a graph so does $\mathcal{G}_{T_f}$. By (7.7), we have

$$\psi_F \pi_{\mathcal{G}_T} = \pi_{(\mathcal{G}_T)_F} \Psi_{\mathcal{G}_T,F} = \pi_{\mathcal{G}_{T_f}} \Psi_{\mathcal{G}_T,F}. \tag{7.8}$$

Let ξ and η be the factor maps given by p and q, respectively. Let ξ_f and η_f be the factor maps given by p_f and q_f, respectively. Then

$$R_T^0 = \{\frac{\xi(z)}{\eta(z)} \mid z \in X_\Gamma\}$$

and

$$R_{T_f}^0 = \{\frac{\xi_f(z)}{\eta_f(z)} \mid z \in X_{\Gamma_f}\}.$$

We can identify R_T^0 with $X_{\mathcal{G}_T}$ and $R_{T_f}^0$ with $X_{\mathcal{G}_{T_f}}$. Let $\check{\omega}_T : X_{\mathcal{G}_T} \to X_G$ and $\hat{\omega}_T : X_{\mathcal{G}_T} \to X_G$ be the projections such that

$$\check{\omega}_T(\frac{\xi(z)}{\eta(z)}) = \xi(z) \quad \text{and} \quad \hat{\omega}_T(\frac{\xi(z)}{\eta(z)}) = \eta(z), \quad z \in X_\Gamma.$$

Let $\check{\omega}_{T_f} : X_{\mathcal{G}_{T_f}} \to X_{G_f}$ and $\hat{\omega}_{T_f} : X_{\mathcal{G}_{T_f}} \to X_{G_f}$ be defined similarly. Clearly we have

$$\psi_f \check{\omega}_T = \check{\omega}_{T_f} \psi_F \tag{7.9}$$

and

$$\psi_f \hat{\omega}_T = \hat{\omega}_{T_f} \psi_F. \tag{7.10}$$

Since $\xi = \check{\omega}_T \pi_{\mathcal{G}_T}$ and $\xi_f = \check{\omega}_{T_f} \pi_{\mathcal{G}_{T_f}}$, by (7.9) and (7.8) we have

$$\begin{aligned} \psi_f \xi &= \psi_f \check{\omega}_T \pi_{\mathcal{G}_T} = \check{\omega}_{T_f} \psi_F \pi_{\mathcal{G}_T} = \check{\omega}_{T_f} \pi_{\mathcal{G}_{T_f}} \Psi_{\mathcal{G}_T, F} \\ &= \xi_f \Psi_{\mathcal{G}_T, F}. \end{aligned}$$

Similarly, since $\eta = \hat{\omega}_T \pi_{\mathcal{G}_T}$ and $\eta_f = \hat{\omega}_{T_f} \pi_{\mathcal{G}_{T_f}}$, by (7.10) and (7.8) we have

$$\psi_f \eta = \eta_f \Psi_{\mathcal{G}_T, F}$$

Therefore defining

$$\Psi_{T,f} = \Psi_{\mathcal{G}_T, F},$$

We have the following lemma.

Lemma 7.14. Let $T = (p, q : \Gamma \to G)$ be a textile system such that $\mathcal{G}_T$ is biresolving. Let f be the bipartite expression defined by either (7.3) or (7.5) (and we assume without loss of generality either (7.4) or (7.6) accordingly). Let $T_f = (p_f, q_f : \Gamma_f \to G_f)$ be the textile system induced by T and f. Then $\mathcal{G}_{T_f}$ is biresolving and T and T_f are strongly topologically conjugate by the following commutative diagram:

$$\begin{array}{ccc} X_G & \xrightarrow{\psi_f} & X_{G_f} \\ \xi \uparrow & & \uparrow \xi_f \\ X_\Gamma & \xrightarrow{\Psi_{T,f}} & X_{\Gamma_f} \\ \eta \downarrow & & \downarrow \eta_f \\ X_G & \xrightarrow[\psi_f]{} & X_{G_f} \end{array}$$

where $\psi_f : X_G \to X_{G_f}$ is the bipartite code induced by f, $\Psi_{T,f}$ is a bipartite code, ξ and η are the factor maps defined by p and q, respectively, and ξ_f and η_f are the factor maps defined by p_f and q_f, respectively. If $\mathcal{G}_T$ is a graphoid, then $\mathcal{G}_{T_f}$ is also a graphoid. In particular if $\mathcal{G}_T$ represents a graph, then so does $\mathcal{G}_{T_f}$.

Lemma 7.15. Let $T = (p, q : \Gamma \to G)$ be a textile system such that $\mathcal{G}_T$ is biresolving. Let f be the bipartite expression defined by (7.3) and assume (7.4). Let $T_f = (p_f, q_f : \Gamma_f \to G_f)$ be the textile system induced by T and f. Let k be a nonnegative integer. Then if T is k resolvable, then T_f is $k+1$ resolvable.

Proof. Using e instead of $\bar{e}$, we have

$$\begin{aligned} V_{G_f} &= \{e(a) \mid a \in A_G\} \quad \text{and} \\ A_{G_f} &= \{ae(b) \mid a, b \in A_G, t_G(a) = i_G(b)\} \end{aligned}$$

and for $ae(b) \in A_{G_f}$ we have

$$i_{G_f}(ae(b)) = e(a) \quad \text{and} \quad t_{G_f}(ae(b)) = e(b).$$

If $(a_1e(b_1))(a_2e(b_2)) \in L_2(G_f)$ with $a_1e(b_1), a_2e(b_2) \in A_{G_f}$ then we have $e(b_1) = e(a_2)$ together with $t_G(a_1) = i_G(b_1)$ and $t_G(a_2) = i_G(b_2)$, so that we have

$$(a_1e(b_1))(a_2e(b_2)) = (a_1e(a_2))(a_2e(b_2)) \quad \text{with } a_1a_2b_2 \in L_3(G)$$

because $i_G(b_1) = i_G(a_2)$ by (7.4). Thus any path in $L_m(G_f)$ with $m \in \mathbf{N}$ can be written in the form

$$(a_1e(a_2))(a_2e(a_3)) \cdots (a_me(a_{m+1})),$$

where

$$a_1 \cdots a_{m+1} \in L_{m+1}(G), \quad a_i \in A_G.$$

We assume that a path in $L_m(G_f)$ with $m \in \mathbf{N}$ is always written in this form.

For $\alpha \in A_\Gamma$, let

$$\bar{E}(\alpha) = (i_\Gamma(\alpha), \frac{e(p(\alpha))}{e(q(\alpha))}).$$

Then we have

$$\begin{aligned} V_{\Gamma_f} &= \{\bar{E}(\alpha) \mid \alpha \in A_\Gamma\} \quad \text{and} \\ A_{\Gamma_f} &= \{\alpha\bar{E}(\beta) \mid \alpha, \beta \in A_\Gamma, t_\Gamma(\alpha) = i_\Gamma(\beta)\}, \end{aligned}$$

and for $\alpha\bar{E}(\beta) \in A_{\Gamma_f}$ we have

$$\begin{aligned} i_{\Gamma_f}(\alpha\bar{E}(\beta)) &= \bar{E}(\alpha), \quad t_{\Gamma_f}(\alpha\bar{E}(\beta)) = \bar{E}(\beta) \\ p_f(\alpha\bar{E}(\beta)) &= p(\alpha)e(p(\beta)) \quad \text{and} \quad q_f(\alpha\bar{E}(\beta)) = q(\alpha)e(q(\beta)) \end{aligned}$$

For a similar reason to that stated above, any path in $L_m(\Gamma_f)$ with $m \in \mathbf{N}$ can be written in the form

$$(\alpha_1 \bar{E}(\alpha_2))(\alpha_2 \bar{E}(\alpha_3)) \cdots (\alpha_m \bar{E}(\alpha_{m+1})),$$

where

$$\alpha_1 \cdots \alpha_{m+1} \in L_{m+1}(\Gamma), \quad \alpha_i \in A_\Gamma.$$

We assume that a path in $L_m(\Gamma_f)$ with $m \in \mathbf{N}$ is always written in this form.

(i) Assume that

$$\bar{C} = (\alpha_{ij} \bar{E}(\alpha_{i,j+1}))_{i \in \mathbf{N}, 1 \leq j \leq k+1}$$

is a curtain of length $k+1$ weaved by T_f. Let $\bar{x}$ be the upper side of $\bar{C}$, that is, $\bar{x} = (p(\alpha_{11})e(p(\alpha_{12})))(p(\alpha_{12})e(p(\alpha_{13}))) \cdots (p(\alpha_{1,k+1})e(p(\alpha_{1,k+2})))$. The left side of $\bar{C}$ is

$$\downarrow (\bar{E}(\alpha_{i1}))_{i \in \mathbf{N}}.$$

Let $l > k+1$ and let

$$\bar{z} = (a_1 e(a_2))(a_2 e(a_3)) \cdots (a_l e(a_{l+1}))$$

be a path in $L(G_f)$ such that

$$(a_1 e(a_2))(a_2 e(a_3)) \cdots (a_{k+1} e(a_{k+2})) = \bar{x}.$$

Let

$$C = (\alpha_{ij})_{i \in \mathbf{N}, 1 \leq j \leq k+1}$$

and let $x = a_1 \cdots a_{k+1}$ and let $z = a_1 \cdots a_{l+1}$. Then C is a curtain weaved by T, x is the upper side of C and $z \in L_{l+1}(G)$. Since T is k resolvable, T weaves a curtain

$$D = (\beta_{ij})_{i \in \mathbf{N}, 1 \leq j \leq l+1}$$

whose upper side is z and whose left side is the same as that of C. Noting the condition (ii) of Definition 7.1, we infer that the initial subcurtain of length 1 of C and D are the same, i.e.,

$$\downarrow (\beta_{i1})_{i \in \mathbf{N}} = \downarrow (\alpha_{i1})_{i \in \mathbf{N}}.$$

Let

$$\bar{D} = (\beta_{ij} \bar{E}(\beta_{i,j+1}))_{i \in \mathbf{N}, 1 \leq j \leq l}.$$

Then $\bar{D}$ is a curtain weaved by T_f, the upper side of $\bar{D}$ is equal to $\bar{z}$ and the left side of $\bar{D}$ is

$$\downarrow (\bar{E}(\beta_{i1}))_{i \in \mathbf{N}} = \downarrow (\bar{E}(\alpha_{i1}))_{i \in \mathbf{N}}$$

which is equal to the left side of $\bar{C}$. Thus we have shown that T_f satisfies the condition (i) of Definition 7.1 for $k+1$.

(ii) Let $l > k+1$ and let

$$\bar{C} = (\alpha_{ij}\bar{E}(\alpha_{i,j+1}))_{i\in\mathbf{N},1\leq j\leq l}$$

and

$$\bar{D} = (\beta_{ij}\bar{E}(\beta_{i,j+1}))_{i\in\mathbf{N},1\leq j\leq l}$$

be two curtains weaved by T_f having the same upper side and the same left side. Let $C = (\alpha_{ij})_{i\in\mathbf{N},1\leq j\leq l}$ and let $D = (\beta_{ij})_{i\in\mathbf{N},1\leq j\leq l}$. Then C and D are curtains weaved by T. Since $\bar{C}$ and $\bar{D}$ have the same upper side, so do C and D. Since $\bar{C}$ and $\bar{D}$ have the same left side, we have

$$\downarrow(\bar{E}(\alpha_{i1}))_{i\in\mathbf{N}} = \downarrow(\bar{E}(\beta_{i1}))_{i\in\mathbf{N}}$$

which implies

$$\downarrow(i_\Gamma(\alpha_{i1}))_{i\in\mathbf{N}} = \downarrow(i_\Gamma(\beta_{i1}))_{i\in\mathbf{N}}$$

so that C and D have the same left side. Therefore since T is k resolvable,

$$(\alpha_{ij})_{i\in\mathbf{N},1\leq j\leq l-k} = (\beta_{ij})_{i\in\mathbf{N},1\leq j\leq l-k},$$

which implies

$$(\alpha_{ij}\bar{E}(\alpha_{i,j+1}))_{i\in\mathbf{N},1\leq j\leq l-k-1} = (\beta_{ij}\bar{E}(\beta_{i,j+1}))_{i\in\mathbf{N},1\leq j\leq l-k-1}.$$

(iii) Let

$$\bar{x} = (a_1e(a_2))(a_2e(a_3))\cdots(a_{k+1}e(a_{k+2}))$$

and

$$\bar{y} = (b_1e(b_2))(b_2e(b_3))\cdots(b_{k+1}e(b_{k+2}))$$

be two distinct compatible paths in $L_{k+1}(G_f)$ with respect to T_f. Then there are $m \in \mathbf{N}$ and two cloths

$$\bar{C}_0 = (\alpha_{ij}\bar{E}(\alpha_{i,j+1}))_{1\leq i\leq m,1\leq j\leq k+1}$$

and

$$\bar{D}_0 = (\beta_{ij}\bar{E}(\beta_{i,j+1}))_{1\leq i\leq m,1\leq j\leq k+1}$$

weaved by T_f, having the same upper side and the same left side and having $\bar{x}$ and $\bar{y}$ as their lower sides, respectively. Let $C_0 = (\alpha_{ij})_{1\leq i\leq m,1\leq j\leq k+1}$ and let $D_0 = (\beta_{ij})_{1\leq i\leq m,1\leq j\leq k+1}$. Then C_0 and D_0 are cloths weaved by T. Let $x = a_1\cdots a_{k+1}$ and let $y = b_1\cdots b_{k+1}$. Since $\bar{C}_0$ and $\bar{D}_0$ have the same left side, we have

$$\bar{E}(\alpha_{i1}) = \bar{E}(\beta_{i1}), \quad 1\leq i\leq m,$$

which implies

$$i_\Gamma(\alpha_{i1}) = i_\Gamma(\beta_{i1}), \quad 1\leq i\leq m,$$

so that C_0 and D_0 have the same left side. Since $\bar{C}_0$ and $\bar{D}_0$ have the same upper side, it is clear that C_0 and D_0 have the same upper side. Moreover, since $\bar{x}$ is the lower side of $\bar{C}_0$, x is that of C_0 and since $\bar{y}$ is the lower side of $\bar{D}_0$, y is that of D_0. Thus x and y are compatible paths in $L_{k+1}(G)$ with respect to T. Assume that T_f weaves a curtain

$$\bar{C}_1 = (\alpha'_{ij}\bar{E}(\alpha'_{i,j+1}))_{i\in\mathbf{N},1\le j\le k+1}$$

whose upper side is $\bar{x}$. Let $C_1 = (\alpha'_{ij})_{i\in\mathbf{N},1\le j\le k+1}$. Then C_1 is a curtain weaved by T and its upper side is x. Since T is k resolvable and x and y are compatible paths with respect to T, T weaves a curtain $D_1 = (\beta'_{ij})_{i\in\mathbf{N},1\le j\le k+1}$ whose upper side is y and whose left side is the same as that of C_1. Let $C = (\alpha_{ij})_{i\in\mathbf{N},1\le j\le k+1}$ and $D = (\beta_{ij})_{i\in\mathbf{N},1\le j\le k+1}$ be the curtains obtained by joining C_1 to C_0 and D_1 to D_0, respectively, that is, $\alpha_{i+m,j} = \alpha'_{ij}$ and $\beta_{i+m,j} = \beta'_{ij}$ for all $i \in \mathbf{N}$ and $1 \le j \le k+1$. Clearly C and D have the same upper side and the same left side and are of length $k+1$. Therefore since T is k resolvable, the initial subcurtains of length 1 of C and D are the same so that we have

$$\downarrow (\alpha'_{i1})_{i\in\mathbf{N}} = \ \downarrow (\beta'_{i1})_{i\in\mathbf{N}}.$$

Since T is k resolvable, the existence of the terminal subcurtain of length k of D_1 implies the existence of a curtain

$$D_2 = (\beta''_{ij})_{i\in\mathbf{N},1\le j\le k+2}$$

such that

$$\downarrow (\beta''_{i1})_{i\in\mathbf{N}} = \ \downarrow (\beta'_{i1})_{i\in\mathbf{N}}$$

and the upper side of D_2 is equal to

$$yb_{k+2} = b_1 \cdots b_{k+1}b_{k+2}.$$

Let

$$\bar{D}_2 = (\beta''_{ij}\bar{E}(\beta''_{i,j+1}))_{i\in\mathbf{N},1\le j\le k+1}.$$

Then the left side of $\bar{D}_2$ is

$$\downarrow (\bar{E}(\beta''_{i1}))_{i\in\mathbf{N}} = \ \downarrow (\bar{E}(\beta'_{i1}))_{i\in\mathbf{N}} = \ \downarrow (\bar{E}(\alpha'_{i1}))_{i\in\mathbf{N}},$$

which is the left side of $\bar{C}_1$, and the upper side of $\bar{D}_2$ is equal to $\bar{y}$. □

Lemma 7.16. Let $T = (p, q : \Gamma \to G)$ be a textile system such that $\mathcal{G}_T$ is biresolving. Let f be the bipartite expression defined by (7.5) and assume (7.6). Let $T_f = (p_f, q_f : \Gamma_f \to G_f)$ be the textile system induced by T and f. Let k be a nonnegative integer. Then if T is k resolvable, then T_f is k resolvable.

Proof. Since (7.6) is assumed, we can use e instead of $\bar{e}$. We have

$$\begin{aligned} V_{G_f} &= \{e(a) \mid a \in A_G\} \text{ and} \\ A_{G_f} &= \{e(a)b \mid a, b \in A_G, t_G(a) = i_G(b)\}, \end{aligned}$$

and for $e(a)b \in A_{G_f}$ we have

$$i_{G_f}(e(a)b) = e(a) \quad \text{and} \quad t_{G_f}(e(a)b) = e(b).$$

If $(e(a_1)b_1)(e(a_2)b_2) \in L_2(G_f)$ with $e(a_1)b_1, e(a_2)b_2 \in A_{G_f}$, then we have $e(b_1) = e(a_2)$ together with $t_G(a_1) = i_G(b_1)$ and $t_G(a_2) = i_G(b_2)$. Hence we have

$$(e(a_1)b_1)(e(a_2)b_2) = (e(a_1)b_1)(e(b_1)b_2) \quad \text{with} \quad a_1b_1b_2 \in L_3(G),$$

because $t_G(b_1) = t_G(a_2)$ by (7.6). Therefore any path in $L_m(G_f)$ with $m \in \mathbf{N}$ can be written in the form

$$(e(a_0)a_1)(e(a_1)a_2)\cdots(e(a_{m-1})a_m),$$

where

$$a_0 \cdots a_m \in L_{m+1}(G), \quad a_i \in A_G.$$

We assume that a path in $L_m(G_f)$ with $m \in \mathbf{N}$ is always written in this form.

For $\alpha \in A_\Gamma$, let

$$\bar{E}(\alpha) = (\frac{e(p(\alpha))}{e(q(\alpha))}, t_\Gamma(\alpha)).$$

Then we have

$$\begin{aligned} V_{\Gamma_f} &= \{\bar{E}(\alpha) \mid \alpha \in A_\Gamma\} \\ A_{\Gamma_f} &= \{\bar{E}(\alpha)\beta \mid \alpha, \beta \in A_\Gamma, t_\Gamma(\alpha) = i_\Gamma(\beta)\} \end{aligned}$$

and for $\bar{E}(\alpha)\beta \in A_{\Gamma_f}$ we have

$$\begin{aligned} i_{\Gamma_f}(\bar{E}(\alpha)\beta) &= \bar{E}(\alpha), \quad t_{\Gamma_f}(\bar{E}(\alpha)\beta) = \bar{E}(\beta) \\ p_f(\bar{E}(\alpha)\beta) &= e(p(\alpha))p(\beta) \quad \text{and} \quad q_f(\bar{E}(\alpha)\beta) = e(q(\alpha))q(\beta). \end{aligned}$$

For a similar reason to that stated above, any path in $L_m(\Gamma_f)$ with $m \in \mathbf{N}$ can be written in the form

$$(\bar{E}(\alpha_0)\alpha_1)(\bar{E}(\alpha_1)\alpha_2)\cdots(\bar{E}(\alpha_{m-1})\alpha_m),$$

where

$$\alpha_0 \cdots \alpha_m \in L_{m+1}(\Gamma), \quad \alpha_i \in A_\Gamma.$$

We assume that a path in $L_m(\Gamma_f)$ with $m \in \mathbf{N}$ is always written in this form.

(i) Assume that $k = 0$. Suppose that

$$\bar{y} = \downarrow (\bar{E}(\alpha_{i0}))_{i \in \mathbf{N}}, \quad \alpha_{i0} \in A_\Gamma,$$

is a curtain of length 0 weaved by T_f. Then the upper side of $\bar{y}$ is $e(p(\alpha_{10}))$. Let

$$\bar{z} = (e(a_0)a_1)(e(a_1)a_2)\cdots(e(a_{l-1})a_l)$$

be any path in $L(G_f)$ with $l \in \mathbf{N}$ such that

$$e(a_0) = e(p(\alpha_{10})).$$

Since $(\bar{E}(\alpha_{i0}))_{i\in\mathbf{N}} \in \tilde{X}_{(G_f)^T}$, we have

$$e(q(\alpha_{i0})) = (q_f)_V(\bar{E}(\alpha_{i0})) = (p_f)_V(\bar{E}(\alpha_{i+1,0})) = e(p(\alpha_{i+1,0}))$$

for all $i \in \mathbf{N}$. Therefore by (7.6), we have

$$t_G(a_0) = t_G(p(\alpha_{1,0}))$$

and

$$t_G(q(\alpha_{i0})) = t_G(p(\alpha_{i+1,0})) \quad \text{for all } i \in \mathbf{N}.$$

Let

$$y = \downarrow (t_\Gamma(\alpha_{i,0}))_{i\in\mathbf{N}}.$$

Then since

$$p_V(t_\Gamma(\alpha_{i,0})) = t_G(p(\alpha_{i,0})) \quad \text{and} \quad q_V(t_\Gamma(\alpha_{i,0})) = t_G(q(\alpha_{i,0})),$$

it follows from the above that y is a curtain of length 0 weaved by T whose upper side is equal to $t_G(a_0)$. Also we have $a_0 \cdots a_l \in L_{l+1}(G)$. Since T is 0 resolvable, T weaves a curtain

$$(\alpha_{ij})_{i\in\mathbf{N},1\le j\le l}, \quad \alpha_{ij} \in A_\Gamma$$

whose upper side is $a_1 \cdots a_l$ and whose left side is y. Since $y = \downarrow (t_\Gamma(\alpha_{i,0}))_{i\in\mathbf{N}}$ is the left side of the curtain $(\alpha_{ij})_{i\in\mathbf{N},1\le j\le l}$, we have $\alpha_{i0}\alpha_{i1} \in L_2(\Gamma)$ for all $i \in \mathbf{N}$. Therefore

$$(\bar{E}(\alpha_{ij})\alpha_{i,j+1})_{i\in\mathbf{N},0\le j\le l-1}$$

is a curtain weaved by T_f, its upper side is $\bar{z}$ and its left side is $\bar{y}$.

Assume that $k > 0$. Suppose that

$$\bar{C} = (\bar{E}(\alpha_{ij})\alpha_{i,j+1})_{i\in\mathbf{N},0\le j\le k-1}, \quad \alpha_{ij} \in A_\Gamma,$$

is a curtain of length k weaved by T_f. Let $\bar{x}$ be the upper side of $\bar{C}$. Let $l > k$ and let

$$\bar{z} = (e(a_0)a_1)(e(a_1)a_2)\cdots(e(a_{l-1})a_l)$$

be a path in $L_l(G_f)$ such that

$$(e(a_0)a_1)\cdots(e(a_{k-1})a_k) = \bar{x}.$$

Then $a_0 \cdots a_l \in L(G)$. Let

$$C = (\alpha_{ij})_{i\in\mathbf{N},1\le j\le k}.$$

Then C is a curtain of length k weaved by T, its upper side is $a_1 \cdots a_k$ and its left side is

$$\downarrow (t_\Gamma(\alpha_{i0}))_{i\in\mathbf{N}}.$$

Since T is k resolvable, T weaves a curtain $D = (\beta_{ij})_{i\in\mathbf{N},1\le j\le l}$ whose upper side is $a_1 \cdots a_l$ and whose left side is $\downarrow (t_\Gamma(\alpha_{i0}))_{i\in\mathbf{N}}$. Put $\alpha_{i0} = \beta_{i0}$ for all $i \in \mathbf{N}$. Then we have a curtain

$$(\bar{E}(\beta_{ij})\beta_{i,j+1})_{i\in\mathbf{N},0\le j\le l-1}$$

weaved by T_f. Its upper side is $\bar{z}$ and its left side is

$$\downarrow (\bar{E}(\beta_{i0}))_{i\in\mathbf{N}} = \downarrow (\bar{E}(\alpha_{i0}))_{i\in\mathbf{N}},$$

which is the left side of $\bar{C}$.

Thus we have proved that T_f satisfies (i) of Definition 7.1 for k.

(ii) Let $l > k$. Let

$$\bar{C} = (\bar{E}(\alpha_{ij})\alpha_{i,j+1})_{i\in\mathbf{N},0\le j\le l-1}$$

and

$$\bar{D} = (\bar{E}(\beta_{ij})\beta_{i,j+1})_{i\in\mathbf{N},0\le j\le l-1}$$

be two curtains weaved by T_f and having the same upper side and the same left side. Then

$$(\bar{E}(\alpha_{10})\alpha_{11})(\bar{E}(\alpha_{11})\alpha_{12})\cdots(\bar{E}(\alpha_{1,l-1})\alpha_{1l}) = (\bar{E}(\beta_{10})\beta_{11})(\bar{E}(\beta_{11})\beta_{12})\cdots(\bar{E}(\beta_{1,l-1})\beta_{1l})$$

and

$$\downarrow (\bar{E}(\alpha_{i0}))_{i\in\mathbf{N}} = \downarrow (\bar{E}(\beta_{i0}))_{i\in\mathbf{N}},$$

so that we have

$$\alpha_{11} \cdots \alpha_{1l} = \beta_{11} \cdots \beta_{1l}$$

and

$$\downarrow (t_\Gamma(\alpha_{i0}))_{i\in\mathbf{N}} = \downarrow (t_\Gamma(\beta_{i0}))_{i\in\mathbf{N}}.$$

Therefore $(\alpha_{ij})_{i\in\mathbf{N},1\le j\le l}$ and $(\beta_{ij})_{i\in\mathbf{N},1\le j\le l}$ are two curtains weaved by T and having the same upper side and the same left side. since T is k resolvable, we have

$$(\alpha_{ij})_{i\in\mathbf{N},1\le j\le l-k} = (\beta_{ij})_{i\in\mathbf{N},1\le j\le l-k}.$$

Therefore, since $\downarrow (\bar{E}(\alpha_{i0}))_{i\in\mathbf{N}} = \downarrow (\bar{E}(\beta_{i0}))_{i\in\mathbf{N}}$, we have

$$(\bar{E}(\alpha_{ij})\alpha_{i,j+1})_{i\in\mathbf{N},0\le j\le l-k-1} = (\bar{E}(\beta_{ij})\beta_{i,j+1})_{i\in\mathbf{N},0\le j\le l-k-1},$$

which means that the initial subcurtains of length $l-k$ of $\bar{C}$ and $\bar{D}$ are the same. Thus we have shown that the condition (ii) of Definition 7.1 is satisfied by T_f for k.

(iii) Let $k > 0$. Let

$$\bar{x} = (e(a_0)a_1)(e(a_1)a_2)\cdots(e(a_{k-1})a_k)$$

and

$$\bar{y} = (e(b_0)b_1)(e(b_1)b_2)\cdots(e(b_{k-1})b_k)$$

be compatible paths in $L_k(G_f)$ with respect to T_f. Then there are $m \geq 1$ and two cloths

$$(\bar{E}(\alpha_{ij})\alpha_{i,j+1})_{1\leq i\leq m, 0\leq j\leq k-1} \quad \text{and} \quad (\bar{E}(\beta_{ij})\beta_{i,j+1})_{1\leq i\leq m, 0\leq j\leq k-1}$$

weaved by T_f, having the same upper side and the same left side, and having $\bar{x}$ and $\bar{y}$ as their lower sides, respectively. Therefore since

$$\downarrow (\bar{E}(\alpha_{i0}))_{1\leq i\leq m} = \downarrow (\bar{E}(\beta_{i0}))_{1\leq i\leq m} \quad \Rightarrow \quad \downarrow (t_\Gamma(\alpha_{i0}))_{1\leq i\leq m} = \downarrow (t_\Gamma(\beta_{i0}))_{1\leq i\leq m},$$

we have two cloths

$$(\alpha_{ij})_{1\leq i\leq m, 1\leq j\leq k} \quad \text{and} \quad (\beta_{ij})_{1\leq i\leq m, 1\leq j\leq k}$$

weaved by T, having the same upper side and the same left side and having $a_1\cdots a_k$ and $b_1\cdots b_k$ as their lower sides, respectively. Therefore $a_1\cdots a_k$ and $b_1\cdots b_k$ are compatible paths in $L_k(G)$ with respect to T. Assume that there is a curtain

$$(\bar{E}(\gamma_{ij})\gamma_{i,j+1})_{i\in\mathbf{N}, 0\leq j\leq k-1}, \quad \gamma_{ij} \in A_\Gamma$$

weaved by T_f and having $\bar{x}$ as its upper side. Then $(\gamma_{ij})_{i\in\mathbf{N}, 1\leq j\leq k}$ is a curtain weaved by T, its upper side is $a_1\cdots a_k$ and its left side is

$$(t_\Gamma(\gamma_{i0}))_{i\in\mathbf{N}} = (i_\Gamma(\gamma_{i1}))_{i\in\mathbf{N}},$$

which is in $\tilde{X}_{G^T}$. Since T is k resolvable, T weaves a curtain

$$(\gamma'_{ij})_{i\in\mathbf{N}, 1\leq j\leq k}, \quad \gamma'_{ij} \in A_\Gamma,$$

whose left side is $(t_\Gamma(\gamma_{i0}))_{i\in\mathbf{N}}$ and whose upper side is $b_1\cdots b_k$. Putting $\gamma_{i0} = \gamma'_{i0}$ for $i \in \mathbf{N}$, we have a curtain

$$(\bar{E}(\gamma'_{ij})\gamma'_{i,j+1})_{i\in\mathbf{N}, 0\leq j\leq k-1}$$

weaved by T_f. Its upper side is $\bar{y}$ and its left side is the same as that of

$$(\bar{E}(\gamma_{ij})\gamma_{i,j+1})_{i\in\mathbf{N}, 0\leq j\leq k-1}.$$

Thus we have shown that the condition (iii) of Definition 7.1 is satisfied by T_f for $k > 0$.

If $k = 0$, then T_f trivially satisfies the condition (iii) of Definition 7.1 for $k = 0$. □

Let T be a textile system over a graph G. For a path x in $L(G)$ and a path w of length greater than 0 in $L(G^T)$, a path x' in $L(G)$ is called a *successor of x with respect to w* if there is a cloth weaved by T such that its upper side is x, its left side is $\downarrow w$ and its lower side is x'.

Lemma 7.17. Let k be a nonnegative integer. Let T be a k-resolvable textile system over a graph G. Let x and x' be in $L_k(G)$ and assume that they are equivalent. Then successors of x and x' with respect to the same path of length greater than 0 in $L(G^T)$ are also equivalent.

Proof. Assume that there are two cloths C and C' such that the upper side of C is x, the upper side of C' is x', and C and C' have the same left side, say $\downarrow w$. Let z and z' be the lower sides of C and C', respectively. We shall show that z and z' are equivalent. Let $y \in \tilde{X}_{G^T}$ and assume that there is a curtain D whose left side is $\downarrow y$ and whose upper side is z. Consider the curtain D_1 obtained by joining C to D (put C on the top of D). Then the upper side of D_1 is x and the left side of D_1 is $\downarrow wy$. (Here, wy is a sequence in $\tilde{X}_{G^T}$ defined as follows: if $w = c_1 \cdots c_m, c_i \in A_{G^T}$, and $y = (b_i)_{i \in \mathbf{N}}, b_i \in A_{G^T}$, then $wy = (d_i)_{i \in \mathbf{N}}$ with $d_i = c_i$ for $i = 1, \cdots, m$ and $d_{m+i} = b_i$ for $i \in \mathbf{N}$.) Since x and x' are equivalent, there is a curtain D_1' whose upper side is x' and whose left side is $\downarrow wy$. Therefore on D_1' there is a path z'' in $L_k(G)$ which is a successor of x' with respect w. Clearly z' and z'' are compatible. Therefore, since T is k resolvable, z' and z'' are equivalent. Since there is a curtain whose upper side is z'' and whose left side is y, there is a curtain whose upper side is z' and whose left side is $\downarrow y$.

Thus we have seen that for any $y \in \tilde{X}_{G^T}$ if there is a curtain whose upper side is z and whose left side is $\downarrow y$, then there is a curtain whose upper side is z' and whose left side is $\downarrow y$. The converse also holds by symmetry. $\square$

For a textile system $T = (p, q : \Gamma \to G)$, let the textile system

$$T^{(-1)} = (p^{(-1)}, q^{(-1)} : \Gamma^{-1} \to G^{-1})$$

be defined by

$$T^{(-1)} = ((T^*)^{-1})^*.$$

Clearly $T^{(-1)}$ is obtained from T by reversing the direction of arcs of Γ and G and leaving p_A and q_A unchanged. Recall that

$$T^{-1} = (q, p : \Gamma \to G).$$

Let $k \in \mathbf{N}$. Let $T = (p, q : \Gamma \to G)$ be a k-resolvable textile system such that $\mathcal{G}_T$ is biresolving. Then $T^{[k]} = (p^{[k]}, q^{[k]} : \Gamma^{[k]} \to G^{[k]})$ is k resolvable (Remark 7.4) and $\mathcal{G}_{T^{[k]}}$ is biresolving. We define a mapping e of $A_{G^{[k]}}$ onto an alphabet such that for $c, d \in A_{G^{[k]}} = L_k(G)$, $e(c) = e(d)$ if and only if the initial subpaths of length $k-1$ of c and d are the same and c and d are equivalent with respect to T (considering c and d as paths in $L_k(G)$). Let f be the bipartite expression of $A_{G^{[k]}}$ defined by

$$f(c) = e(c)c, \quad c \in A_{G^{[k]}}.$$

We define the *k-resolvent* of T by

$$\mathrm{res}_k(T) = (T^{[k]})_f.$$

By Lemma 7.14, $\mathcal{G}_{\mathrm{res}_k(T)}$ is biresolving. If $\mathrm{res}_k(T) = (p', q' : \Gamma' \to G')$ and the factor map given by p is onto, then the factor map given by p' is onto, by Lemma 7.14. If T is nondegenerate, then so is $\mathrm{res}_k(T)$. Clearly we have

$$((T^{-1})^{[k]})_f = (\mathrm{res}_k(T))^{-1}.$$

For $c \in A_{G^{[k]}} = L_k(G)$ with $c = a_1 \cdots a_k, a_i \in A_G$, let c^t denote the element of $A_{(G^{-1})^{[k]}} = L_k(G^{-1})$ given by

$$c^t = a_k a_{k-1} \cdots a_1.$$

Let $\bar{f}$ be the bipartite expression of $A_{(G^{-1})^{[k]}}$ defined by

$$\bar{f}(c^t) = c^t e'(c^t), \quad c \in A_{G^{[k]}}$$

where $e'(c^t) = e(c)$ for all $c \in A_{G^{[k]}}$. Suppose that $T^{(-1)}$ is l resolvable with $l \geq 0$ and consider $((T^{(-1)})^{[k]})_{\bar{f}}$. Then it follows from Remark 7.4 and Lemma 7.16 that $((T^{(-1)})^{[k]})_{\bar{f}}$ is l resolvable. Therefore since $((T^{(-1)})^{[k]})_{\bar{f}} = ((T^{[k]})_f)^{(-1)}$, we know that $(\mathrm{res}_k(T))^{(-1)}$ is l resolvable.

Similarly, if $(T^{(-1)})^{-1}$ is l resolvable with $l \geq 0$, then $((\mathrm{res}_k(T))^{(-1)})^{-1}$ is l resolvable, because

$$\begin{aligned} ((\mathrm{res}_k(T))^{(-1)})^{-1} &= ((\mathrm{res}_k(T))^{-1})^{(-1)} = (((T^{-1})^{[k]})_f)^{(-1)} = (((T^{-1})^{(-1)})^{[k]})_{\bar{f}} \\ &= (((T^{(-1)})^{-1})^{[k]})_{\bar{f}}. \end{aligned}$$

Lemma 7.18. Let $k \in \mathbf{N}$. Let $T = (p, q : \Gamma \to G)$ be a k-resolvable textile system such that $\mathcal{G}_T$ is biresolving and the factor map ξ given by p is onto. Then $\mathrm{res}_k(T)$ is $k-1$ resolvable and $\mathcal{G}_{\mathrm{res}_k(T)}$ is biresolving. If $\mathrm{res}_k(T) = (p', q' : \Gamma' \to G')$, then the factor map given by p' is onto. If T is nondegenerate, then so is $\mathrm{res}_k(T)$. If $T^{(-1)}$ is l resolvable with $l \geq 0$, then $(\mathrm{res}_k(T))^{(-1)}$ is also l resolvable. If $(T^{(-1)})^{-1}$ is l resolvable with $l \geq 0$, then $((\mathrm{res}_k(T))^{(-1)})^{-1}$ is also l resolvable.

Proof. By the above, it suffices to show that $\mathrm{res}_k(T)$ is $k-1$ resolvable. Let $T^{[k]} = (p^{[k]}, q^{[k]} : \Gamma^{[k]} \to G^{[k]})$. Let $\mathrm{res}_k(T) = (T^{[k]})_f = ((p^{[k]})_f, (q^{[k]})_f : (\Gamma^{[k]})_f \to (G^{[k]})_f)$ and let

$$\bar{E}(\gamma) = (i_{\Gamma^{[k]}}(\gamma), \frac{e(p^{[k]}(\gamma))}{e(q^{[k]}(\gamma))}), \quad \gamma \in A_{\Gamma^{[k]}},$$

where e and f are the same as in the above.

(i) (1) Assume that $k = 1$. Let

$$\downarrow (\bar{E}(\alpha_{i1}))_{i \in \mathbf{N}}, \quad \alpha_{i1} \in A_\Gamma,$$

be a curtain of length 0 weaved by $\mathrm{res}_1(T) = T_f$. Then its upper side is $e(p(\alpha_{11}))$. Let

$$\bar{x} = (a_1 e(a_2))(a_2 e(a_3)) \cdots (a_m e(a_{m+1}))$$

be any path in $L(G_f)$ with $m \in \mathbf{N}$ such that

$$e(a_1) = e(p(\alpha_{11})).$$

Since $\downarrow (\bar{E}(\alpha_{i1}))_{i \in \mathbf{N}}$ is a curtain of length 0 weaved by T_f, we have

$$e(q(\alpha_{i1})) = e(p(\alpha_{i+1,1}))$$

for all $i \in \mathbf{N}$. Therefore we have

$$i_G(q(\alpha_{i1})) = i_G(p(\alpha_{i+1,1})) \quad \text{all } i \in \mathbf{N}$$

so that

$$(i_\Gamma(\alpha_{i1}))_{i \in \mathbf{N}} \in \tilde{X}_{GT}.$$

Moreover,

(†) a_1 and $p(\alpha_{11})$ are equivalent with respect to T and $q(\alpha_{i1})$ and $p(\alpha_{i+1,1})$ are equivalent with respect to T for all $i \in \mathbf{N}$.

We claim that

(‡) there is a curtain

$$D = (\beta_{i1})_{i \in \mathbf{N}}, \quad \beta_{i1} \in A_\Gamma,$$

weaved by T such that the upper side of D is a_1, the left side of D is $\downarrow (i_\Gamma(\alpha_{i1}))_{i \in \mathbf{N}}$ and $q(\beta_{i1})$ is equivalent to $q(\alpha_{i1})$ for all $i \in \mathbf{N}$.

We shall prove this claim by induction.

Since ξ is onto, there is a curtain $\downarrow (\alpha_{l1}^{(0)})_{l \in \mathbf{N}}$ with $\alpha_{11}^{(0)} = \alpha_{11}$. Since a_1 and $p(\alpha_{11})$ are equivalent with respect to T and since there is the curtain $\downarrow (\alpha_{l1}^{(0)})_{l \in \mathbf{N}}$, there is a curtain $\downarrow (\beta_{l1}^{(0)})_{l \in \mathbf{N}}$ whose upper side is a_1 and whose left side is the same as that of $(\alpha_{l1}^{(0)})_{i \in \mathbf{N}}$. Define $\beta_{11} = \beta_{11}^{(0)}$. Then, for the cloth $D_1 = (\beta_{11})$, its upper side is a_1, its left side is $i_\Gamma(\alpha_{11})$, and its lower side $q(\beta_{11})$ is equivalent to $q(\alpha_{11})$ by Lemma 7.17.

Assume that T weaves a cloth $D_i = (\beta_{l1})_{1 \le l \le i}$, $\beta_{l1} \in A_\Gamma$, such that the left side of D_i is $\downarrow i_\Gamma(\alpha_{11}) \cdots i_\Gamma(\alpha_{i1})$ and the lower side $q(\beta_{i1})$ is equivalent to $q(\alpha_{i1})$. Then $q(\beta_{i1})$ is equivalent to $p(\alpha_{i+1,1})$ by (†). Since ξ is onto, there is a curtain $\downarrow (\alpha_{l1}^{(i)})_{l \in \mathbf{N}}$ such that $\alpha_{11}^{(i)} = \alpha_{i+1,1}$. Since $q(\beta_{i1})$ and $p(\alpha_{i+1,1})$ are equivalent, there is a curtain $(\beta_{l1}^{(i)})_{l \in \mathbf{N}}$ whose upper side is $q(\beta_{i1})$ and whose left side is the same as that of $\downarrow (\alpha_{l1}^{(i)})_{l \in \mathbf{N}}$. Define $\beta_{i+1,1} = \beta_{11}^{(i)}$. Then $i_\Gamma(\beta_{i+1,1}) = i_\Gamma(\beta_{11}^{(i)}) = i_\Gamma(\alpha_{11}^{(i)}) = i_\Gamma(\alpha_{i+1,1})$. Then $D_{i+1} = (\beta_{l1})_{1 \le l \le i+1}$ is a cloth whose upper side is a_1 and whose left side is $\downarrow i_\Gamma(\alpha_{11}) \cdots i_\Gamma(\alpha_{i+1,1})$. Since $q(\beta_{i1}) = p(\beta_{i+1,1})$ is equivalent to $p(\alpha_{i+1,1})$ and since $q(\beta_{i+1,1})$ and $q(\alpha_{i+1,1})$ are their successors with respect to $i_\Gamma(\alpha_{i+1,1})$, it follows from Lemma 7.17 that $q(\alpha_{i+1,1})$ and $q(\beta_{i+1,1})$ are equivalent.

We have proved (‡). For $D = (\beta_{i1})_{i\in\mathbf{N}}$ in (‡), $p(\beta_{i1})$ and $p(\alpha_{i1})$ are equivalent for all $i \in \mathbf{N}$, by (†). Since T is 1 resolvable and there is the curtain D of length 1, there is a curtain

$$D' = (\beta'_{ij})_{i\in\mathbf{N},1\leq j\leq m+1}, \quad \beta'_{ij} \in A_\Gamma,$$

whose left side is $\downarrow y = \downarrow (i_\Gamma(\alpha_{i1}))_{i\in\mathbf{N}}$ and whose upper side is $a_1 \cdots a_{m+1}$. Since $p(\beta'_{i1})$ and $p(\beta_{i1})$ are compatible with respect to T for all $i \in \mathbf{N}$ and T is 1 resolvable, $p(\beta'_{i1})$ and $p(\beta_{i1})$ are equivalent for all $i \in \mathbf{N}$. Therefore, it follows that $p(\beta'_{i1})$ and $p(\alpha_{i1})$ are equivalent and $q(\beta'_{i1})$ and $q(\alpha_{i1})$ are equivalent for all $i \in \mathbf{N}$. Thus, noting that the left side of D' is $\downarrow (i_\Gamma(\alpha_{i1}))_{i\in\mathbf{N}}$, we have

$$\bar{E}(\beta'_{i1}) = (i_\Gamma(\beta'_{i1}), \frac{e(p(\beta'_{i1}))}{e(q(\beta'_{i1}))}) = (i_\Gamma(\alpha_{i1}), \frac{e(p(\alpha_{i1}))}{e(q(\alpha_{i1}))}) = \bar{E}(\alpha_{i1})$$

for all $i \in \mathbf{N}$. Clearly T_f can weave the curtain

$$(\beta'_{ij}\bar{E}(\beta'_{i,j+1}))_{i\in\mathbf{N},1\leq j\leq m},$$

whose upper side is $\bar{x}$ and whose left side is

$$\downarrow (\bar{E}(\beta'_{i1}))_{i\in\mathbf{N}} = \downarrow (\bar{E}(\alpha_{i1}))_{i\in\mathbf{N}}.$$

(2) Assume that $k > 1$. Let

$$\bar{C} = \downarrow (\gamma_{ij}\bar{E}(\gamma_{i,j+1}))_{i\in\mathbf{N},1\leq j\leq k-1}$$

be a curtain weaved by $\mathrm{res}_k(T) = (T^{[k]})_f$, where

$$\gamma_{ij} = \alpha_{ij} \cdots \alpha_{i,j+k-1}$$

with

$$\alpha_{i1} \cdots \alpha_{i,2k-1} \in L_{2k-1}(\Gamma), \quad \alpha_{ij} \in A_\Gamma,$$

for all $i \in \mathbf{N}$. Let $m > k$ and let

$$\bar{x} = (c_1e(c_2))(c_2e(c_3)) \cdots (c_me(c_{m+1}))$$

be any path in $L((G^{[k]})_f)$ such that

$$c_j = a_j \cdots a_{j+k-1}$$

with

$$a_1 \cdots a_{m+k} \in L_{m+k}(G), \quad a_j \in A_G,$$

and that $(c_1e(c_2))(c_2e(c_3)) \cdots (c_{k-1}e(c_k))$ is equal to the upper side of $\bar{C}$. Let

$$C = (\alpha_{ij})_{i\in\mathbf{N},1\leq j\leq 2k-2}.$$

Then C is a curtain weaved by T and its upper side is equal to $a_1 \cdots a_{2k-2}$. Since

$$\downarrow (\bar{E}(\gamma_{ik}))_{i \in \mathbf{N}}$$

is a curtain of length 0 weaved by $\operatorname{res}_k(T)$ whose upper side is $e(a_k \cdots a_{2k-1})$, we have

$$e(p(\alpha_{1k}) \cdots p(\alpha_{1,2k-1})) = e(a_k \cdots a_{2k-1})$$

and

$$e(q(\alpha_{ik}) \cdots q(\alpha_{i,2k-1})) = e(p(\alpha_{i+1,k}) \cdots p((\alpha_{i+1,2k-1})) \quad \text{for all } i \in \mathbf{N}.$$

Therefore

(†) $p(\alpha_{1k}) \cdots p(\alpha_{1,2k-1})$ and $a_k \cdots a_{2k-1}$ are equivalent with respect to T and $q(\alpha_{ik}) \cdots q(\alpha_{i,2k-1})$ and $p(\alpha_{i+1,k}) \cdots p(\alpha_{i+1,2k-1})$ are equivalent with respect to T.

The following (‡) is proved by induction using (†) in a similar way to that in (1).

(‡) There is a curtain

$$D = (\beta_{ij})_{i \in \mathbf{N}, k \le j \le 2k-1}$$

weaved by T such that the upper side of D is $a_k \cdots a_{2k-1}$, the left side of D is $\downarrow (i_\Gamma(\alpha_{ik}))_{i \in \mathbf{N}}$, and $q(\beta_{ik}) \cdots q(\beta_{i,2k-1})$ is equivalent to $q(\alpha_{ik}) \cdots q(\alpha_{i,2k-1})$ with respect to T for all $i \in \mathbf{N}$.

Since T is k resolvable, there is a curtain

$$D' = (\beta'_{ij})_{i \in \mathbf{N}, k \le j \le k+m}$$

whose left side is $\downarrow (i_\Gamma(\alpha_{ik}))_{i \in \mathbf{N}}$ and whose upper side is $a_k \cdots a_{k+m}$. We can extend D' to the curtain

$$(\beta'_{ij})_{i \in \mathbf{N}, 1 \le j \le k+m}$$

by setting $\beta'_{ij} = \alpha_{ij}$ for $i \in \mathbf{N}$ and $1 \le j \le k-1$. Clearly $(T^{[k]})_f$ can weave the curtain

$$(\delta_{ij} \bar{E}(\delta_{i,j+1}))_{i \in \mathbf{N}, 1 \le j \le m}$$

with

$$\delta_{ij} = \beta'_{ij} \cdots \beta'_{i,j+k-1} \quad \text{for } i \in \mathbf{N} \text{ and } 1 \le j \le m+1.$$

Its upper side is equal to $\bar{x}$. Its left side is the same as that of $\bar{C}$. In fact, we have

$$\bar{E}(\delta_{i1}) = (i_{\Gamma^{[k]}}(\delta_{i1}), \frac{e(p^{[k]}(\delta_{i1}))}{e(q^{[k]}(\delta_{i1}))}) = (i_{\Gamma^{[k]}}(\gamma_{i1}), \frac{e(p^{[k]}(\gamma_{i1}))}{e(q^{[k]}(\gamma_{i1}))}) = \bar{E}(\gamma_{i1}), \quad \text{all } i \in \mathbf{N}$$

because $\beta'_{i1} \cdots \beta'_{i,k-1} = \alpha_{i1} \cdots \alpha_{i,k-1}$ for all $i \in \mathbf{N}$ and $p(\beta'_{i1}) \cdots p(\beta'_{ik})$ and $p(\alpha_{i1}) \cdots p(\alpha_{ik})$ are equivalent with respect to T for all $i \in \mathbf{N}$ (this follows since T is k resolvable and $p(\beta'_{i1}) \cdots p(\beta'_{ik})$ and $p(\alpha_{i1}) \cdots p(\alpha_{ik})$ are compatible with respect to T for all $i \in \mathbf{N}$).

Thus we have proved that $\operatorname{res}_k(T)$ satisfies the condition (i) of Definition 7.1 for $k-1$.

(ii) Since ξ is onto and T is resolvable, it follows from Remark 7.5(2) that ξ is right closing. Hence the factor map given by $p^{[k]}$ is right closing. Therefore it follows from

Lemma 7.14 that the factor map given by $(p^{[k]})_f$ is right closing. Thus it follows from (i) above and Remark 7.5 (1) that $\mathrm{res}_k(T) = (T^{[k]})_f$ satisfies the condition (ii) of Definition 7.1 for $k-1$.

(iii) If $k=1$, then the condition (iii) of Definition 7.1 is trivially satisfied by T_f for $k-1$. Hence we assume that $k \geq 2$. Let $m \in \mathbf{N}$. Assume that there are cloths

$$\bar{C}_0 = (\gamma_{ij}\bar{E}(\gamma_{i,j+1}))_{1\leq i\leq m, 1\leq j\leq k-1}$$

and

$$\bar{D}_0 = (\delta_{ij}\bar{E}(\delta_{i,j+1}))_{1\leq i\leq m, 1\leq j\leq k-1}$$

weaved by $\mathrm{res}_k(T)$ and having the same upper side and the same left side, where

$$\gamma_{i1}\cdots\gamma_{ik}, \delta_{i1}\cdots\delta_{ik} \in L_k(\Gamma^{[k]}), \quad \gamma_{ij}, \delta_{ij} \in A_{\Gamma^{[k]}},$$

for $1 \leq i \leq m$. Let

$$\bar{C}_1 = (\gamma_{i+m,j}\bar{E}(\gamma_{i+m,j+1}))_{i\in\mathbf{N}, 1\leq j\leq k-1}$$

be a curtain of length $k-1$ weaved by $\mathrm{res}_k(T)$ whose upper side is equal to the lower side of $\bar{C}_0$. Then clearly $(\gamma_{ij}\bar{E}(\gamma_{i,j+1}))_{i\in\mathbf{N}, 1\leq j\leq k-1}$ is a curtain weaved by $\mathrm{res}_k(T)$.

To see that $\mathrm{res}_k(T)$ satisfies the condition (iii) of Definition 7.1 for $k-1$, it suffices to show that there is a curtain $\bar{D}_1$ weaved by $\mathrm{res}_k(T)$ whose upper side is equal to the lower side of $\bar{D}_0$ and whose left side is the same as that of $\bar{C}_1$.

There are paths

$$\alpha_{i1}\cdots\alpha_{i,2k-1} \in L_{2k-1}(\Gamma), \quad \alpha_{ij} \in A_\Gamma, \quad i \in \mathbf{N},$$

such that

$$\gamma_{ij} = \alpha_{ij}\cdots\alpha_{i,j+k-1}, \quad 1 \leq j \leq k,$$

and there are paths

$$\beta_{i1}\cdots\beta_{i,2k-1} \in L_{2k-1}(\Gamma), \quad \beta_{ij} \in A_\Gamma, \quad 1 \leq i \leq m,$$

such that

$$\delta_{ij} = \beta_{ij}\cdots\beta_{i,j+k-1}, \quad 1 \leq j \leq k.$$

Since $\bar{C}_0$ and $\bar{D}_0$ have the same upper side, we have

$$p(\alpha_{11})\cdots p(\alpha_{1,2k-2}) = p(\beta_{11})\cdots p(\beta_{1,2k-2}) \tag{7.11}$$

and

$$e(p(\alpha_{1k})\cdots p(\alpha_{1,2k-1})) = e(p(\beta_{1k})\cdots p(\beta_{1,2k-1})). \tag{7.12}$$

Clearly $(\alpha_{ij})_{i\in\mathbf{N}, 1\leq j\leq 2k-2}$ is a curtain weaved by T and $(\beta_{ij})_{1\leq i\leq m, 1\leq j\leq 2k-2}$ is a cloth weaved by T. Since $\bar{C}_0$ and $\bar{D}_0$ have the same left side, we have $\bar{E}(\gamma_{i1}) = \bar{E}(\delta_{i1})$ for $1 \leq i \leq m$, so that we have

$$\alpha_{i1}\cdots\alpha_{i,k-1} = \beta_{i1}\cdots\beta_{i,k-1} \quad \text{for } 1 \leq i \leq m. \tag{7.13}$$

Since $\downarrow(\bar{E}(\gamma_{i,k}))_{i\in\mathbf{N}}$ is a curtain of length 0 weaved by $\mathrm{res}_k(T)$, we have

$$e(q(\alpha_{ik})\cdots q(\alpha_{i,2k-1})) = e(p(\alpha_{i+1,k})\cdots p(\alpha_{i+1,2k-1}))$$

for $i \in \mathbf{N}$. Therefore the same argument that proves (‡) in (2) of (i) (a similar argument to that used to prove (‡) in (1) of (i)) shows that there is a curtain

$$C' = (\alpha'_{ij})_{i\in\mathbf{N},k\leq j\leq 2k-1}$$

weaved by T such that its upper side is equal to

$$p(\alpha_{1k})\cdots p(\alpha_{1,2k-1}),$$

its left side is equal to

$$\downarrow(i_\Gamma(\alpha_{ik}))_{i\in\mathbf{N}},$$

and $q(\alpha'_{ik})\cdots q(\alpha'_{i,2k-1})$ are equivalent to $q(\alpha_{ik})\cdots q(\alpha_{i,2k-1})$ with respect to T for all $i \in \mathbf{N}$. Hence, in particular, $q(\alpha'_{mk})\cdots q(\alpha'_{m,2k-1})$ and $q(\alpha_{mk})\cdots q(\alpha_{m,2k-1})$ are equivalent. Therefore since there is a curtain whose upper side is $q(\alpha'_{mk})\cdots q(\alpha'_{m,2k-1})$ and whose left side is

$$\downarrow(i_\Gamma(\alpha_{m+i,k}))_{i\in\mathbf{N}},$$

there is a curtain C'' whose upper side is

$$q(\alpha_{mk})\cdots q(\alpha_{m,2k-1})$$

and whose left side is

$$\downarrow(i_\Gamma(\alpha_{m+i,k}))_{i\in\mathbf{N}}.$$

Since $\downarrow(\bar{E}(\delta_{i,k}))_{1\leq i\leq m}$ is the right side of $\bar{D}_0$, we have

$$e(q(\beta_{ik})\cdots q(\beta_{i,2k-1})) = e(p(\beta_{i+1,k})\cdots p(\beta_{i+1,2k-1}))$$

for $1 \leq i \leq m-1$. Therefore, the same argument that proves (‡) in (2) of (i) (a similar argument to that used to prove (‡) in (1) of (i)) shows that there is a cloth

$$D' = (\beta'_{ij})_{1\leq i\leq m,k\leq j\leq 2k-1}$$

weaved by T such that its upper side is

$$p(\beta_{1k})\cdots p(\beta_{1,2k-1}),$$

its left side is

$$\downarrow(i_\Gamma(\beta_{ik}))_{1\leq i\leq m},$$

and $q(\beta'_{mk})\cdots q(\beta'_{m,2k-1})$ and $q(\beta_{mk})\cdots q(\beta_{m,2k-1})$ are equivalent. Since (7.12) implies the equivalence of $p(\alpha_{1k})\cdots p(\alpha_{1,2k-1})$ and $p(\beta_{1k})\cdots p(\beta_{1,2k-1})$ and (7.13) implies that

$$\downarrow(i_\Gamma(\alpha_{ik}))_{1\leq i\leq m} = \downarrow(i_\Gamma(\beta_{ik}))_{1\leq i\leq m},$$

the existence of the cloths $(\alpha'_{ij})_{1\le i\le m,k\le j\le 2k-1}$ and D' implies that $q(\alpha'_{mk})\cdots q(\alpha'_{m,2k-1})$ and $q(\beta'_{mk})\cdots q(\beta'_{m,2k-1})$ are equivalent, by Lemma 7.17 (because T is k resolvable). Therefore, since $q(\alpha'_{mk})\cdots q(\alpha'_{m,2k-1})$ and $q(\alpha_{mk})\cdots q(\alpha_{m,2k-1})$ are equivalent and so are $q(\beta'_{mk})\cdots q(\beta'_{m,2k-1})$ and $q(\beta_{mk})\cdots q(\beta_{m,2k-1})$, it follows that $q(\alpha_{mk})\cdots q(\alpha_{m,2k-1})$ and $q(\beta_{mk})\cdots q(\beta_{m,2k-1})$ are equivalent. Thus the existence of the curtain C'' implies that there is a curtain

$$D'' = (\beta''_{i+m,j})_{i\in\mathbf{N},k\le j\le 2k-1}$$

whose upper side is $q(\beta_{mk})\cdots q(\beta_{m,2k-1})$ and whose left side is $\downarrow(i_\Gamma(\alpha_{m+i,k}))_{i\in\mathbf{N}}$. Let

$$D_1 = (\beta_{m+i,j})_{i\in\mathbf{N},1\le j\le 2k-1}$$

be defined by

$$\beta_{m+i,j} = \alpha_{m+i,j} \quad \text{for } i\in\mathbf{N} \text{ and } 1\le j\le k-1 \tag{7.14}$$

and

$$\beta_{m+i,j} = \beta''_{m+i,j} \quad \text{for } i\in\mathbf{N} \text{ and } k\le j\le 2k-1.$$

Then D_1 is a curtain weaved by T. The upper side of D_1 is equal to $q(\beta_{m1})\cdots q(\beta_{m,2k-1})$ (see(7.13)). Since the curtains $(\alpha_{ij})_{i\in\mathbf{N},1\le j\le k}$ and $(\beta_{ij})_{i\in\mathbf{N},1\le j\le k}$ have the same upper side and the same left side by (7.11), (7.13), and (7.14), $q(\alpha_{i1})\cdots q(\alpha_{ik})$ and $q(\beta_{i1})\cdots q(\beta_{ik})$ are compatible with respect to T for all $i\in\mathbf{N}$. Therefore, $p(\alpha_{m+i,1})\cdots p(\alpha_{m+i,k})$ and $p(\beta_{m+i,1})\cdots p(\beta_{m+i,k})$ are equivalent with respect to T for all $i\in\mathbf{N}$, because T is k resolvable. Hence noting (7.14) again, we have

$$e(p(\alpha_{m+i,1})\cdots p(\alpha_{m+i,k})) = e(p(\beta_{m+i,1})\cdots p(\beta_{m+i,k})) \quad \text{all } i\in\mathbf{N}. \tag{7.15}$$

Let

$$\bar{D}_1 = (\delta_{m+i,j}\bar{E}(\delta_{m+i,j+1}))_{i\in\mathbf{N},1\le j\le k-1}$$

with

$$\delta_{m+i,j} = \beta_{m+i,j}\cdots\beta_{m+i,j+k-1} \quad \text{all } i\in\mathbf{N} \text{ and } 1\le j\le k.$$

Then $\bar{D}_1$ is a curtain weaved by $\mathrm{res}_k(T)$. Its upper side is equal to the lower side of $\bar{D}_0$, because the upper side of D_1 is equal to $q(\beta_{m1})\cdots q(\beta_{m,2k-1})$. It follows from (7.14) and (7.15) that

$$\downarrow(\bar{E}(\delta_{m+i,1}))_{i\in\mathbf{N}} = \downarrow(\bar{E}(\gamma_{m+i,1}))_{i\in\mathbf{N}}$$

so that the left side of $\bar{D}_1$ is equal to that of $\bar{C}_1$. $\square$

Let $T=(p,q:\Gamma\to G)$ and $T'=(p',q':\Gamma'\to G')$ be a textile system. A topological conjugacy $\phi:(X_T,\sigma_T)\to(X_{T'},\sigma_{T'})$ is said to be *relation-preserving with respect to R_T and $R_{T'}$* if for $x,y\in X_T$,

$$\frac{x}{y}\in R_T \quad\Leftrightarrow\quad \frac{\phi(x)}{\phi(y)}\in R_{T'}.$$

A topological conjugacy $\phi_0 : (X_G, \sigma_G) \to (X_{G'}, \sigma_{G'})$ is said to be *relation-preserving with respect to* R_T^0 *and* $R_{T'}^0$ if for $x, y \in X_G$

$$\frac{x}{y} \in R_T^0 \quad \Leftrightarrow \quad \frac{\phi_0(x)}{\phi_0(y)} \in R_{T'}^0.$$

It is clear that if $\phi_0 : (X_G, \sigma_G) \to (X_{G'}, \sigma_{G'})$ is relation-preserving with respect to R_T^0 and $R_{T'}^0$, then the restriction $\phi : (X_T, \sigma_T) \to (X_{T'}, \sigma_{T'})$ of ϕ_0 is relation-preserving with respect to R_T and $R_{T'}$. It is also clear that if $\phi_0 : (X_G, \sigma_G) \to (X_{G'}, \sigma_{G'})$ is relation-preserving with respect to R_T^0 and $R_{T'}^0$, then so is $\phi_0 \sigma_G^l$ for all $l \in \mathbf{Z}$.

Theorem 7.19. Let $T_0 = (p_0, q_0 : \Gamma_0 \to G_0)$ is a textile system such that $\mathcal{G}_{T_0}$ is an extension of a graphoid and p_0 is right resolving. Let $T = (p, q : \Gamma \to G)$ is a textile system such that $\mathcal{G}_T$ is an extension of a graphoid and either the factor map given by p is right closing or $\mathcal{G}_T$ is right resolving. Assume that there is a topological conjugacy $\phi : (X_{G_0}, \sigma_{G_0}) \to (X_G, \sigma_G)$ which is relation-preserving with respect to $R_{T_0}^0$ and R_T^0. Then $T^{[n]}$ is resolvable for some $n \in \mathbf{N}$.

Proof. Since p_0 is right resolving, $\mathcal{G}_{T_0}$ is right resolving. Since $\mathcal{G}_{T_0}$ is an extension of a graphoid, there is $l \in \mathbf{N}$ and a state-equivalence E_0 of $(\mathcal{G}_{T_0^{[l]}})_+ = \mathcal{G}_{T_0^{[l]}}$ such that $(\mathcal{G}_{T_0^{[l]}})_{E_0}$ is a graphoid. Let $T_1 = (p_1, q_1 : \Gamma_1 \to G_1)$ be the textile system such that $\mathcal{G}_{T_1} = (\mathcal{G}_{T_0^{[l]}})_{E_0}$. Then $G_1 = G_0^{[l]}$ and it follows from Remark 7.4 and Lemma 7.7 that T_1 is 0 resolvable. Moreover the topological conjugacy $\rho_{X_{G_0}, l} : (X_{G_0}, \sigma_{G_0}) \to (X_{G_1}, \sigma_{G_1})$ is relation-preserving with respect to $R_{T_0}^0$ and $R_{T_1}^0$ (the definition of $\rho_{X,k}$ for a subshift (X, σ) and $k \in \mathbf{N}$, was given in the paragraph preceding Theorem 5.1′). Since $\mathcal{G}_T$ is an extension of a graphoid, there is $m \in \mathbf{N}$ and a state-equivalence E of $(\mathcal{G}_{T^{[m]}})_+$ such that $((\mathcal{G}_{T^{[m]}})_+)_E$ is a graphoid. Let $T' = (p', q' : \Gamma' \to G')$ be the textile system such that $\mathcal{G}_{T'} = ((\mathcal{G}_{T^{[m]}})_+)_E$. Then $G' = G^{[m]}$ and the topological conjugacy $\rho_{X_G, m} : (X_G, \sigma_G) \to (X_{G'}, \sigma_{G'})$ is relation-preserving with respect to R_T^0 and $R_{T'}^0$. Let

$$\phi_1 = \rho_{X_G, m} \phi (\rho_{X_{G_0}, l})^{-1}$$

Then since ϕ is a topological conjugacy which is relation-preserving with respect to $R_{T_0}^0$ and R_T^0, it follows from the above that ϕ_1 is a topological conjugacy of (X_{G_1}, σ_{G_1}) onto $(X_{G'}, \sigma_{G'})$ which is relation-preserving with respect to $R_{T_1}^0$ and $R_{T'}^0$. By Theorem 5.1′, there are $m', m'' \in \mathbf{N}$ and a sequence of bipartite codes $\psi_1, \cdots, \psi_{m''}$ such that

$$\rho_{X_{G'}, m'} \phi_1 = \psi_{m''} \cdots \psi_2 \psi_1 \tag{7.16}$$

and each ψ_i is induced by a bipartite expression, say f_i, either of the form (7.3) or of the form (7.5). Let

$$T_2 = (\cdots ((T_1)_{f_1})_{f_2} \cdots)_{f_{m''}}. \tag{7.17}$$

Then since $\mathcal{G}_{T_1}$ is a graphoid, it follows from Lemma 7.14 that $\mathcal{G}_{T_2}$ is a graphoid and $\psi_{m''}\cdots\psi_1$ is relation-preserving with respect to $R^0_{T_1}$ and $R^0_{T_2}$. Let

$$T'' = (T')^{[m']}.$$

Since $\mathcal{G}_{T'}$ is a graphoid and $\mathcal{G}_{T''} = \mathcal{G}_{(T')^{[m']}} = (\mathcal{G}_{T'})^{[m']}$, it follows that $\mathcal{G}_{T''}$ is a graphoid. Of course, $\rho_{X_{G'},m'}$ is relation-preserving with respect to $R^0_{T'}$ and $R^0_{T''}$. Hence $\rho_{X_{G'},m'}\phi_1$ is relation preserving with respect to $R^0_{T_1}$ and $R^0_{T''}$. Therefore, since $\psi_{m''}\cdots\psi_1$ is relation preserving with respect to $R^0_{T_1}$ and $R^0_{T_2}$, it follows from (7.16) that $R^0_{T''} = R^0_{T_2}$. Since $\mathcal{G}_{T_2}$ and $\mathcal{G}_{T''}$ are graphoids and a graphoid is canonical (by Proposition 7.24 given at the end of this section), we have $\mathcal{G}_{T_2} = \mathcal{G}_{T''}$. Thus we have

$$T_2 = T''.$$

Since T_1 is resolvable and T_2 is defined by (7.17), it follows from Lemmas 7.15 and 7.16 that T_2 is resolvable. Since $\mathcal{G}_{T_2} = \mathcal{G}_{T''} = (\mathcal{G}_{T'})^{[m']}$ and $\mathcal{G}_{T'} = (((\mathcal{G}_T)^{[m]})_+)_E$, it follows from Lemma 7.11 that there is a state-equivalence E' of $((\mathcal{G}_T)^{[m+m']})_+$ such that

$$\mathcal{G}_{T_2} = (((\mathcal{G}_T)^{[m+m']})_+)_{E'} = ((\mathcal{G}_{T^{[m+m']}})_+)_{E'}$$

Put $m+m'=n$. If the factor map given by p is right closing, then the factor map given by $p^{[n]}$ is right closing, where $T^{[n]} = (p^{[n]}, q^{[n]} : \Gamma^{[n]} \to G^{[n]})$. If $\mathcal{G}_T$ is right resolving, then so is $\mathcal{G}_{T^{[n]}}$ so that

$$\mathcal{G}_{T_2} = (\mathcal{G}_{T^{[n]}})_{E'}.$$

Therefore, since T_2 is resolvable, it follows from Lemmas 7.6 and 7.7 that $T^{[n]}$ is resolvable. □

Theorem 7.20. Let T be a textile system. Then the following statements are valid.

(1) If $T^{[n]}$ is resolvable for some $n \in \mathbf{N}$ and $\pi_{\mathcal{G}_T}$ is left closing, then there are a textile system $T_0 = (p_0, q_0 : \Gamma_0 \to G_0)$ with p_0 right resolving and a topological conjugacy $\phi : (X_T, \sigma_T) \to (X_{T_0}, \sigma_{T_0})$ which is relation-preserving with respect to R_T and R_{T_0}. If in addition T is nondegenerate, then T_0 can be nondegenerate.

(2) If $(T^{[n]})^{-1}$ and $(T^{[n]})^{(-1)}$ are resolvable for some $n \in \mathbf{N}$, then there are an LR textile system T_0 and a topological conjugacy $\phi : (X_T, \sigma_T) \to (X_{T_0}, \sigma_{T_0})$ which is relation-preserving with respect to R_T and R_{T_0}.

(3) If $(T^{[n]})^{-1}$ and $((T^{[n]})^{(-1)})^{-1}$ are resolvable for some $n \in \mathbf{N}$, then there are a textile system $T_0 = (p_0, q_0 : \Gamma_0 \to G_0)$ with q_0 biresolving and a topological conjugacy $\phi : (X_T, \sigma_T) \to (X_{T_0}, \sigma_{T_0})$ which is relation-preserving with respect to R_T and R_{T_0}. If in addition T is nondegenerate, then T_0 can be nondegenerate.

Proof. (1) Suppose that $T^{[n]}$ is k resolvable with $k \geq 1$. (If $T^{[n]}$ is 0 resolvable, then let $T_0 = \overline{\mathrm{trim}}(T^{[n]})$). Let $\bar{T} = (\bar{p}, \bar{q} : \bar{\Gamma} \to \bar{G})$ be defined by

$$\bar{T} = \overline{\mathrm{trim}}((T^{[n]})^{[k]}).$$

Then by Remarks 7.4 and 7.3(3), $\bar{T}$ is resolvable. Since $\pi_{\mathcal{G}_T}$ is left closing, so is $\pi_{\mathcal{G}_{(T^{[n]})^{[k]}}} = \pi_{\mathcal{G}_{T^{[n+k-1]}}}$ so that $\pi_{\mathcal{G}_{\bar{T}}}$ is left closing. By Remark 7.5 (3)(2), the factor map $\bar{\xi}$ given by $\bar{p}$ is onto and right closing. Hence $\pi_{\mathcal{G}_T}$ is right closing. If we put $\bar{\phi} = \rho_{X_T,n+k-1}$, then $\bar{\phi} : (X_T, \sigma_T) \to (X_{\bar{T}}, \sigma_{\bar{T}})$ is a topological conjugacy which is relation preserving with respect to R_T and $R_{\bar{T}}$. Since $\pi_{\mathcal{G}_{\bar{T}}}$ is biclosing, it follows from Lemma 7.8 that there are $m \in \mathbf{N}$ and a state-equivalence E of $(\mathcal{G}_{\bar{T}^{[m]}})_+$ such that $((\mathcal{G}_{\bar{T}^{[m]}})_+)_E$ is biresolving. Let $T_1 = (p_1, q_1 : \Gamma_1 \to G_1)$ be the textile system which is represented by $((\mathcal{G}_{\bar{T}^{[m]}})_+)_E$. Let $\phi_1 = \rho_{X_{\bar{T}},m}$. Then $\phi_1 : (X_{\bar{T}}, \sigma_{\bar{T}}) \to (x_{T_1}, \sigma_{T_1})$ is a topological conjugacy which is relation preserving with respect to $R_{\bar{T}}$ and R_{T_1}. Since $\bar{\xi}$ is right closing and $\bar{T}$ is resolvable, it follows from Lemmas 7.6 and 7.7 that T_1 is resolvable. Since $\bar{\xi}$ is onto, the factor map ξ_1 given by p_1 is onto.

Suppose that T_1 is k_1 resolvable with $k_1 \geq 0$. If $k_1 > 0$, then take the k_1-resolvent of T_1. The resultant is $k_1 - 1$ resolvable by Lemma 7.18. Note that the properties required to apply Lemma 7.18, are preserved under the operation of taking a resolvent. If $k_1 - 1 > 0$, then take its $(k_1 - 1)$-resolvent, which is $k_1 - 2$ resolvable by Lemma 7.18. Continuing in this way, we eventually obtain a textile system $T_0 = (p_0, q_0 : \Gamma_0 \to G_0)$ which is 0 resolvable. The factor map ξ_0 given by p_0 is onto. Thus, by Remark 7.5(2), p_0 is right resolving. By Lemma 7.14 (and by the fact that for any textile system T and $l \in \mathbf{N}$, $\rho_{X_T,l}$ is a topological conjugacy which is relation-preserving with respect to R_T and $R_{T^{[l]}}$), we see that there is a topological conjugacy $\phi_0 : (X_{T_1}, \sigma_{T_1}) \to (X_{T_0}, \sigma_{T_0})$ which is relation-preserving with respect to R_{T_1} and R_{T_0}. Let $\phi = \phi_0 \phi_1 \bar{\phi}$. Then $\phi : (X_T, \sigma_T) \to (X_{T_0}, \sigma_{T_0})$ is relation-preserving with respect to R_T and R_{T_0}. Since all of the transformations $T \to T^{[n]}, T^{[n]} \to \bar{T}, \bar{T} \to T_1, T_1 \to T_0$ in the above preserve nondegeneracy for textile systems, we see that if T is nondegenerate, then T_0 is nondegenerate.

(2) By Remark 7.2, we may assume that $(T^{[n]})^{-1}$ and $(T^{[n]})^{(-1)}$ are both k resolvable with $k \geq 1$. Let $\bar{T} = (\bar{p}, \bar{q} : \bar{\Gamma} \to \bar{G})$ be defined by

$$\bar{T} = \overline{\operatorname{trim}}((T^{[n]})^{[k]}).$$

Then by Remarks 7.4 and 7.3(3), $\bar{T}^{-1}$ and $\bar{T}^{(-1)}$ are resolvable. Since $(T^{[n]})^{-1}$ is k resolvable, it follows from Remark 7.5 that the factor map $\bar{\eta}$ given by $\bar{q}$ is right closing and onto. Since $(T^{[n]})^{(-1)}$ is k resolvable, it also follows from Remark 7.5 that the factor map $\bar{\xi}$ given by $\bar{p}$ is left closing and onto. Therefore $\pi_{\mathcal{G}_{\bar{T}}}$ is biclosing and $\bar{T}$ is nondegenerate. As seen in the proof of (1), there is a topological conjugacy $\bar{\phi} : (X_T, \sigma_T) \to (X_{\bar{T}}, \sigma_{\bar{T}})$ which is relation-preserving with respect to R_T and $R_{\bar{T}}$. Since $\pi_{\mathcal{G}_{\bar{T}}}$ is biclosing, it follows from Lemma 7.8 that there are $m \in \mathbf{N}$, a state-equivalence E of $(\mathcal{G}_{\bar{T}^{[m]}})_+$ and a state-equivalence E' of $((\mathcal{G}_{\bar{T}^{[m]}})_-)^{-1}$ such that $((\mathcal{G}_{\bar{T}^{[m]}})_+)_E$ and $((((\mathcal{G}_{\bar{T}^{[m]}})_-)^{-1})_{E'})^{-1}$ are the same biresolving λ-graph. Let $T_1 = (p_1, q_1 : \Gamma_1 \to G_1)$ be the textile system represented by this biresolving λ-graph. Since $\bar{T}$ is nondegenerate, it follows that T_1 is nondegenerate. Since

$$\mathcal{G}_{T_1^{-1}} = ((\mathcal{G}_{(\bar{T}^{-1})^{[m]}})_+)_E$$

and $\bar{\eta}$ is right closing, it follows from Lemmas 7.6 and 7.7 that T_1^{-1} is resolvable. Since

$$\mathcal{G}_{T_1^{(-1)}} = ((\mathcal{G}_{(\bar{T}^{(-1)})^{[m]}})_+)_{E'}$$

and $\bar{\xi}$ is left closing, it follows from Lemmas 7.6 and 7.7 that $T_1^{(-1)}$ is resolvable.

We suppose that T_1^{-1} is k_1 resolvable and $T_1^{(-1)}$ is l_1 resolvable with $k_1, l_1 \geq 0$. Starting from T_1^{-1} and repeating the operation of taking a resolvent k_1 times we obtain a nondegenerate textile system $\tilde{T}_0^{-1}$ which is 0 resolvable, by Lemma 7.18. Since $((T_1^{-1})^{(-1)})^{-1} = T_1^{(-1)}$ is l_1 resolvable, $\tilde{T}_0^{(-1)} = ((\tilde{T}_0^{-1})^{(-1)})^{-1}$ is l_1 resolvable, by Lemma 7.18. Starting from $\tilde{T}_0^{(-1)}$ and repeating the operation of taking a resolvent l_1 times, we obtain a nondegenerate textile system $T_0^{(-1)}$ which is 0 resolvable, by Lemma 7.18. It also follows from Lemma 7.18 that T_0^{-1} is 0 resolvable (because $\tilde{T}_0^{-1} = ((\tilde{T}_0^{(-1)})^{(-1)})^{-1}$ and $T_0^{-1} = ((T_0^{(-1)})^{(-1)})^{-1}$). Let $T_0 = (p_0, q_0 : \Gamma_0 \to G_0)$. Then, since $T_0^{(-1)}$ and T_0^{-1} are 0 resolvable and T_0 is nondegenerate, p_0 is left resolving and q_0 is right resolving, by Remark 7.5(2). Applying Lemma 7.14 to the transformations of taking resolvents above, we know that there is a topological conjugacy $\phi_0 : (X_{\bar{G}}, \sigma_{\bar{G}}) \to (X_{G_0}, \sigma_{G_0})$ which is relation-preserving with respect to $R^0_{\bar{T}}$ and $R^0_{T_0}$. Let $\phi = \phi_0\bar{\phi}$. Then $\phi : (X_T, \sigma_T) \to (X_{T_0}, \sigma_{T_0})$ is relation preserving with respect to R_T and R_{T_0}.

(3) The proof of (3) of the theorem is similar to that of (1) and (2). As in (1), the transformations made to obtain T_0 from T are easily seen to preserve nondegeneracy for textile systems. Therefore if T is nondegenerate, then T_0 is nondegenerate. □

Theorem 7.21. Let $T = (p, q : \Gamma \to G)$ be a one-sided 1-1 and nondegenerate textile system. Then the following statements are valid.

(1) There are a one-sided 1-1 and nondegenerate textile system $T_0 = (p_0, q_0 : \Gamma_0 \to G_0)$ with p_0 left resolving and a topological conjugacy $\phi : (X_G, \sigma_G) \to (X_{G_0}, \sigma_{G_0})$ with $\varphi_T = \phi^{-1}\varphi_{T_0}\phi$ if and only if $(T^{[n]})^{(-1)}$ is resolvable for some $n \in \mathbf{N}$.

(2) There are a one-sided 1-1 LR textile system $T_0 = (p_0, q_0 : \Gamma_0 \to G_0)$ and a topological conjugacy $\phi : (X_G, \sigma_G) \to (X_{G_0}, \sigma_{G_0})$ with $\varphi_T = \phi^{-1}\varphi_{T_0}\phi$ if and only if $(T^{[n]})^{-1}$ and $(T^{[n]})^{(-1)}$ are resolvable for some $n \in \mathbf{N}$.

(3) There are a one-sided 1-1 and nondegenerate textile system $T_0 = (p_0, q_0 : \Gamma_0 \to G_0)$ with q_0 biresolving and a topological conjugacy $\phi : (X_G, \sigma_G) \to (X_{G_0}, \sigma_{G_0})$ with $\varphi_T = \phi^{-1}\varphi_{T_0}\phi$ if and only if $(T^{[n]})^{-1}$ and $((T^{[n]})^{(-1)})^{-1}$ are resolvable for some $n \in \mathbf{N}$.

Proof. We describe the proof of (2) in detail.

Let ξ and η be the factor maps given by p and q, respectively. Since T is one-sided 1-1 and nondegenerate, ξ is a topological conjugacy, η is onto, and $\varphi_T = \eta\xi^{-1}$.

Assume that there are a one-sided 1-1 LR textile system $T_0 = (p_0, q_0 : \Gamma_0 \to G_0)$ and a topological conjugacy $\phi : (X_G, \sigma_G) \to (X_{G_0}, \sigma_{G_0})$ with

$$\varphi_T = \phi^{-1}\varphi_{T_0}\phi.$$

Let ξ_0 and η_0 be the factor maps given by p_0 and q_0, respectively. Since T_0 is LR, T_0 is nondegenerate (Fact 3.3). Since q_0 is right resolving, η_0 is right closing. Therefore since ξ_0 is a topological conjugacy and $\varphi_{T_0} = \eta_0\xi_0^{-1}$, φ_{T_0} is right closing so that φ_T is right closing. Since $\eta\xi^{-1}$ is right closing, η is right closing. Of course ξ is left closing.

Since ξ is a topological conjugacy, the sofic cover defined by $\mathcal{G}_T$ is a topological conjugacy. Therefore, by Lemma 7.8, $\mathcal{G}_T$ is an extension of a graphoid (a graph). Similarly $\mathcal{G}_{T_0}$ is an extension of a graphoid (a graph). Since $\varphi_T = \phi^{-1}\varphi_{T_0}\phi, \phi : (X_G, \sigma_G) \to (X_{G_0}, \sigma_{G_0})$ is relation preserving with respect to R_T^0 and $R_{T_0}^0$. Therefore, since ξ is left closing, it follows from Theorem 7.19 that $(T^{(-1)})^{[n_1]}$ is resolvable for some $n_1 \in \mathbf{N}$, and since η is right closing, it also follows from Theorem 7.19 that $(T^{-1})^{[n_2]}$ is resolvable for some $n_2 \in \mathbf{N}$. Put $n = \max\{n_1, n_2\}$. By Remark 7.4, $(T^{(-1)})^{[n]}$ and $(T^{-1})^{[n]}$ are resolvable so that $(T^{[n]})^{(-1)}$ and $(T^{[n]})^{-1}$ are resolvable.

Conversely assume that $(T^{[n]})^{(-1)}$ and $(T^{[n]})^{-1}$ are resolvable for some $n \in \mathbf{N}$. Then, by Theorem 7.20, there are an LR textile system T_0 and a topological conjugacy ϕ : $(X_T, \sigma_T) \to (X_{T_0}, \sigma_{T_0})$ which is relation preserving with respect to R_T and R_{T_0}. Since T and T_0 are nondegenerate $(X_T, \sigma_T) = (X_G, \sigma_G)$ and $(X_{T_0}, \sigma_{T_0}) = (X_{G_0}, \sigma_{G_0})$. Hence we have the topological conjugacy $\phi : (X_G, \sigma_G) \to (X_{G_0}, \sigma_{G_0})$ with $\varphi_T = \phi^{-1}\varphi_{T_0}\phi$.

Similarly (1) and (3) follow from Theorems 7.19 and 7.20. □

Recall that an endomorphism φ of a topological Markov shift is called an LR endomorphism if there is a one-sided 1-1 LR textile system such that $\varphi_T = \varphi$.

Theorem 7.22. Let $\varphi : (X_G, \sigma_G) \to (X_G, \sigma_G)$ be an onto endomorphism of a topological Markov shift. Let $k \in \mathbf{N}$. Then the following statements are valid.

(1) If there is a one-sided 1-1 and nondegenerate textile system $T_0 = (p_0, q_0 : \Gamma_0 \to G)$ such that p_0 is left resolving and $\varphi_{T_0} = \varphi^k$, then there are a one-sided 1-1 and nondegenerate textile system $T_0' = (p_0', q_0' : \Gamma_0' \to G')$ with p_0' left resolving and a topological conjugacy $\phi : (X_G, \sigma_G) \to (X_{G'}, \sigma_{G'})$ such that $\varphi = \phi^{-1}\varphi_{T_0'}\phi$.

(2) If φ^k is an LR endomorphism, then there are an LR endomorphism $\varphi' : (X_{G'}, \sigma_{G'}) \to (X_{G'}, \sigma_{G'})$ of a topological Markov shift and a topological conjugacy $\phi : (X_G, \sigma_G) \to (X_{G'}, \sigma_{G'})$ such that $\varphi = \phi^{-1}\varphi'\phi$.

(3) If there is a one-sided 1-1 and nondegenerate textile system $T_0 = (p_0, q_0 : \Gamma_0 \to G)$ such that q_0 is biresolving and $\varphi_{T_0} = \varphi^k$ then there are a one-sided 1-1 and nondegenerate textile system $T_0' = (p_0', q_0' : \Gamma_0' \to G')$ with q_0' biresolving and a topological conjugacy $\phi : (X_G, \sigma_G) \to (X_{G'}, \sigma_{G'})$ such that $\varphi = \phi^{-1}\varphi_{T_0'}\phi$.

Proof. We describe the proof of (2) in detail.

Since φ^k is an LR endomorphism, there is a textile system $T_0 = (p_0, q_0 : \Gamma_0 \to G)$ such that p_0 is left resolving, q_0 is right resolving and $\varphi_{T_0} = \varphi^k$. Let ξ_0 and η_0 be the factor maps given by p_0 and q_0, respectively. Then $\varphi_{T_0} = \eta_0\xi_0^{-1}$. Since η_0 is right closing and ξ_0^{-1} is a topological conjugacy, φ_{T_0} is right closing. Therefore, since $\varphi_{T_0} = \varphi^k$, φ

is right closing. Since φ is an onto endomorphism, there is a one-sided 1-1, nondegenerate textile system $T = (p, q : \Gamma \to G)$ with $\varphi_T = \varphi$ (Fact 2.2). Let ξ and η be the factor maps given by p and q, respectively. Then ξ is a conjugacy, η is onto and $\varphi = \eta\xi^{-1}$. Since φ is right closing, so is η. Since $\varphi_{T^k} = \varphi^k$ and φ^k is an LR endomorphism, it follows from Theorem 7.21 that there is $n \in \mathbf{N}$ such that $((T^k)^{[n]})^{-1}$ and $((T^k)^{[n]})^{(-1)}$ are resolvable. We can identify $(T^k)^{[n]}$ with $(T^{[n]})^k$ (because ξ is a conjugacy). Hence $((T^{[n]})^k)^{-1}$ and $((T^{[n]})^k)^{(-1)}$ are resolvable, so that $((T^{[n]})^{-1})^k$ and $((T^{[n]})^{(-1)})^k$ are resolvable. Put $(T^{[n]})^{-1} = (p', q' : \Gamma^{[n]} \to G^{[n]})$ and $(T^{[n]})^{(-1)} = (p'', q'' : (\Gamma^{[n]})^{-1} \to (G^{[n]})^{-1})$. Let ξ', η', ξ'', and η'' be the factor maps given by p', q', p'', and q'', respectively. Since η is right closing, so is ξ'. Since ξ is onto, so is η'. Since ξ is left closing, ξ'' is right closing. Since η is onto, so is η''. Therefore using Lemma 7.9 we know that $(T^{n]})^{-1}$ and $(T^{[n]})^{(-1)}$ are resolvable. Thus by Theorem 7.21, we get the conclusion of (2).

The proofs of (1) and (3) are similar. □

Theorem 7.23. Let $\varphi : (X_G, \sigma_G) \to (X_G, \sigma_G)$ be an onto endomorphism of a topological Markov shift. Let $k \in \mathbf{N}$. Let $\psi : (X_{G^k}, \sigma_{G^k}) \to (X_{G_0}, \sigma_{G_0})$ be a conjugacy between topological Markov shifts. Then the following statements are valid.

(1) If there is a one-sided 1-1 and nondegenerate textile system $T_0 = (p_0, q_0 : \Gamma_0 \to G_0)$ such that p_0 is left resolving and $\varphi_{T_0} = \psi\varphi^{(k)}\psi^{-1}$, then there are a one-sided 1-1 and nondegenerate textile system $T_0' = (p_0', q_0' : \Gamma_0' \to G')$ with p_0' left resolving and a topological conjugacy $\phi : (X_G, \sigma_G) \to (X_{G'}, \sigma_{G'})$ such that $\varphi = \phi^{-1}\varphi_{T_0'}\phi$.

(2) If $\psi\varphi^{(k)}\psi^{-1}$ is an LR endomorphism of (X_{G_0}, σ_{G_0}), then there are an LR endomorphism $\varphi' : (X_{G'}, \sigma_{G'}) \to (X_{G'}, \sigma_{G'})$ of a topological Markov shift and a topological conjugacy $\phi : (X_G, \sigma_G) \to (X_{G'}, \sigma_{G'})$ such that $\varphi = \phi^{-1}\varphi'\phi$.

(3) If there is a one-sided 1-1 and nondegenerate textile system $T_0 = (p_0, q_0 : \Gamma_0 \to G_0)$ such that q_0 is biresolving and $\varphi_{T_0} = \psi\varphi^{(k)}\psi^{-1}$, then there are a one-sided 1-1 and nondegenerate textile system $T_0' = (p_0', q_0' : \Gamma_0' \to G')$ with q_0' biresolving and a topological conjugacy $\phi : (X_G, \sigma_G) \to (X_{G'}, \sigma_{G'})$ such that $\varphi = \phi^{-1}\varphi'\phi$.

Proof. We only prove (2), because the proofs of (1) and (3) are similar.

There is a one-sided 1-1 and nondegenerate textile system T with $\varphi_T = \varphi$. Since $\varphi_{T^{(k)}} = \varphi^{(k)}$ and $\psi\varphi^{(k)}\psi^{-1}$ is an LR endomorphism, it follows from Theorem 7.21 that there is $n \in \mathbf{N}$ such that $((T^{(k)})^{[n]})^{-1}$ and $((T^{(k)})^{[n]})^{(-1)}$ are resolvable. It is seen that $(T^{(k)})^{[n]}$ is identified with $(T^{[m]})^{(k)}$, where $m = kn - k + 1$. Since $\varphi^{(k)}$ is right closing, so φ. Hence by a similar argument to that in the proof of Theorem 7.22, but using Lemma 7.10 instead of Lemma 7.9, we get the conclusion of (2). □

In concluding this section, we shall prove the following proposition used in the proof of Theorem 7.19 and give its applications.

Proposition 7.24. A graphoid is canonical.

Proof. Let $\mathcal{G} = (G, \lambda)$ be a graphoid. Put $(Y, \sigma) = (X_{\mathcal{G}}, \sigma_{\mathcal{G}})$. Then by symmetry, it suffices to show that $(\mathcal{K}_Y^+)_- = \mathcal{G}$. Let $A = \lambda(A_G)$. Let $\mathcal{K} = (K, \lambda_{\mathcal{K}})$ be the right resolving λ-graph defined as follows: V_K is the family of all nonempty subsets U of V_G such that there exists a left infinite sequence $\cdots a_{-2}a_{-1}a_0 \in Y_-, a_i \in A$, such that for infinitely many $i \in \mathbf{N}, S_+(V_G, a_{-i} \cdots a_0) = U$, where Y_- and S_+ are the same as in Section 4;

$$A_K = \{(U, a) \mid U \in V_K, a \in A, S_+(U, a) \neq \emptyset\};$$

for $(U, a) \in A_K$,

$$i_K((U, a)) = U, \quad t_K((U, a)) = S_+(U, a) \text{ and } \lambda_{\mathcal{K}}((U, a)) = a.$$

Since every vertex of $\mathcal{G}$ has a tracing magic word for itself, it follows that $\mathcal{K}$ is reduced. Therefore, by the construction described in Appendix of [N3], we know that $\mathcal{K} = \mathcal{K}_Y^+$. For any vertex v of $\mathcal{G}$, if w is a homing magic word for v, then $S_+(V_G, w) = \{v\}$. Hence every singleton subset of V_G is an element of V_K. If $\alpha \in A_G$, then $S_+(\{i_G(\alpha)\}, \lambda(\alpha)) = \{t_G(\alpha)\}$ because $\mathcal{G}$ is right resolving. Thus $\mathcal{G}$ is a sub-λ-graph of $\mathcal{K}$ if we identify $u \in V_G$ with $\{u\} \in V_K$ and $\alpha \in A_G$ with $(\{i_G(\alpha)\}, \lambda(\alpha)) \in A_K$. Assume that there are $U_1, \cdots, U_l, U' \in V_K$ and $x \in L(\mathcal{G})$ such that

$$S_+(U_1, x) = S_+(U_2, x) = \quad \cdots \quad = S_+(U_l, x) = U'.$$

Let $u' \in U'$. Then there is $u \in \bigcap_{i=1}^{l} U_i$ such that $S_+(\{u\}, x) = \{u'\}$, because $\mathcal{G}$ is biresolving. If y is a tracing magic word for u', then

$$S_+(U_1, xy) = S_+(U_2, xy) = \quad \cdots \quad = S_+(U_l, xy) = S_+(\{u\}, xy).$$

Therefore we know that if $\{U_1, \cdots, U_m\}$ with $U_i \in V_K$ is a maximal left-compatible set for $\mathcal{K}$, then it contains $\{u\}$ such that $u \in \bigcap_{i=1}^{m} U_i$. For $u \in V_G$, let $F(u)$ denote the family of all subsets U in V_K such that $U \ni u$. For $u \in V_G$ if w is a tracing magic word for u, then $S_+(U, w) = S_+(\{u\}, w)$ for all $U \in F(u)$. Thus for any $u \in V_G, F(u)$ is a maximal left-compatible set for $\mathcal{K}$, and every maximal left-compatible set for $\mathcal{K}$ is equal to $F(u)$ for some $u \in V_G$.

Let $u, v \in V_G$ and $a \in A_G$ with $S_+(\{u\}, a) = \{v\}$. Then it follows that for $U \in V_K, S_+(U, a) \in F(v)$ if and only if $U \in F(u)$. Therefore $F(u)$ is the a-predecessor of $F(v)$ in $\mathcal{K}_-$. Thus there is a λ-graph-isomorphism between $\mathcal{G}$ and $\mathcal{K}_-$. $\square$

Corollary 7.25. If φ is an LR endomorphism of a topological Markov shift, then a one-sided 1-1 LR textile system T with $\varphi_T = \varphi$, is unique.

Proof. Let $T = (p, q : \Gamma \to G)$ be a one-sided 1-1 LR textile system with $\varphi_T = \varphi$. Since T is LR, $\mathcal{G}_T$ is biresolving. For the same reason T is nondegenerate. Therefore, since T is one-sided 1-1, the factor map given by p is 1-1. Hence, it follows that every vertex of $\mathcal{G}_T$ has a tracing magic word and a homing magic word for itself. Thus $\mathcal{G}_T$ is a graphoid. Since a graphoid is canonical, $\mathcal{G}_T$ is uniquely determined by R_T^0, which is equal to R_T. Therefore $\mathcal{G}_T$ is uniquely determined by R_T so that T is uniquely determined by $\varphi_T = \varphi$. □

Corollary 7.26. Let $T = (p, q : \Gamma \to G)$ and $T' = (p', q' : \Gamma' \to G)$ be a one-sided 1-1 and nondegenerate textile system with $\varphi_T = \varphi_{T'}$. If both q and q' are biresolving, then $T = T'$.

Proof. The proof is similar to that of Corollary 7.25. □

In connection with Corollary 6.11, we have the following result. A related result is found in [KMT] as that of J.B. Wagoner.

Corollary 7.27. Let φ be a forward automorphism of a topological shift. Then the following statements are valid.

(1) A specified equivalence associated with φ is unique for φ.

(2) For every nonnegative integer n, a specified shift equivalence of lag n associated with φ is unique for φ and n, if any.

Proof. Since a forward automorphism is LR, (1) is a direct consequence of Corollary 7.25.

To show (2), assume that there is a specified shift equivalence

$$\tilde{M}\tilde{P} \overset{k}{\simeq} \tilde{P}\tilde{M}, \ \tilde{M}\tilde{Q} \overset{k'}{\simeq} \tilde{Q}\tilde{M}, \ \tilde{M}^n \overset{l}{\simeq} \tilde{P}\tilde{Q}, \ \tilde{Q}\tilde{P} \overset{l'}{\simeq} \tilde{M}^n$$

associated with φ. Then by (1), $\tilde{M}\tilde{P} \overset{k}{\simeq} \tilde{P}\tilde{M}$ is unique for φ and $\tilde{M}\tilde{Q} \overset{k'}{\simeq} \tilde{Q}\tilde{M}$ is unique for $\varphi^{-1}\sigma_M^n$. Therefore we know that $\begin{pmatrix} 0 & \tilde{P} \\ \tilde{Q} & 0 \end{pmatrix}$ is unique for φ and n and (2) is valid for $n = 0$. Assume that $n \geq 1$. Let G be the bipartite graph represented by $\begin{pmatrix} 0 & \tilde{P} \\ \tilde{Q} & 0 \end{pmatrix}$, let $\acute{A}_G$ be the set of the arcs represented by the symbols in $\tilde{P}$ and $\grave{A}_G$ the set of the arcs represented by the symbols in $\tilde{Q}$. Let H be the graph represented by $\tilde{M}^n$. Let $T = (p, q : \Gamma \to H)$ be the 1-1 LR textile system associated with the specified equivalence induced by the 1-step specified strong shift equivalence

$$\tilde{M}^n \overset{l}{\simeq} \tilde{P}\tilde{Q}, \ \tilde{Q}\tilde{P} \overset{l'}{\simeq} \tilde{M}^n.$$

Then, Γ, p, and q are given as follows:

$$V_\Gamma = \acute{A}_G, \quad A_\Gamma = \{rsr' \in L_3(G) \mid r, r' \in \acute{A}_G, \ s \in \grave{A}_G\}$$

and for $rsr' \in A_\Gamma$

$$i_\Gamma(rsr') = r, \quad t_\Gamma(rsr') = r', \quad p(rsr') = l^{-1}(rs), \quad q(rsr') = l'(sr').$$

Assume that there is another 1-step specified strong shift equivalence

$$\tilde{M}^n \overset{m}{\simeq} \tilde{P}\tilde{Q}, \; \tilde{Q}\tilde{P} \overset{m'}{\simeq} \tilde{M}^n$$

which gives a κ-ζ factorization of $\varphi^{(n)}$. Then if we define a textile system $\bar{T} = (\bar{p}, \bar{q} : \Gamma \to H)$ by

$$\bar{p}(rsr') = m^{-1}(rs), \quad \bar{q}(rsr') = m'(sr'), \quad rsr' \in A_\Gamma,$$

then $\bar{T}$ is 1-1 and LR and we have

$$\varphi_{\bar{T}} = \varphi^{(n)} = \varphi_T.$$

Therefore, by the proof of Corollary 7.25, $\mathcal{G}_T$ and $\mathcal{G}_{\bar{T}}$ are λ-graph-isomorphic, so that there is a graph-automorphism $g : \Gamma \to \Gamma$ which preserves the label of each arc, that is, $p(\alpha)/q(\alpha) = \bar{p}(g(\alpha))/\bar{q}(g(\alpha))$ for $\alpha \in A_\Gamma$. Therefore if $g(rsr') = r_1 s_1 r_1'$ with $rsr' \in A_\Gamma$, then

$$r_1 s_1 = m(l^{-1}(rs)) \quad \text{and} \quad s_1 r_1' = (m')^{-1} l'(sr').$$

Let $\hat{g} : A_\Gamma \to \grave{A}_G$ be defined by

$$\hat{g}(rsr') = s_1, \quad rsr' \in A_\Gamma,$$

where s_1 is the arc of $\grave{A}_G$ such that $g(rsr') = r_1 s_1 r_1'$. If $rsr', \bar{r}s\bar{r}' \in A_\Gamma$, then $rs\bar{r}' \in A_\Gamma$ and hence, by the above,

$$\hat{g}(rsr') = \hat{g}(rs\bar{r}') = \hat{g}(\bar{r}s\bar{r}').$$

Thus we can define a graph-automorphism $h : G \to G$ as follows: if $r \in \acute{A}_G$, then $h(r) = g_V(r)$, and if $s \in \grave{A}_G$, then $h(s) = \hat{g}(rsr')$ for $r, r' \in \acute{A}_G$ with $rsr' \in A_\Gamma$. (It is clear that h is a graph-homomorphism and onto.) By the above, we have

$$l^{-1}(rs) = m^{-1}(h(r)h(s))$$

for any $rs \in L_2(G)$ with $r \in \acute{A}_G$ and $s \in \grave{A}_G$ and we also have

$$l'(sr) = m'(h(s)h(r))$$

for any $sr \in L_2(G)$ with $s \in \grave{A}_G$ and $r \in \acute{A}_G$. Therefore it follows that h fixes the vertices i.e., $h_V = \mathrm{id}_{V_G}$ and that if we rename the arcs of G according to h, then the specified shift equivalence

$$\tilde{M}\tilde{P} \overset{k}{\simeq} \tilde{P}\tilde{M}, \; \tilde{M}\tilde{Q} \overset{k'}{\simeq} \tilde{Q}\tilde{M}, \; \tilde{M}^n \overset{l}{\simeq} \tilde{P}\tilde{Q}, \; \tilde{Q}\tilde{P} \overset{l'}{\simeq} \tilde{M}^n$$

becomes the same as the specified shift equivalence

$$\tilde{M}\tilde{P} \overset{k}{\simeq} \tilde{P}\tilde{M}, \; \tilde{M}\tilde{Q} \overset{k'}{\simeq} \tilde{Q}\tilde{M}, \; \tilde{M}^n \overset{m}{\simeq} \tilde{P}\tilde{Q}, \; \tilde{Q}\tilde{P} \overset{m'}{\simeq} \tilde{M}^n.$$

Thus both the specified shift equivalences are the same up to a permutation of the symbols in $\tilde{P}$ and $\tilde{Q}$ which leaves $\tilde{P}$ and $\tilde{Q}$ unchanged. $\square$

It is clear from the proofs of Corollaries 7.25 and 7.26 that these results are generalized to factor maps with respect to textile relation systems (see the last paragraph of Section 6). It is also clear from the proof of Corollary 7.27 that Corollary 7.27 is generalized to forward conjugacies between topological Markov shifts, like Corollary 6.11 is generalized to forward conjugacies between topological Markov shifts (see the last paragraph of Section 6).

8. ESSENTIALLY LR AUTOMORPHISMS

Let P be a property of endomorphisms or automorphisms of subshifts. We say that an endomorphism or automorphism φ of a dynamical system (X, σ) *essentially have* the property P if there is a subshift (X', σ') and a topological conjugacy ψ : $(X, \sigma) \to (X', \sigma')$ such that $\psi\varphi\psi^{-1}$ have the property P. For example, if φ is an automorphism of a dynamical system (X, σ) and there are a subshift (X', σ') and a conjugacy $\psi : (X, \sigma) \to (X', \sigma')$ such that $\psi\varphi\psi^{-1}$ is a forward automorphism of (X', σ'), then φ is an *essentially forward automorphism* of (X, σ).

By Theorem 7.22, we know that if φ is an endomorphism of a topological Markov shift (X, σ) and φ^k is an essentially LR endomorphism, then φ is an essentially LR endomorphism of (X, σ). By Theorem 7.23 we also know that if φ is an endomorphism of a topological Markov shift (X, σ) and $\varphi^{(k)}$ is an essentially LR endomorphism of $(X^{(k)}, \sigma^{(k)})$, then φ is an essentially LR endomorphism of (X, σ). Therefore by Theorem 6.29 we have the following theorem.

Theorem 8.1. Let φ be an automorphism of a topological Markov shift (X, σ) and let $k \in \mathbf{N}$. Then the following statements are valid.

(1) If φ^k is an essentially forward automorphism of (X, σ), then so is φ.

(2) If $\varphi^{(k)}$ is an essentially forward automorphism of $(X^{(k)}, \sigma^{(k)})$, then φ is an essentially forward automorphism of (X, σ).

The result that an automorphism of finite order of a subshift is an essentially symbolic automorphism, appears in [BLR] being attributed to J. Franks. For completeness, we give a proof.

Proposition 8.2. If an automorphism of a subshift has finite order, then it is an essentially symbolic automorphism.

Proof. Let (X, σ) be a subshift over an alphabet A. Let φ be an automorphism of (X, σ). Assume that $\varphi^k = \mathrm{id}_X$ with $k \in \mathbf{N}$. If $k = 1$, then the proposition is trivial. Assume that $k \geq 2$. Let B be the set of all elements of the form

$$\downarrow a_{1j} \cdots a_{kj}$$

with $a_{ij} \in A$, and let X' be the set of all bisequences over B of the form

$$(\downarrow a_{1j} \cdots a_{kj})_{j \in \mathbf{Z}},$$

where $a_{ij} \in A$, $(a_{ij})_{j\in\mathbf{Z}} \in X$ and

$$\varphi((a_{ij})_{j\in\mathbf{Z}}) = (a_{i+1,j})_{j\in\mathbf{Z}} \quad \text{for} \quad i = 1, \cdots, k-1.$$

Let $\psi : X \to X'$ be defined by

$$\psi((a_j)_{j\in\mathbf{Z}}) = (\downarrow a_{1j} \cdots a_{kj})_{j\in\mathbf{Z}}, \quad (a_j)_{j\in\mathbf{Z}} \in X,$$

where $(a_{1j})_{j\in\mathbf{Z}} = (a_j)_{j\in\mathbf{Z}}$ and $\varphi((a_{ij})_{j\in\mathbf{Z}}) = (a_{i+1,j})_{j\in\mathbf{Z}}$ for $i = 1, \cdots k-1$, with $a_j, a_{ij} \in A$. Clearly ψ is a topological conjugacy of (X, σ) onto (X', σ'). Let $f : L_1(X') \to L_1(X')$ be defined by

$$f(\downarrow a_1 \cdots a_k) = \downarrow a_2 \cdots a_k a_1, \quad \downarrow a_1 \cdots a_k \in L_1(X'), \ a_i \in A.$$

Then f gives an automorphism φ' of (X', σ') such that $\psi\varphi = \varphi'\psi$. Thus φ is an essentially symbolic automorphism of (X, σ). □

By the proposition above, Corollary 6.12, Theorem 6.23, and Theorem 6.10, we have the following proposition.

Proposition 8.3. Let (X, σ) be a subshift with $h(\sigma) > 0$. Let φ be an automorphism of finite order of (X, σ). Let $k, n \in \mathbf{Z}$. Then $(X, \varphi^k\sigma^n)$ is expansive if and only if $n \neq 0$. If (X, σ) is a sofic system and $n \neq 0$, then $(X, \varphi^k\sigma^n)$ is topologically conjugate to a sofic system. If (X, σ) is a topological Markov shift and $n \neq 0$, then $(X, \varphi^k\sigma^n)$ is topologically conjugate to a topological Markov shift.

Problem 8.a. For a subshift (X, σ) and an automorphism of φ of (X, σ), classify the dynamical systems $(X, \varphi^k\sigma^n)$, $k, n \in \mathbf{Z}$.

The preceding proposition gives a solution of this problem for the case when φ has finite order. Throughout the remainder of this section, we study further the problem for the case when (X, σ) is a topological Markov shift.

Let $T = (p, q : \Gamma \to G)$ be an LR textile system. Then T^k is LR for all $k \in \mathbf{N}$. (Recall that T^k denote the kth composition-power of T.) It follows from Lemma 6.25 that T is 1-1 if and only if p is a definite left resolving graph-homomorphism and q is a definite right resolving graph-homomorphism. For an integer $n \geq 0$, let us say that T is n *definite* if both p and q are n definite. Let H be the graph such that T^* is over H, that is, $H = G^T$. By definition it is clear that for $n \geq 1$ and $k \geq 1$, T^k is n definite if and only if there are mappings $f : L_n(G) \to L_k(H)$ and $g : L_n(G) \to L_k(H)$ such that for any cloth $C = (\alpha_{ij})_{1\leq i\leq k, 1\leq j\leq n}$, $\alpha_{ij} \in A_\Gamma$, weaved by T, if the upper side of C is x, then the left side of C is $f(x)$, and if the lower side of C is x', then the right side of C is $g(x)$. Since T is LR, T^* is also LR.

Lemma 8.4. Let $T = (p, q : \Gamma \to G)$ be a 1-1 LR textile system. Let $k, n \in \mathbf{N}$. If T^k is n definite, then $\sigma_T^n \varphi_T^{-k}$ is an essentially LR automorphism of $(X_T, \sigma_T) = (X_G, \sigma_G)$.

Proof. Let H be the graph such that T^* is over H. Since T^k is n definite, there are mappings $f : L_n(G) \to L_k(H)$ and $g : L_n(G) \to L_k(H)$ such that for any cloth $C = (\alpha_{ij})_{1 \le i \le k, 1 \le j \le n}, \alpha_{ij} \in A_\Gamma$, weaved by T such that if the upper side of C is x, then the left side of C is $f(x)$, and if the lower side of C is x', then the right side of C is $g(x')$. Let $T_0 = (T^k)^{(n)}$ with $T_0 = (p_0, q_0 : \Gamma_0 \to G_0)$ (recall that $T^{(n)}$ denotes the nth product-power of T). Then T_0 is 1-1 and LR. Every $\gamma \in A_{\Gamma_0}$ is a cloth weaved by T and written in the form $\gamma = (\alpha_{ij})_{1 \le i \le k, 1 \le j \le n}, \alpha_{ij} \in A_\Gamma$. Clearly $i_{\Gamma_0}(\gamma), t_{\Gamma_0}(\gamma), (p_0)_A(\gamma)$ and $(q_0)_A(\gamma)$ are the left, right, upper, and lower sides of the cloth γ. We define a textile system

$$T_1 = (p_1, q_1 : G_0^{[2]} \to \Gamma_0)$$

as follows: for $xx' \in A_{G_0^{[2]}} = L_2(G_0)$ with $x, x' \in A_{G_0}, (p_1)_A(xx')$ is the arc γ in A_{Γ_0} such that

$$(p_0)_A(\gamma) = x \quad \text{and} \quad t_{\Gamma_0}(\gamma) = f(x'),$$

and $(q_1)_A(xx')$ is the arc γ' in A_{Γ_0} such that

$$i_{\Gamma_0}(\gamma') = g(x) \quad \text{and} \quad (q_0)_A(\gamma') = x'.$$

Since T_0 is LR, these γ and γ' exist and unique.

To see that T_1 is LR, let $x \in A_{G_0} = V_{G_0^{[2]}}$. Then it is seen that $(p_1)_V(x) = f(x)$. Let $\gamma \in A_{\Gamma_0}$ such that $t_{\Gamma_0}(\gamma) = f(x)$. Let $w = (p_0)_A(\gamma)$. Then $wx \in A_{G_0^{[2]}}$ and $(p_1)_A(wx) = \gamma$. If $w'x \in A_{G_0^{[2]}}$ with $w' \in A_{G_0}$ and $(p_1)_A(w'x) = \gamma$, then $w' = w$. Thus p_1 is left resolving. It is also seen that $(q_1)_V(x) = g(x)$. Let $\gamma' \in A_{\Gamma_0}$ such that $i_{\Gamma_0}(\gamma') = g(x)$. Let $y = (q_0)_A(\gamma')$. Then $xy \in A_{G_0^{[2]}}$ and $(q_1)_A(xy) = \gamma'$. If $xy' \in A_{G_0^{[2]}}$ with $y' \in A_{G_0}$ and $(q_1)_A(xy') = \gamma'$, then $y' = y$. Thus q_1 is right resolving.

Let ξ_0, η_0, ξ_1 and η_1 be the factor maps given by p_0, q_0, p_1 and q_1, respectively. Since T_0 is 1-1 and LR, ξ_0 and η_0 are 1-1 and onto. Let $(x_j)_{j \in \mathbf{Z}} \in X_{G_0}$ with $x_j \in A_{G_0}$. Put

$$\xi_0^{-1}((x_j)_{j \in \mathbf{Z}}) = (\gamma_j)_{j \in \mathbf{Z}}, \quad \gamma_j \in A_{\Gamma_0}$$

and

$$\eta_0^{-1}((x_j)_{j \in \mathbf{Z}}) = (\gamma'_j)_{j \in \mathbf{Z}}, \quad \gamma'_j \in A_{\Gamma_0}.$$

Then we have

$$(p_1)_A(x_j x_{j+1}) = \gamma_j$$

and

$$(q_1)_A(x_j x_{j+1}) = \gamma'_{j+1}$$

for all $j \in \mathbf{Z}$. Therefore it follows that

$$\xi_1 \rho_{X_{G_0},2} = \xi_0^{-1}$$

and

$$\eta_1 \rho_{X_{G_0},2} = \sigma_{\Gamma_0} \eta_0^{-1}.$$

Hence ξ_1 and η_1 are 1-1 and onto. We have

$$\begin{aligned} \varphi_{T_1} = \eta_1 \xi_1^{-1} &= \eta_1 \rho_{X_{G_0},2}(\xi_1 \rho_{X_{G_0},2})^{-1} \\ &= \sigma_{\Gamma_0} \eta_0^{-1} \xi_0 = \eta_0^{-1}(\sigma_{G_0} \xi_0 \eta_0^{-1}) \eta_0 \\ &= \eta_0^{-1}(\sigma_{G_0} \varphi_{T_0}^{-1}) \eta_0 \end{aligned}$$

Therefore, $\sigma_{G_0}\varphi_{T_0}^{-1}$ is an essentially LR automorphism of (X_{G_0}, σ_{G_0}). Therefore since $(\sigma_G^n \varphi_T^{-k})^{(n)} = \sigma_{G_0}\varphi_{T_0}^{-1}$, $\sigma_G^n \varphi_T^{-k}$ is an essentially LR automorphism of (X_G, σ_G), by Theorem 7.23. □

Lemma 8.5. Let T be a 1-1 LR textile system over a graph G. Let $k, n \in \mathbf{N}$. If $\varphi_T^{-k}\sigma_T^n$ is an LR automorphism of $(X_T, \sigma_T) = (X_G, \sigma_G)$, then T^k is n definite.

Proof. Since $\sigma_T^n \varphi_T^{-k}$ is an LR automorphism of (X_G, σ_G), there is a 1-1 LR textile system T' over G such that $\varphi_{T'} = \sigma_T^n \varphi_T^{-k}$. Let

$$T_1 = (T^k \circ T')^{(n)}$$

and let

$$T_2 = (T' \circ T^k)^{(n)}.$$

Then T_1 and T_2 are 1-1 LR textile systems, by Corollary 3.17. We have

$$\varphi_{T_1} = (\varphi_T^k(\sigma_T^n \varphi_T^{-k}))^{(n)} = \sigma_{G^n}$$

and we also have

$$\varphi_{T_2} = \sigma_{G^n}.$$

Let T_σ be the textile system such that

$$T_\sigma = (p_\sigma, q_\sigma : G^{[2]} \to G),$$

where p_σ and q_σ are defined by

$$p_\sigma(ab) = a \quad \text{and} \quad q_\sigma(ab) = b, \quad ab \in A_{G^{[2]}} = L_2(G), \quad a, b \in A_G.$$

Then T_σ is 1-1 and LR and $\varphi_{T_\sigma} = \sigma_G$. Therefore $((T_\sigma)^{(n)})^n$ is 1-1 and LR, and we have $\varphi_{((T_\sigma)^{(n)})^n} = \sigma_{G^n}$.

Therefore, by Corollary 7.25, we have

$$T_1 = T_2 = ((T_\sigma)^{(n)})^n.$$

Note that if we let

$$((T_\sigma)^{(n)})^n = (\tilde{p}, \tilde{q} : (G^n)^{[2]} \to G^n),$$

then for $w = (a_1 \cdots a_n)(a_{n+1} \cdots a_{2n}) \in A_{(G^n)^{[2]}}$ with $a_i \in A_G$,

$$\tilde{p}_A(w) = i_{(G^n)^{[2]}}(w) = a_1 \cdots a_n$$

and

$$\tilde{q}_A(w) = t_{(G^n)^{[2]}}(w) = a_{n+1} \cdots a_{2n}.$$

Let $T_1 = (p_1, q_1 : \Gamma_1 \to G^n)$. Then $V_{\Gamma_1} = A_{H^k K}$, where H is the graph such that T^* is over H and K is the graph such that $(T')^*$ is over K. Since $T_1 = ((T_\sigma)^{(n)})^n$, it follows from the above that for $\alpha \in A_{\Gamma_1}$,

$$p_1(\alpha) = a_1 \cdots a_n \quad \text{with } a_i \in A_G$$

and

$$i_{\Gamma_1}(\alpha) = b_1 \cdots b_k c \quad \text{with } b_j \in A_H \text{ and } c \in A_K$$

uniquely correspond to each other. This implies that if we let

$$(T^k)^{(n)} = (\hat{p}, \hat{q} : \hat{\Gamma} \to G^n),$$

then for $\hat{\alpha} \in A_{\hat{\Gamma}}$, $\hat{p}(\hat{\alpha})$ in A_{G^n} uniquely determines $i_{\hat{\Gamma}}(\hat{\alpha})$ in A_{H^k}. A similar argument for T_2 shows that for $\hat{\alpha} \in A_{\hat{\Gamma}}$, $\hat{q}(\hat{\alpha})$ in A_{G^n} uniquely determines $t_{\hat{\Gamma}}(\hat{\alpha})$ in A_{H^k}. Thus T^k is n definite. □

Lemma 8.6. Let $T = (p, q : \Gamma \to G)$ be a 1-1 LR textile system with T^* 1-1. Let $k, n \in \mathbf{N}$. If $(T^*)^n$ is k definite, then $\varphi_T^k \sigma_T^{-n}$ is an essentially LR automorphism of $(X_T, \sigma_T) = (X_G, \sigma_G)$.

Proof. Let H be the graph such that T^* is over H. Since $(T^*)^n$ is k definite, there are mappings $f : L_k(H) \to L_n(G)$ and $g : L_k(H) \to L_n(G)$ such that for any cloth $C = (\alpha_{ij})_{1 \le i \le k, 1 \le j \le n}, \alpha_{ij} \in A_\Gamma$, weaved by T, if the left side of C is $\downarrow w, w \in L_k(H)$, then the upper side of C is $f(w)$, and if the right side of C is $\downarrow w', w' \in L_k(H)$, then the lower side of C is $g(w')$.

Let $T_0 = (T^k)^{(n)}$ with $T_0 = (p_0, q_0 : \Gamma_0 \to G_0)$. Then T_0 is 1-1 and LR. Every $\gamma \in A_{\Gamma_0}$ is a cloth weaved by T and written in the form $\gamma = (\alpha_{ij})_{1 \le i \le k, 1 \le j \le n}, \alpha_{ij} \in A_\Gamma$. Clearly $i_{\Gamma_0}(\gamma), t_{\Gamma_0}(\gamma), p_0(\gamma)$ and $q_0(\gamma)$ are the left, right, upper and lower sides of the cloth γ. We define a textile system

$$\bar{T}_0 = (\bar{p}_0, \bar{q}_0 : \bar{\Gamma}_0 \to \bar{G}_0)$$

as follows: $\bar{G}_0 = G_0^{[2]}$. $A_{\bar{\Gamma}_0}$ and $V_{\bar{\Gamma}_0}$ are copies of A_{Γ_0} and V_{Γ_0} such that

$$A_{\bar{\Gamma}_0} = \{\bar{\gamma} \mid \gamma \in A_{\Gamma_0}\}$$

and

$$V_{\bar{\Gamma}_0} = \{\bar{w} \mid w \in V_{\Gamma_0}\}.$$

For $\bar{\gamma} \in A_{\bar{\Gamma}_0}$ with $\gamma \in A_{\Gamma_0}$, we define

$$\begin{aligned} i_{\bar{\Gamma}_0}(\bar{\gamma}) &= \overline{i_{\Gamma_0}(\gamma)}, \quad t_{\bar{\Gamma}_0}(\bar{\gamma}) = \overline{t_{\Gamma_0}(\gamma)} \\ (\bar{p}_0)_A(\bar{\gamma}) &= (p_0)_A(\gamma)f(t_{\Gamma_0}(\gamma)) \\ (\bar{q}_0)_A(\bar{\gamma}) &= g(i_{\Gamma_0}(\gamma))(q_0)_A(\gamma). \end{aligned}$$

Note that for $\bar{w} \in V_{\bar{\Gamma}_0}$ with $w \in V_{\Gamma_0}$,

$$(\bar{p}_0)_V(\bar{w}) = f(w) \quad \text{and} \quad (\bar{q}_0)_V(\bar{w}) = g(w).$$

We shall show that $\bar{T}_0$ is LR. Let $\bar{w} \in V_{\bar{\Gamma}_0}$ with $w \in V_{\Gamma_0}$. Then $(\bar{p}_0)_V(\bar{w}) = f(w)$. Let $x \in A_{G_0}$ such that $xf(w) \in L_2(G_0)$. Since w and $f(w)$ have the same initial vertex, we have $(p_0)_V(w) = t_{G_0}(x)$. Since p_0 is left resolving, there is unique $\gamma \in A_{\Gamma_0}$ such that $t_{\Gamma_0}(\gamma) = w$ and $(p_0)_A(\gamma) = x$, and hence

$$(\bar{p}_0)_A(\bar{\gamma}) = (p_0)_A(\gamma)f(t_{\Gamma_0}(\gamma)) = xf(w)$$

and

$$t_{\bar{\Gamma}_0}(\bar{\gamma}) = \bar{w}.$$

Thus $\bar{p}_0$ is left resolving. Similarly since q_0 is right resolving, it is seen that $\bar{q}_0$ is right resolving.

Let ξ_0 and η_0 be the factor maps given by p_0 and q_0, respectively, and let $\bar{\xi}_0$ and $\bar{\eta}_0$ be the factor maps given by $\bar{p}_0$ and $\bar{q}_0$, respectively. Let $(\bar{\gamma}_j)_{j\in\mathbf{Z}} \in X_{\bar{\Gamma}_0}$ with $\gamma_j \in A_{\Gamma_0}$. Then $(\gamma_j)_{j\in\mathbf{Z}} \in X_{\Gamma_0}$. We have

$$\begin{aligned} \xi_0((\gamma_j)_{j\in\mathbf{Z}}) &= (p_0(\gamma_j))_{j\in\mathbf{Z}} \\ \bar{\xi}_0((\bar{\gamma}_j)_{j\in\mathbf{Z}}) &= (p_0(\gamma_j)f(t_{\Gamma_0}(\gamma_j)))_{j\in\mathbf{Z}} \\ &= (p_0(\gamma_j)p_0(\gamma_{j+1}))_{j\in\mathbf{Z}} \\ \eta_0((\gamma_j)_{j\in\mathbf{Z}}) &= (q_0(\gamma_j))_{j\in\mathbf{Z}} \\ \bar{\eta}_0((\bar{\gamma}_j)_{j\in\mathbf{Z}}) &= (g(i_{\Gamma_0}(\gamma_j))q_0(\gamma_j))_{j\in\mathbf{Z}} \\ &= (q_0(\gamma_{j-1})q_0(\gamma_j))_{j\in\mathbf{Z}}. \end{aligned}$$

Since T_0 is 1-1, ξ_0 and η_0 are 1-1 so that $\bar{\xi}_0$ and $\bar{\eta}_0$ are 1-1. Therefore $\bar{T}_0$ is 1-1. Let $\rho = \rho_{X_{G_0},2}$. Then

$$\varphi_{\bar{T}_0} = \rho\varphi_{T_0}\sigma_{G_0}^{-1}\rho^{-1}.$$

Thus, $\varphi_{T_0}\sigma_{G_0}^{-1}$ is an essentially LR automorphism of $(X_{G_0}, \sigma_{G_0}) = (X_T^{(n)}, \sigma_T^{(n)})$. Since $\varphi_{T_0}\sigma_{G_0}^{-1} = (\varphi_T^k\sigma_T^{-n})^{(n)}$, $\varphi_T^k\sigma_T^{-n}$ is an essentially LR automorphism of (X_T, σ_T), by Theorem 7.23. □

Lemma 8.7. Let T be a 1-1 LR textile system over a graph G. Let $k, n \in \mathbf{N}$. If $\varphi_T^k \sigma_T^{-n}$ is an LR automorphism of $(X_T, \sigma_T) = (X_G, \sigma_G)$, then $(T^*)^n$ is k definite.

Proof. Since $\varphi_T^k \sigma_T^{-n}$ is an LR automorphism of $(X_T, \sigma_T) = (X_G, \sigma_G)$, there is a 1-1 LR automorphism T' over G such that $\varphi_{T'} = \varphi_T^k \sigma_T^{-n}$. Let $T_1 = (T')^{(n)}$ with $T_1 = (p_1, q_1 : \Gamma_1 \to G^n)$. Then T_1 is a 1-1 LR textile system with $\varphi_{T_1} = (\varphi_T^k \sigma_T^{-n})^{(n)}$. Let $T_0 = (T_\sigma^n)^{(n)}$, where T_σ is the same as in the proof of Lemma 8.5. Then $\varphi_{T_0} = (\sigma_T^n)^{(n)}$. Let

$$T_2 = T_0 \circ T_1$$

and let

$$T_2' = T_1 \circ T_0.$$

Then T_2 and T_2' are 1-1 and LR, by Corollary 3.17. We have $\varphi_{T_2} = \varphi_{T_1} \varphi_{T_0} = (\varphi_T^k)^{(n)}$ and $\varphi_{T_2'} = (\varphi_T^k)^{(n)}$. Let $\bar{T} = (T^k)^{(n)}$ with $\bar{T} = (\bar{p}, \bar{q} : \bar{\Gamma} \to G^n)$. Then $\bar{T}$ is 1-1 and LR and $\varphi_{\bar{T}} = (\varphi_T^k)^{(n)}$. Thus, by Corollary 7.25, we have

$$\bar{T} = T_2 = T_2'.$$

Let $T_0 = (p_0, q_0 : \Gamma_0 \to G^n)$. Then for $\gamma \in A_{\Gamma_0}$, $p_0(\gamma) = i_{\Gamma_0}(\gamma)$ and $q_0(\gamma) = t_{\Gamma_0}(\gamma)$. Since $\bar{T} = T_2 = T_0 \circ T_1$, each arc in $A_{\bar{\Gamma}}$ is written in the form

$$\downarrow (\gamma, \delta),$$

where $\gamma \in A_{\Gamma_0}$ and $\delta \in A_{\Gamma_1}$ with $q_0(\gamma) = p_1(\delta)$. We have

$$\bar{p}(\downarrow (\gamma, \delta)) = p_0(\gamma)$$

and

$$i_{\bar{\Gamma}}(\downarrow (\gamma, \delta)) = \downarrow (i_{\Gamma_0}(\gamma), i_{\Gamma_1}(\delta)) = \downarrow (p_0(\gamma), i_{\Gamma_1}(\delta)).$$

Therefore, it follows that for every $\beta \in A_{\bar{\Gamma}}, \bar{p}(\beta)$ is uniquely determined by $i_{\bar{\Gamma}}(\beta)$. A similar argument for T_2' shows that for every $\beta \in A_{\bar{\Gamma}}, \bar{q}(\beta)$ is uniquely determined by $t_{\bar{\Gamma}}(\beta)$. Thus, since $\bar{T} = (T^k)^{(n)}$, we conclude that $(T^*)^n$ is k definite. □

Proposition 8.8. If φ is an essentially LR expansive automorphism of topological Markov shift (X, σ), then there is $m \in \mathbf{N}$ such that for all $n \in \mathbf{N}$ with $n \geq m$, φ^n is an LR automorphism of (X, σ).

Proof. Assume that φ is an essentially LR automorphism of (X, σ). Then there is a topological Markov shift (X', σ') and a topological conjugacy $\psi : (X, \sigma) \to (X', \sigma')$ such that $\psi\varphi\psi^{-1}$ is an LR automorphism of (X', σ'). Put $\psi\varphi\psi^{-1} = \varphi'$. Then there is a 1-1 LR textile system T such that $\varphi_T = \varphi'$. Since φ' is expansive, T^* is 1-1, by Theorem 2.5. Therefore T^* is k definite for some $k \in \mathbf{N}$. Thus, by Lemma 8.6, $(\varphi')^k(\sigma')^{-1}$ is essentially LR automorphism of (X', σ'). There are a topological Markov shift (X'', σ'')

and a topological conjugacy $\psi' : (X', \sigma') \to (X'', \sigma'')$ such that $\psi'(\varphi')^k(\sigma')^{-1}(\psi')^{-1}$ is an LR automorphism of (X'', σ''). Put

$$\psi'(\varphi')^k(\sigma')^{-1}(\psi')^{-1} = \varphi''.$$

We may assume that ψ' is a forward conjugacy. Assume that ψ' is of lag l' with $l' \geq 0$. Then $(\psi')^{-1}(\sigma'')^{l'}$ is a forward conjugacy. Since

$$(\varphi')^k = (\psi')^{-1}\varphi''\psi'\sigma' = (\psi')^{-1}\sigma''\varphi''\psi',$$

we have

$$\begin{aligned}(\varphi')^{k(l'+1)} &= (\psi')^{-1}(\sigma'')^{l'+1}(\varphi'')^{l'+1}\psi' \\ &= \sigma'\tilde{\varphi},\end{aligned}$$

where $\tilde{\varphi} = (\psi')^{-1}(\sigma'')^{l'}(\varphi'')^{l'+1}\psi'$. It follows that $\tilde{\varphi}$ is a forward automorphism of (X', σ'), since φ'' is a forward automorphism by Theorem 6.29. We may assume that ψ is a forward conjugacy. Assume that ψ is of lag l with $l \geq 0$. Then $\psi^{-1}(\sigma')^l$ is a forward conjugacy. Let

$$m = k(l'+1)l$$

and let i be any nonnegative integer. Then

$$\begin{aligned}\varphi^{m+i} &= \psi^{-1}(\varphi')^{m+i}\psi = \psi^{-1}(\varphi')^{k(l'+1)l}(\varphi')^i\psi \\ &= \psi^{-1}(\sigma')^l(\tilde{\varphi})^l(\varphi')^i\psi,\end{aligned}$$

which is a forward automorphism of (X, σ). Thus, again by Theorem 6.29, we conclude that φ^{m+i} is an LR automorphism of (X, σ). □

Let T be a 1-1 LR textile system with $h(\sigma_T) > 0$. Then T^k is a 1-1 LR textile system for all $k \in \mathbf{N}$. Hence for all $k \in \mathbf{N}$, there is $n \geq 0$ such that T^k is n definite. We can see that for integers $k, l > 0$ and $m, n \geq 0$, if T^k is m definite and T^l is n definite, then T^{k+l} is $m+n$ definite.

To see this, let $T = (p, q : \Gamma \to G)$ and consider a cloth $C = (\alpha_{ij})_{1 \leq i \leq k+l, 1 \leq j \leq m+n}$ weaved by T. Since T^k is m definite, $p(\alpha_{1,n+1}) \cdots p(\alpha_{1,m+n})$ uniquely determines $\downarrow i_\Gamma(\alpha_{1,n+1}) \cdots i_\Gamma(\alpha_{k,n+1}) = \downarrow t_\Gamma(\alpha_{1,n}) \cdots t_\Gamma(\alpha_{k,n})$. This and $p(\alpha_{11}) \cdots p(\alpha_{1n})$ uniquely determine the subcloth $\bar{C} = (\alpha_{ij})_{1 \leq i \leq k, 1 \leq j \leq n}$ of C, because p is left resolving. Hence the upper side of C, or $p(\alpha_{11}) \cdots p(\alpha_{1,m+n})$, uniquely determines the left and lower sides of $\bar{C}$. Since T^l is n definite, the lower side of $\bar{C}$ uniquely determines $\downarrow i_\Gamma(\alpha_{k+1,1}) \cdots i_\Gamma(\alpha_{k+l,1})$. Since this and the left side of $\bar{C}$ constitute the left side of C, the left side of C is uniquely determined by the upper side of C. An analogous argument shows that the right side of C is uniquely determined by the lower side of C.

For $k \in \mathbf{N}$, let n_k be the nonnegative integer such that T^k is *strictly* n_k *definite*, that is, T^k is n_k definite but not $n_k - 1$ definite. Then by the above, we have

$$n_k + n_l \geq n_{k+l} \quad \text{for all } k, l \in \mathbf{N}.$$

Therefore $\lim_{k\to\infty} n_k/k$ exists and equals $\inf_k n_k/k$ (see, e.g., Theorem 4.9 of [Wal]). We define

$$\kappa(T) = \lim_{k\to\infty} \frac{n_k}{k}.$$

We also define $\bar{\kappa}(T)$ as follows: if T^* is not 1-1 (i.e., φ_T is not expansive), then

$$\bar{\kappa}(T) = 0;$$

if T^* is 1-1 (i.e., φ_T is expansive), then

$$\bar{\kappa}(T) = \frac{1}{\kappa(T^*)},$$

that is, if for $n \in \mathbf{N}$, k_n is the nonnegative integer such that $(T^*)^n$ is strictly k_n definite, then

$$\bar{\kappa}(T) = \lim_{n\to\infty} \frac{n}{k_n} = \sup_n \frac{n}{k_n}.$$

We note that $k_n \neq 0$ for $n \in \mathbf{N}$. For otherwise, $\varphi_{(T^*)^n}$ would be a symbolic automorphism of (X_{T^*}, σ_{T^*}), which is impossible because $h(\varphi_{(T^*)^n}) = h(\sigma_T)^n > 0$ by our hypothesis.

Proposition 8.9. Let T be a 1-1 LR textile system with $h(\sigma_T) > 0$. Then the following statements are valid:

(1) for integers $k, n > 0$, $\varphi_T^{-k}\sigma_T^n$ is an essentially LR expansive automorphism of (X_T, σ_T) if and only if $n/k > \kappa(T)$;

(2) if $\varphi_T^{-k}\sigma_T^n$ is an essentially LR nonexpansive automorphism of (X_T, σ_T) with integers $k, n > 0$, then $\kappa(T) = n/k$;

(3) $\varphi_T^k\sigma_T^n$ is an expansive LR automorphism of (X_T, σ_T) for all integers $k \geq 0$ and $n > 0$;

(4) for integers $k > 0$ and $n \geq 0$, $\varphi_T^k\sigma_T^{-n}$ is an essentially LR expansive automorphism of (X_T, σ_T) if and only if $n/k < \bar{\kappa}(T)$;

(5) if $\varphi_T^k\sigma_T^{-n}$ is an essentially LR nonexpansive automorphism of (X_T, σ_T) with integers $k > 0$ and $n \geq 0$, then $\bar{\kappa}(T) = n/k$.

Proof. (1) Let $k, n \in \mathbf{N}$. Assume that $n/k > \kappa(T)$. Then there is $l \in \mathbf{N}$ such that $n/k > n_l/l$, where n_l is the nonnegative integer such that T^l is strictly n_l definite. Since T^{lk} is $n_l k$ definite, it follows from Lemma 8.4 that $\varphi_T^{-lk}\sigma_T^{n_l k}$ is an essentially LR automorphism of (X_T, σ_T). Therefore, since

$$\varphi_T^{-lk}\sigma_T^{ln} = \varphi_T^{-lk}\sigma_T^{n_l k}\sigma_T^{nl-kn_l}$$

with $nl - kn_l > 0$, $\varphi_T^{-lk}\sigma_T^{ln}$ is expansive, by Corollary 6.5, and it is an essentially LR automorphism of (X_T, σ_T), by Corollary 3.18. Therefore, by Theorem 7.22(2), $\varphi_T^{-k}\sigma_T^n$ is an essentially LR expansive automorphism of (X_T, σ_T).

Conversely assume that $\varphi_T^{-k}\sigma_T^n$ is an essentially LR expansive automorphism of (X_T, σ_T). Then by Proposition 8.8, there is $l \in \mathbf{N}$ such that $\varphi_T^{-kl}\sigma_T^{nl}$ is an LR automorphism of (X_T, σ_T). Let $\tilde{T}$ be the 1-1 LR textile system with $\varphi_{\tilde{T}} = \varphi_T^{-kl}\sigma_T^{nl}$. Since $\varphi_{\tilde{T}}$ is expansive, $\tilde{T}^*$ is 1-1 (by Lemma 2.5) so that there is $m \in \mathbf{N}$ such that $(\tilde{T}^*)^2$ is m definite. Hence, by Lemma 8.6, $\varphi_{\tilde{T}}^m\sigma_{\tilde{T}}^{-2}$ is an essentially LR automorphism of $(X_{\tilde{T}}, \sigma_{\tilde{T}}) = (X_T, \sigma_T)$ so that $\varphi_T^{-klm}\sigma_T^{nlm-2}$ is an essentially LR automorphism of (X_T, σ_T). Therefore, by Corollary 6.5, $\varphi_T^{-klm}\sigma_T^{nlm-1}$ is an essentially LR expansive automorphism of (X_T, σ_T). Thus , by Proposition 8.8, there is $l' \in \mathbf{N}$ such that $\varphi_T^{-klml'}\sigma_T^{(nlm-1)l'}$ is an LR automorphism of (X_T, σ_T). By Lemma 8.5, $T^{klml'}$ is $(nlm-1)l'$ definite so that

$$n/k > (nlm-1)l'/klml' \geq \kappa(T).$$

(2) Assume that $\varphi_T^{-k}\sigma_T^n$ is an essentially LR nonexpansive automorphism of (X_T, σ_T) with $k, n \in \mathbf{N}$. There is a 1-1 LR textile system $\bar{T}$ and a topological conjugacy ψ : $(X_T, \sigma_T) \to (X_{\bar{T}}, \sigma_{\bar{T}})$ such that

$$\psi\varphi_T^{-k}\sigma_T^n\psi^{-1} = \varphi_{\bar{T}}.$$

Let $k', n' \in \mathbf{N}$. Assume that $n'/k' > n/k$. Then $\varphi_{\bar{T}}^{k'}\sigma_{\bar{T}}^{kn'-k'n}$ is an expansive LR automorphism of $(X_{\bar{T}}, \sigma_{\bar{T}})$ by Corollary 6.5 and Corollary 3.18. Therefore, since

$$\varphi_{\bar{T}}^{k'}\sigma_{\bar{T}}^{kn'-k'n} = \psi\varphi_T^{-kk'}\sigma_T^{kn'}\psi^{-1},$$

using Theorem 7.22(2) we conclude that $\varphi_T^{-k'}\sigma_T^{n'}$ is an essentially LR expansive automorphism of (X_T, σ_T).

Assume that $n'/k' \leq n/k$ and that $\varphi_T^{-k'}\sigma_T^{n'}$ were an essentially LR expansive automorphism of (X_T, σ_T). Then $(\varphi_T^{-k}\sigma_T^n)^{k'} = (\varphi_T^{-k'}\sigma_T^{n'})^k\sigma_T^{nk'-n'k}$ would have to be expansive by Corollary 6.5. But this contradicts the hypothesis that $\varphi_T^{-k}\sigma_T^n$ is nonexpansive.

Thus we have proved that $\varphi_T^{-k'}\sigma_T^{n'}$ is an essentially LR expansive automorphism of (X_T, σ_T) if and only if $n'/k' > n/k$. Therefore by (1), we conclude that $\kappa(T) = n/k$.

(3) By Corollaries 6.5 and 3.18, (3) follows.

(4) Let n, k be integers with $n \geq 0$ and $k > 0$. Assume that $n/k < \bar{\kappa}(T)$. Then $\bar{\kappa}(T) > 0$. Hence T^* is 1-1 and there is $m \in \mathbf{N}$ such that $n/k < m/k_m$, where k_m is the positive integer such that $(T^*)^m$ is strictly k_m definite. Since $(T^*)^{km}$ is kk_m definite, it follows from Lemma 8.6 that $\varphi_T^{kk_m}\sigma_T^{-km}$ is an essentially LR automorphism of (X_T, σ_T). Since

$$\varphi_T^{kk_m}\sigma_T^{-nk_m} = (\varphi_T^{kk_m}\sigma_T^{-km})\sigma_T^{km-k_m n}$$

with $km - k_m n > 0$, $\varphi_T^k\sigma_T^{-n}$ is an essentially LR expansive automorphism of (X_T, σ_T), by Corollary 6.5, Corollary 3.18 and Theorem 7.22(2) as above.

The proof of the converse is given in the same way as in the proof of (2) by using Lemma 8.7 instead of Lemma 8.5.

(5) Suppose that $\varphi_T^k \sigma_T^{-n}$ is an essentially LR nonexpansive automorphism of (X_T, σ_T) with integers $k > 0$ and $n \geq 0$. If $n = 0$, then T^* is not 1-1 (by Corollary 6.5) so that $\bar{\kappa}(T) = 0 = n/k$. If $n > 0$, then an argument which is similar to that in the proof of (2) but uses (4) instead of (1), shows that $\bar{\kappa}(T) = n/k$. □

Proposition 8.10. Let φ_1 and φ_2 be essentially LR automorphisms of topological Markov shift (X, σ). If φ_2 is expansive and $\varphi_1\varphi_2 = \varphi_2\varphi_1$, then $\varphi_1\varphi_2$ is an essentially LR expansive automorphism of (X, σ).

Proof. Since φ_1 is an essentially LR automorphism of (X, σ), there are a topological Markov shift (X', σ') and a topological conjugacy $\psi : (X, \sigma) \to (X', \sigma')$ such that $\psi\varphi_1\psi^{-1}$ is an LR automorphism of (X', σ'). Let $\varphi_1' = \psi\varphi_1\psi^{-1}$ and let $\varphi_2' = \psi\varphi_2\psi^{-1}$. Then φ_2' is an essentially LR expansive automorphism of (X', σ'). By Proposition 8.8, there is $m \in \mathbf{N}$ such that $(\varphi_2')^m$ is an LR automorphism of (X', σ'). Since $(\varphi_2')^m$ is an expansive LR automorphism of (X', σ'), there is a 1-1 LR textile system T such that T^* is 1-1, $(X_T, \sigma_T) = (X', \sigma')$, and $\varphi_T = (\varphi_2')^m$ (Theorem 2.5). Since $(T^*)^2$ is 1-1 and LR, there is $k > 0$ such that $(T^*)^2$ is k definite. Therefore, by Lemma 8.6, $\varphi_T^k \sigma_T^{-2}$ is an essentially LR automorphism of (X_T, σ_T). Thus, by Corollaries 3.18 and 6.5, $(\varphi_2')^{mk}(\sigma')^{-1} = (\varphi_T^k \sigma_T^{-2})\sigma_T$ is an essentially LR expansive automorphism of (X', σ'). Again by Proposition 8.8, there is $l \in \mathbf{N}$ such that $((\varphi_2')^{mk}(\sigma')^{-1})^l$ is an LR automorphism of (X', σ'). Put $(\varphi_2')^{mkl}(\sigma')^{-l} = \bar{\varphi}$. Then since $\varphi_1'\varphi_2' = \varphi_2'\varphi_1'$, we have

$$(\varphi_1'\varphi_2')^{klm} = (\varphi_1')^{klm}\bar{\varphi}(\sigma')^l.$$

By Corollary 3.18, $(\varphi_1')^{klm}\bar{\varphi}$ is an LR automorphism of (X', σ') and hence $(\varphi_1'\varphi_2')^{klm}$ is an expansive LR automorphism of (X', σ') by Corollaries 6.5 and 3.18. Therefore $\varphi_1'\varphi_2'$ is an essentially LR expansive automorphism of (X', σ'), by Theorem 7.22(2), so that $\varphi_1\varphi_2$ is an essentially LR expansive automorphism of (X, σ). □

Proposition 8.11. Let (X, σ) be an irreducible topological Markov shift. Then the following statements are valid.

(1) If φ_1 and φ_2 are LR endomorphisms of (X, σ), then $h(\varphi_2\varphi_1) = h(\varphi_1) + h(\varphi_2)$.

(2) If φ_1 and φ_2 are essentially LR automorphisms of (X, σ) with at least one of them expansive, then for all sufficiently large m, $h(\varphi_2^m\varphi_1^m) = m(h(\varphi_1) + h(\varphi_2))$.

(3) If φ_1 and φ_2 are essentially LR automorphisms of (X, σ) with at least one of them expansive and $\varphi_1\varphi_2 = \varphi_2\varphi_1$, then $h(\varphi_2\varphi_1) = h(\varphi_1) + h(\varphi_2)$.

(4) If φ_1, φ_2, and $\varphi_1\varphi_2$ are essentially LR automorphisms of (X, σ) with $\varphi_1\varphi_2 = \varphi_2\varphi_1$, then $h(\varphi_2\varphi_1) = h(\varphi_1) + h(\varphi_2)$.

Proof. (1) Let T_i be a one-sided 1-1 LR textile system with $\varphi_{T_i} = \varphi_i$ for $i = 1,2$. Let $\tilde{M}\tilde{N}_i \overset{k_i}{\simeq} \tilde{N}_i\tilde{M}$ be the specified equivalence associated with T_i for $i = 1,2$, where M is an irreducible nonnegative matrix with $(X,\sigma) = (X_M,\sigma_M)$. Let λ, λ_1 and λ_2 be the maximal eigenvalues of M, N_1 and N_2, respectively. Since M is irreducible and $MN_i = N_iM$ for $i = 1,2$, N_1 and N_2 have the same positive eigenvector, so that N_1N_2 has $\lambda_1\lambda_2$ as its maximal eigenvalue (cf. [BP], Chapter 2). By Proposition 3.5 and Observation 6.9, we have $h(\varphi_2\varphi_1) = \log\lambda_1\lambda_2 = \log\lambda_1 + \log\lambda_2 = h(\varphi_1) + h(\varphi_2)$.

(2) We assume without loss of generality that φ_2 is expansive. There are a topological Markov shift (X',σ') and a topological conjugacy $\psi : (X,\sigma) \to (X',\sigma')$ such that $\psi\varphi_1\psi^{-1}$ is an LR automorphism of (X',σ'). Let $\varphi_i' = \psi\varphi_i\psi^{-1}$ for $i = 1.2$. Then by Proposition 8.8, there is $m_0 \in \mathbf{N}$ such that for $m \geq m_0$ $(\varphi_2')^m$ is an LR automorphism of (X',σ'). Using (1) we have $h(\varphi_1^m\varphi_2^m) = h((\varphi_1')^m(\varphi_2')^m) = m(h(\varphi_1') + h(\varphi_2')) = m(h(\varphi_1) + h(\varphi_2))$.

(3) From (2), (3) of the proposition follows.

(4) Assume that φ_1, φ_2, and $\varphi_1\varphi_2$ are essentially LR automorphisms of (X,σ) and $\varphi_1\varphi_2 = \varphi_2\varphi_1$. Then using (3) we have $h(\varphi_2\varphi_1) + h(\sigma) = h(\varphi_2\varphi_1\sigma) = h(\varphi_2) + h(\varphi_1\sigma) = h(\varphi_2) + h(\varphi_1) + h(\sigma)$. □

Remark 8.12. Let T be a 1-1 LR textile system with $h(\sigma_T) > 0$. Then $\bar{\kappa}(T) \leq \kappa(T)$. Moreover if (X_T,σ_T) is an irreducible topological Markov shift, then $\bar{\kappa}(T) \leq h(\varphi_T)/h(\sigma_T) \leq \kappa(T)$.

Proof. If there were $k, n \in \mathbf{N}$ such that $\kappa(T) < n/k < \bar{\kappa}(T)$, then $\varphi_T^{-k}\sigma_T^n$ and $\varphi_T^k\sigma_T^{-n}$ would be essentially LR, expansive automorphisms of (X_T,σ_T), by Proposition 8.9, so that by Proposition 8.10, the identity map would be expansive, which contradicts the hypothesis that $h(\sigma_T) > 0$.

Assume that (X_T,σ_T) is an irreducible topological Markov shift. Let $k,n \in \mathbf{N}$ with $n/k > \kappa(T)$. Let $\varphi = \varphi_T^{-k}\sigma_T^n$. Then by Proposition 8.9, φ is an essentially LR automorphism of (X_T,σ_T). Since $\varphi\varphi_T^k = \varphi_T^k\varphi = \sigma_T^n$, it follows from Proposition 8.11 that $h(\varphi)+kh(\varphi_T) = nh(\sigma_T)$ so that $0 \leq h(\varphi) = nh(\sigma_T)-kh(\varphi_T)$. Hence $n/k \geq h(\varphi_T)/h(\sigma_T)$ so that $\kappa(T) \geq h(\varphi_T)/h(\sigma_T)$.

Similarly it follows from Propositions 8.9 and 8.11 that $\bar{\kappa}(T) \leq h(\varphi_T)/h(\sigma_T)$. □

Proposition 8.13. Let T be a 1-1 LR textile system with $h(\sigma_T) > 0$. Let $\varphi_1 = \varphi_T^{-k}\sigma_T^n$ and $\varphi_2 = \varphi_T^{k'}\sigma_T^{-n'}$ with integers $k,k' > 0$ and $n,n' \geq 0$. Assume that φ_1 and φ_2 are essentially LR nonexpansive automorphisms of (X_T,σ_T). Then $n/k \geq n'/k'$. If $n/k > n'/k'$, then $\varphi_1\varphi_2$ is an essentially LR expansive automorphism of (X_T,σ_T). If $n/k = n'/k'$, then φ_1 and φ_2 are essentially symbolic automorphisms.

Proof. By Proposition 8.9 and Remark 8.12, we have $n/k \geq n'/k'$.

Suppose that $n/k > n'/k'$. Then it is not the case that $n \leq n'$ and $k \geq k'$. If $n > n'$ and $k \leq k'$, then

$$\varphi_1\varphi_2 = \varphi_T^{k'-k}\sigma_T^{n-n'}$$

is an expansive LR automorphism of (X_T, σ_T). If $n > n'$ and $k > k'$, then $(n-n')/(k-k') > n/k = \kappa(T)$, and if $n \leq n'$ and $k < k'$, then $(n'-n)/(k'-k) < n'/k' = \bar{\kappa}(T)$. Hence $\varphi_1\varphi_2$ is an essentially LR expansive automorphism of (X_T, σ_T), by Proposition 8.9.

Assume that $n/k = n'/k'$. Then we have $\varphi_1^{n'} = \varphi_2^{-n}$. Put $\varphi_1^{n'} = \varphi$. Since φ_1 and φ_2 are essentially LR automorphisms of (X_T, σ_T), so are φ and φ^{-1}. There is a 1-1 and nondegenerate textile system T_1 such that $\varphi_{T_1} = \varphi$ (Fact 2.2). Then it follows from Theorem 7.21 (2) and Remark 7.4 that there is $m \in \mathbf{N}$ such that $T_1^{[m]}$ and $(T_1^{[m]})^{(-1)}$ are resolvable. Hence it follows from Theorem 7.21 (3) that there are a 1-1 and nondegenerate textile system $T_0 = (p_0, q_0 : \Gamma_0 \to G_0)$ with p_0 biresolving and a topological conjugacy $\psi : (X_T, \sigma_T) \to (X_{G_0}, \sigma_{G_0})$ with $\varphi = \psi^{-1}\varphi_{T_0}\psi$. Since p_0 is biresolving and the factor map given by p_0 is a conjugacy, we infer that p_0 is a graph-isomorphism. Since the graph-homomorphism $q_0p_0^{-1} : G_0 \to G_0$ gives the automorphism φ_{T_0} of (X_{G_0}, σ_{G_0}), $q_0p_0^{-1}$ is a graph-automorphism and hence φ_{T_0} is a symbolic automorphism. Thus φ has finite order. Therefore, by Proposition 8.2, we conclude that φ_1 and φ_2 are essentially symbolic automorphisms. □

Let $\sigma : X \to X$ be a homeomorphism such that (X, σ) is topologically conjugate to a topological Markov shift. Let $\varphi : X \to X$ be a homeomorphism such that $\sigma\varphi = \varphi\sigma$. Then φ is an *essentially LR automorphism* of (X, σ) if there are a topological Markov shift (X', σ') and a topological conjugacy $\psi : (X, \sigma) \to (X', \sigma')$ such that $\psi\varphi\psi^{-1}$ is an LR automorphism of (X', σ').

Proposition 8.14. Let (X, σ) be a dynamical system which is topologically conjugate to a topological Markov shift with $h(\sigma) > 0$. Let φ be an automorphism of (X, σ). Then the following statements are valid.

(1) The homeomorphism σ is an essentially LR automorphism of (X, σ).

(2) If φ is an essentially LR expansive automorphism of (X, σ), then σ is an essentially LR (expansive) automorphism of (X, φ).

(3) If there are $k, n \in \mathbf{Z}$ such that $\varphi^k\sigma^n$ is an essentially LR expansive automorphism of (X, σ) and φ is an essentially LR (not necessarily expansive) automorphism of $(X, \varphi^k\sigma^n)$, then φ is an essentially LR automorphism of (X, σ).

Proof. (1) Since there is a topological Markov shift (X', σ') to which (X, σ) is topologically conjugate and σ' is an LR automorphism of (X', σ'), σ is an essentially LR automorphism of (X, σ).

(2) Since φ is an essentially LR expansive automorphism of (X, σ), there are a 1-1 LR textile system T with T^* 1-1 and a topological conjugacy $\psi : (X, \sigma) \to (X_T, \sigma_T)$ such that $\psi\varphi\psi^{-1} = \varphi_T$. Let $\chi : X_T \to X_{T^*}$ be the homeomorphism defined by

$$\chi(\theta_T(t)) = \theta_T^*(t), \quad t \in U_T.$$

Then by Theorem 2.5, the following diagrams commute:

$$\begin{array}{ccc} X & \xrightarrow{\varphi} & X \\ \psi\downarrow & & \downarrow\psi \\ X_T & \xrightarrow{\varphi_T} & X_T \\ \chi\downarrow & & \downarrow\chi \\ X_{T^*} & \xrightarrow[\sigma_{T^*}]{} & X_{T^*} \end{array} \qquad \begin{array}{ccc} X & \xrightarrow{\sigma} & X \\ \psi\downarrow & & \downarrow\psi \\ X_T & \xrightarrow{\sigma_T} & X_T \\ \chi\downarrow & & \downarrow\chi \\ X_{T^*} & \xrightarrow[\varphi_{T^*}]{} & X_{T^*} \end{array} .$$

Thus $\chi\psi : (X, \varphi) \to (X_{T^*}, \sigma_{T^*})$ is a topological conjugacy and

$$\varphi_{T^*} = (\chi\psi)\sigma(\chi\psi)^{-1}$$

is an LR automorphism of (X_{T^*}, σ_{T^*}). Therefore, σ is an essentially LR automorphism of (X, φ).

(3) Assume that there are $k, n \in \mathbf{Z}$ such that $\varphi^k\sigma^n$ is an essentially LR expansive automorphism of (X, σ) and φ is an essentially LR automorphism of $(X, \varphi^k\sigma^n)$. It is not the case that $n = 0$ and $k = 0$ because id_X is not expansive. If $n = 0$ and $k < 0$, then φ^{-k} would be an essentially LR automorphism of (X, φ^k), which is impossible because, otherwise, $\mathrm{id}_X = \varphi^{-k}\varphi^k$ would be expansive by (1) and Proposition 8.10. Assume that $n < 0$ and $k \leq 0$. By (2), σ is an essentially LR expansive automorphism of $(X, \varphi^k\sigma^n)$. Therefore it follows from Proposition 8.10 that $\varphi^{-k}\sigma^{-n}$ is an essentially LR automorphism of $(X, \varphi^k\sigma^n)$, which is also impossible for the same reason as above. Thus it is not the case that $n \leq 0$ and $k \leq 0$.

Put $\varphi^k\sigma^n = \tau$. Assume that $n \leq 0$ and $k > 0$. Then τ is an essentially LR expansive automorphism of (X, σ) and hence so is $\tau\sigma^{-n}$. Therefore φ^k is essentially LR automorphism of (X, σ). Thus, by Theorem 7.22(2), φ is an essentially LR automorphism of (X, σ).

Now we assume that $n > 0$. Since φ is an essentially LR automorphism of (X, τ), there are a 1-1 LR textile system T_0 and a topological conjugacy $\psi : (X, \tau) \to (X_{T_0}, \sigma_{T_0})$ such that

$$\psi\varphi\psi^{-1} = \varphi_{T_0}.$$

Since σ is an essentially LR expansive automorphism of $(X, \tau), \psi\sigma\psi^{-1}$ is an essentially LR expansive automorphism of (X_{T_0}, σ_{T_0}). Therefore, by Proposition 8.8, there is $m \in \mathbf{N}$ such that $(\psi\sigma\psi^{-1})^m$ is an expansive LR automorphism of (X_{T_0}, σ_{T_0}). Thus there is a 1-1 LR textile system T with T^* 1-1 such that $(X_T, \sigma_T) = (X_{T_0}, \sigma_{T_0})$ and

$$\psi\sigma^m\psi^{-1} = \varphi_T$$

(by Theorem 2.5). Since

$$\sigma_T^m = \sigma_{T_0}^m = \psi\tau^m\psi^{-1} = (\psi\varphi^{km}\psi^{-1})(\psi\sigma^{nm}\psi^{-1}) = \varphi_{T_0}^{km}\varphi_T^n,$$

we have

$$\varphi_{T_0}^{km} = \varphi_T^{-n}\sigma_T^m.$$

Since $\varphi_{T_0}^{km}$ is an LR automorphism of (X_T, σ_T), it follows from Lemma 8.5 that T^n is m definite. Since T^* is a 1-1 LR textile system with $(T^*)^*$ 1-1, it follows from Lemma 8.6 that $\varphi_{T^*}^m\sigma_{T^*}^{-n}$ is an essentially LR automorphism of (X_{T^*}, σ_{T^*}). Let χ_T be the homeomorphism defined by

$$\chi_T(\theta_T(t)) = \theta_T^*(t), \quad t \in U_T.$$

Then, by Theorem 2.5, $\chi_T : (X_T, \varphi_T) \to (X_{T^*}, \sigma_{T^*})$ is a topological conjugacy so that $\chi_T\psi : (X, \sigma^m) \to (X_{T^*}, \sigma_{T^*})$ is a topological conjugacy. Using Theorem 2.5, we have

$$(\chi_T\psi)\varphi^{km}(\chi_T\psi)^{-1} = \chi_T\varphi_{T_0}^{km}\chi_T^{-1} = \chi_T\varphi_T^{-n}\sigma_T^m\chi_T^{-1} = \varphi_{T^*}^m\sigma_{T^*}^{-n}.$$

Therefore φ^{km} is an essentially LR automorphism of (X, σ^m). Thus it follows from Theorem 7.22(2) that φ is an essentially LR automorphism of (X, σ^m). Thus it follows from Theorem 7.23(2) that φ is an essentially LR automorphism of (X, σ). □

Corollary 8.15. Let (X, σ) be a topological Markov shift with $h(\sigma) > 0$. Let φ be an automorphism of (X, σ). Let K_0 be the set of elements $(n, k) \in \mathbf{Z}^2$ such that $(X, \varphi^k\sigma^n)$ is topologically conjugate to a topological Markov shift. Let ELR be the relation on K_0 such that for $(n, k), (n', k') \in K_0, (n, k)ELR(n', k')$ means that $\varphi^k\sigma^n$ is an essentially LR automorphism of $(X, \varphi^{k'}\sigma^{n'})$. Then ELR is an equivalence relation on K_0.

Proof. By the preceding proposition, ELR is reflexive and symmetric. Suppose that $(n, k)ELR(n', k')$ and $(n', k')ELR(n'', k'')$ with $(n, k), (n', k'), (n'', k'') \in K_0$. Then $\varphi^k\sigma^n$ and $\varphi^{k''}\sigma^{n''}$ are essentially LR expansive automorphisms of $(X, \varphi^{k'}\sigma^{n'})$. (But in the remainder of the proof, we do not assume the expansiveness of $\varphi^k\sigma^n$ for the purpose of another use of the proof.)

Assume that $k''n = kn''$. If $k = 0$ and $n = 0$, then $\varphi^k\sigma^n = \mathrm{id}_X$ is an essentially LR automorphism of $(X, \varphi^{k''}\sigma^{n''})$.

If $k = 0$ and $n \neq 0$, then $k'' = 0$ and $n'' \neq 0$ so that σ^n is an essentially LR automorphism of $(X, \varphi^{k'}\sigma^{n'})$ and $\sigma^{n''}$ is an essentially LR expansive automorphism of $(X, \varphi^{k'}\sigma^{n'})$. If $n > 0$ and $n'' < 0$, then by Proposition 8.10, $(\sigma^n)^{-n''}(\sigma^{n''})^n = \mathrm{id}_X$ would be expansive, which is not the case. Similarly it is not the case that $n < 0$ and $n'' > 0$. If $nn'' > 0$, then it follows from Theorem 7.22(2) that σ^n is essentially LR automorphism of $(X, \sigma^{n''})$ because $(\sigma^n)^{|n''|} = (\sigma^{n''})^{|n|}$.

If $k \neq 0$, then $k'' \neq 0$ because

$$(\varphi^k\sigma^n)^{k''} = (\varphi^{k''}\sigma^{n''})^k$$

and $\varphi^{k''}\sigma^{n''}$ is expansive. Since

$$(\varphi^k\sigma^n)^{k''}(\varphi^{k''}\sigma^{n''})^{-k} = \mathrm{id}_X,$$

it follows from Proposition 8.10 that neither the case where $k < 0$ and $k'' > 0$ nor the case where $k < 0$ and $k'' > 0$, can occur. If $kk'' > 0$, then it follows from Theorem 7.22(2) that $\varphi^k\sigma^n$ is an essentially LR automorphism of $(X, \varphi^{k''}\sigma^{n''})$, because

$$(\varphi^k\sigma^n)^{|k''|} = (\varphi^{k''}\sigma^{n''})^{|k|}.$$

Suppose that $k''n \neq kn''$. Then there are $m \in \mathbf{N}$ and $\bar{n}, \bar{k} \in \mathbf{Z}$ such that

$$(\varphi^{k'}\sigma^{n'})^m = (\varphi^k\sigma^n)^{\bar{k}}(\varphi^{k''}\sigma^{n''})^{\bar{n}}.$$

Since $\varphi^k\sigma^n$ and $\varphi^{k''}\sigma^{n''}$ are essentially LR automorphisms of $(X, \varphi^{k'}\sigma^{n'})$, they are essentially LR automorphisms of $(X, (\varphi^{k'}\sigma^{n'})^m)$. (This follows from the fact that if T is an LR textile system, then so is $T^{(m)}$ for $m \in \mathbf{N}$.) Hence $\varphi^k\sigma^n$ and $\varphi^{k''}\sigma^{n''}$ are essentially LR automorphisms of $(X, (\varphi^k\sigma^n)^{\bar{k}}(\varphi^{k''}\sigma^{n''})^{\bar{n}})$. Thus by the preceding proposition, $\varphi^k\sigma^n$ is an essentially LR automorphism of $(X, \varphi^{k''}\sigma^{n''})$. □

Let (X, σ) be a topological Markov shift with $h(\sigma) > 0$ and let φ be an automorphism of (X, σ). For an automorphism $\tau = \varphi^{k_0}\sigma^{n_0}$ of (X, σ) with (n_0, k_0) in $\mathbf{Z}^2$ such that (X, τ) is topologically conjugate to a topological Markov shift, define

$$K_\varphi(\tau) = \{(n, k) \in \mathbf{Z}^2 \mid \varphi^k\sigma^n \text{ is an essentially LR automorphism of } (X, \tau)\}$$

and define $CK_\varphi(\tau)$ to be the convex hull of $K_\varphi(\tau)$ in $\mathbf{R}^2$. It follows from Theorem 6.29 that there is $l \in \mathbf{N}$ such that all of $\sigma\tau^l, \sigma^{-1}\tau^l, \varphi\tau^l$ and $\varphi^{-1}\tau^l$ are essentially LR automorphisms of (X,τ). Therefore there is $(n_1,k_1)\in\mathbf{Z}^2$ such that $n_0k_1-k_0n_1>0$ and $\varphi^{k_1}\sigma^{n_1}$ is an essentially LR automorphism of (X,τ). In fact, we can set

$$(n_1,k_1)=\begin{cases}(ln_0,1) & \text{if } k_0=0 \text{ and } n_0>0\\ (ln_0,-1) & \text{if } k_0=0 \text{ and } n_0<0\\ (ln_0-1,lk_0) & \text{if } k_0>0\\ (ln_0+1,lk_0) & \text{if } k_0<0.\end{cases}$$

There are a 1-1 LR textile system T and a topological conjugacy $\psi:(X,\tau)\to(X_T,\sigma_T)$ such that $\varphi_T=\psi(\varphi^{k_1}\sigma^{n_1})\psi^{-1}$. Then for any $(\hat{n},\hat{k})\in\mathbf{Z}^2$, if

$$(n,k)=(\hat{n},\hat{k})\begin{pmatrix}n_0 & k_0\\ n_1 & k_1\end{pmatrix},$$

then $\varphi^k\sigma^n=\psi^{-1}\varphi_T^{\hat{k}}\sigma_T^{\hat{n}}\psi$ and hence $\varphi_T^{\hat{k}}\sigma_T^{\hat{n}}$ is an essentially LR automorphism of (X_T,σ_T) if and only if $\varphi^k\sigma^n$ is an essentially LR automorphism of (X,τ). For any $(n,k)\in\mathbf{Z}^2$, there is $(\hat{n},\hat{k})\in\mathbf{Z}^2$ such that

$$(n_0k_1-k_0n_1)(n,k)=(\hat{n},\hat{k})\begin{pmatrix}n_0 & k_0\\ n_1 & k_1\end{pmatrix}$$

and hence $\varphi_T^{\hat{k}}\sigma_T^{\hat{n}}$ is an essentially LR automorphism of (X_T,σ_T) if and only if $\varphi^k\sigma^n$ is an essentially LR automorphism of (X,τ) (by Theorem 7.22(2)). Therefore it follows from Proposition 8.9 and Remark 8.12 that $CK_\varphi(\tau)$ is a convex cone in $\mathbf{R}^2$ which is either *pointed* (i.e., $CK_\varphi(\tau)\cap(-CK_\varphi(\tau))=\{(0,0)\}$) or equal to a closed half-plane (i.e., $CK_\varphi(\tau)=\{(x,y)\in\mathbf{R}^2 \mid ax+by\geq 0\}$ for some $(a,b)\in\mathbf{R}^2$ with $(a,b)\neq(0,0)$). Clearly $K_\varphi(\tau)$ is the set of all lattice points contained in $CK_\varphi(\tau)$. We call $CK_\varphi(\tau)$ the *ELR cone* (essentially-LR cone) *of* τ *for* the automorphism $\varphi:(X,\sigma)\to(X,\sigma)$ or an *ELR cone for* φ. It follows from Proposition 8.9 that $CK_\varphi(\tau)$ always has its interior and that for $(n,k)\in\mathbf{Z}^2$, $\varphi^k\sigma^n$ is an essentially LR expansive automorphism of (X,τ) if and only if (n,k) is a lattice point in the interior of $CK_\varphi(\tau)$. It also follows from Proposition 8.9 that for $(n,k)\in\mathbf{Z}^2$, $\varphi^k\sigma^n$ is an essentially LR nonexpansive automorphism of (X,τ) if and only if (n,k) is a lattice point which is in $CK_\varphi(\tau)$ and on the boundary of $CK_\varphi(\tau)$. Hence if $\varphi^k\sigma^n$ is an essentially symbolic automorphism of (X,τ), then $CK_\varphi(\tau)$ is the closed half-plane containing (n_0,k_0) and having the line passing through (n,k) as its boundary. (This readily follows since a symbolic automorphism of a topological Markov shift with positive entropy and its inverse are nonexpansive LR automorphisms of the shift.) Therefore, by Proposition 8.13, we conclude that $CK_\varphi(\tau)$ is pointed if and only if there is no essentially symbolic automorphism of (X,τ) of the form $\varphi^k\sigma^n$ with $(n,k)\in\mathbf{Z}^2$ $-\{(0,0)\}$. By Corollary 8.15, we know that no two distinct ELR cones share interior points. Moreover, it follows from the proof of Corollary 8.15 that for $(n,k)\in\mathbf{Z}^2$ and for $(n',k')\in\mathbf{Z}^2$ such that $(X,\varphi^{n'}\sigma^{k'})$ is topologically conjugate to a topological Markov shift, $\varphi^k\sigma^n$ is an essentially LR (not necessarily expansive) automorphism of $(X,\varphi^{k'}\sigma^{n'})$ if and only if (n,k) and (n',k') belong to the same ELR cone.

Assume that the topological Markov shift (X, σ) is irreducible. Let $(n, k), (n', k') \in CK_\varphi(\tau)$. Let $\omega = \varphi^k \sigma^n$ and let $\omega' = \varphi^{k'} \sigma^{n'}$. We shall show that

$$h(\omega\omega') = h(\omega) + h(\omega').$$

Let (X, σ) have period $s \in \mathbf{N}$. Then, by a result of Boyle and Krieger (Lemma 2.14 of [BK]), we know that for some l dividing s, X is the union of l pairwise disjoint closed open sets $X_1, \cdots, X_l$ such that $\sigma(X_i) = X_{i+1 (\text{mod } l)}$ and $\tau(X_i) = X_i$ for $1 \le i \le l$, and $(X_i, \tau_i), 1 \le i \le l$, are topologically conjugate to an irreducible topological Markov shift (with period s/l), where $\tau_i = \tau | X_i$, $1 \le i \le l$. Since ω and ω' are automorphisms of (X, τ) with (X_i, τ_i) conjugate to an irreducible topological Markov shift for $1 \le i \le l$, there are permutations r, r' on $\{1, \cdots, l\}$ such that $\omega(X_i) = X_{r(i)}$ and $\omega'(X_i) = X_{r'(i)}$ for $1 \le i \le l$. Therefore there is $m \in \mathbf{N}$ such that $\omega^m(X_i) = X_i$ and $(\omega')^m(X_i) = X_i$ for $1 \le i \le l$. Let $v_i = \omega^m | X_i$ and let $v'_i = (\omega')^m | X_i$ for $1 \le i \le l$. Then for $1 \le i, j \le l$, we have $(\sigma^{j-i}|X_i) v_i = v_j (\sigma^{j-i}|X_i)$, so that (X_i, v_i) and (X_j, v_j) are topologically conjugate. Similarly for $1 \le i, j \le l$, (X_i, v'_i) and (X_j, v'_j) are topologically conjugate and so are $(X_i, v_i v'_i)$ and $(X_j, v_j v'_j)$. Hence we have $h(\omega^m) = h(v_1), h((\omega')^m) = h(v'_1)$, and $h(\omega^m (\omega')^m) = h(v_1 v'_1)$. Since $\omega^m, (\omega')^m$, and $\omega^m(\omega')^m$ are essentially LR automorphisms of (X, τ), it follows that v_1, v'_1 and $v_1 v'_1$ are essentially LR automorphisms of (X_1, τ_1). Since (X_1, τ_1) is topologically conjugate to an irreducible topological Markov shift, it follows from Proposition 8.11 that $h(v_1 v'_1) = h(v_1) + h(v'_1)$. Hence $h(\omega^m(\omega')^m) = h(\omega^m) + h((\omega')^m)$ so that $h(\omega\omega') = h(\omega) + h(\omega')$.

We summarize those above in the following theorem.

Theorem 8.16. Let (X, σ) be a topological Markov shift with $h(\sigma) > 0$ and φ an automorphism of (X, σ). Then the ELR cones for φ have the following properties.

(i) Each ELR cone is a convex cone which is either pointed or equal to a closed half-plane.

(ii) Each ELR cone has its interior.

(iii) The interiors of any two distinct ELR cones are disjoint.

(iv) For any lattice point $(n, k) \in \mathbf{Z}^2, (X, \varphi^k \sigma^n)$ is topologically conjugate to a topological Markov shift if and only if (n, k) is an interior point of some ELR cone.

(v) If a lattice point (n, k) in an ELR cone is on the boundary of the cone, then $\varphi^k \sigma^n$ is nonexpansive.

(vi) For $(n, k) \in \mathbf{Z}^2$ and for $(n', k') \in \mathbf{Z}^2$ such that $(X, \varphi^{k'} \sigma^{n'})$ is topologically conjugate to a topological Markov shift, $\varphi^k \sigma^n$ is an essentially LR automorphism of $(X, \varphi^{k'} \sigma^{n'})$ if and only if (n, k) and (n', k') belong to the same ELR cone.

(vii) Every (some) ELR cone is pointed if and only if there is no lattice point $(n, k) \in \mathbf{Z}^2 - \{(0, 0)\}$ such that $\varphi^k \sigma^n$ is an essentially symbolic automorphism of (X, σ).

(viii) If the topological Markov shift (X, σ) is irreducible, then entropy function is additive on each ELR cone, that is, for $(n, k), (n', k') \in \mathbf{Z}^2$ if (n, k) and (n', k') are in the same ELR cone, then

$$h((\varphi^k \sigma^n)(\varphi^{k'} \sigma^{n'})) = h(\varphi^k \sigma^n) + h(\varphi^{k'} \sigma^{n'}).$$

Question 8.b. Is an ELR cone is always closed?

Question 8.c. What about the number of ELR cones for an automorphism of an irreducible topological Markov shift with positive entropy ? Is it bounded, or can it be infinite ?

There is an automorphism of a topological Markov shift such that distinct pointed ELR cones for it share a boundary half-line. But there is also an automorphism φ of a topological Markov shift (X, σ) such that there are $\alpha, \beta \in \boldsymbol{Q}$ with $\alpha < \beta$ such that for all integers $k \neq 0$ and n with $\alpha \leq n/k \leq \beta$, $\varphi^k \sigma^n$ is nonexpansive. An example of such an automorphism will be given in Section 10. Such an example has also been given by Mike Boyle and Doug Lind (private communication).

We conclude this section by adding the following proposition.

Proposition 8.17. Let T be an LR textile system. Let $k, l \in \mathbf{N}$. Then for all integers $m, n \geq 0$, the automorphism $\psi_{T,(k,l)}^{(m,n)}$ of $(\check{X}_T^{(k,l)}, \check{\sigma}_T^{(k,l)})$ defined in Proposition 2.4, is an LR automorphism. For all integers $m, n \geq 0$, $\boldsymbol{\sigma}_T^{(m,n)}$ is an essentially LR automorphism of $(U_T, \boldsymbol{\sigma}_T^{(k,l)})$.

Proof. Let $\tilde{M}\tilde{N} \overset{k}{\simeq} \tilde{N}\tilde{M}$ be the specified equivalence associated with T. Then $(\check{X}_T^{(k,l)}, \check{\sigma}_T^{(k,l)}) = (X_{N^k M^l}, \sigma_{N^k M^l})$ and $\check{\theta}_T^{(k,l)} : (U_T, \boldsymbol{\sigma}_T^{(k,l)}) \to (\check{X}_T^{(k,l)}, \check{\sigma}_T^{(k,l)})$ is a topological conjugacy (Lemma 6.2). Since $\psi_{T,(k,l)}^{(m,n)} = \check{\theta}_T^{(k,l)} \boldsymbol{\sigma}_T^{(m,n)} (\check{\theta}_T^{(k,l)})^{-1}$ by definition, it suffices to show that $\psi_{T,(k,l)}^{(m,n)}$ is an LR automorphism of $(\check{X}_T^{(k,l)}, \check{\sigma}_T^{(k,l)})$. Since $\psi_{T,(k,l)}^{(m,n)} = (\psi_{T,(k,l)}^{(1,0)})^m (\psi_{T,(k,l)}^{(0,1)})^n$, it suffices to show that $\psi_{T,(k,l)}^{(1,0)}$ and $\psi_{T,(k,l)}^{(0,1)}$ are LR automorphisms (by Corollary 3.18). Let

$$(\tilde{N}^k \tilde{M}^l)\tilde{N} \overset{k'}{\simeq} \tilde{N}(\tilde{N}^k \tilde{M}^l)$$

be the specified equivalence which is naturally induced by $\tilde{M}\tilde{N} \overset{k}{\simeq} \tilde{N}\tilde{M}$. Let T' be the LR textile system associated with it. Let

$$(\tilde{N}^k \tilde{M}^l)\tilde{M} \overset{k''}{\simeq} \tilde{M}(\tilde{N}^k \tilde{M}^l)$$

be the specified equivalence which is naturally induced by $\tilde{N}\tilde{M} \overset{k^{-1}}{\simeq} \tilde{M}\tilde{N}$. Let T'' be the LR textile system associated with it. Then T' and T'' are 1-1 (because $\check{\theta}_T^{(k,l)}$ is 1-1) and

$$\varphi_{T'} = \psi_{T,(k,l)}^{(1,0)} \quad \text{and} \quad \varphi_{T''} = \psi_{T,(k,l)}^{(0,1)}.$$

□

9. SIMILARITY

In this section, we introduce the notions of "similarity" and "weak similarity" for topological conjugacies for sofic systems. Closely related materials for topological Markov shifts were studied by Wagoner in [Wag1],[Wag2], and [Wag3]. In particular, results closely related with (3) of Corollary 9.5, Proposition 9.16, and Remark 9.17 were presented in [Wag3] and [Wag1] (see Theorem 1.7 of [Wag3] and Section 4 of [Wag1]). Our approaches are different. Similar treatment to ours for "weak similarity" is found in [KMT] (see Section 3 of [KMT]). Our intention is to introduce the classifications by "similarity" and "weak similarity" to automorphisms of sofic systems and to consider their relation to the dynamics of the automorphisms.

Let ϕ be a forward conjugacy of a sofic system $(X_{\mathcal{M}}, \sigma_{\mathcal{M}})$ onto a sofic system $(X_{\mathcal{N}}, \sigma_{\mathcal{N}})$, where $\mathcal{M}$ and $\mathcal{N}$ are representation matrices. Let $n \in \mathbf{N}$ and let

$$\mathcal{M} \simeq \mathcal{P}_1\mathcal{Q}_1,\ \mathcal{Q}_1\mathcal{P}_1 \simeq \mathcal{P}_2\mathcal{Q}_2,\ \cdots,\ \mathcal{Q}_{n-1}\mathcal{P}_{n-1} \simeq \mathcal{P}_n\mathcal{Q}_n,\ \mathcal{Q}_n\mathcal{P}_n \simeq \mathcal{N}$$

be a strong shift equivalence of representation matrices from $\mathcal{M}$ to $\mathcal{N}$. We say that ϕ is *associated with* this strong shift equivalence if there are specifications $k_0, \cdots, k_n$ such that the specified strong shift equivalence

$$\mathcal{M} \overset{k_0}{\simeq} \mathcal{P}_1\mathcal{Q}_1,\ \mathcal{Q}_1\mathcal{P}_1 \overset{k_1}{\simeq} \mathcal{P}_2\mathcal{Q}_2,\ \cdots,\ \mathcal{Q}_{n-1}\mathcal{P}_{n-1} \overset{k_{n-1}}{\simeq} \mathcal{P}_n\mathcal{Q}_n,\ \mathcal{Q}_n\mathcal{P}_n \overset{k_n}{\simeq} \mathcal{N}$$

gives a κ-ζ factorization of ϕ (cf. Theorem 5.3).

Let ϕ be a forward conjugacy of a topological Markov shift (X_M, σ_M) onto (X_N, σ_N), where M and N are square nonnegative integral matrices. Let $n \in \mathbf{N}$ and let

$$M = P_1Q_1,\ Q_1P_1 = P_2Q_2,\ \cdots,\ Q_{n-1}P_{n-1} = P_nQ_n,\ Q_nP_n = N$$

be a strong shift equivalence from M to N. We say that ϕ is *associated with* this strong shift equivalence if there are specifications $k_0, \cdots, k_n$ such that the specified strong shift equivalence

$$\tilde{M} \overset{k_0}{\simeq} \tilde{P}_1\tilde{Q}_1,\ \tilde{Q}_1\tilde{P}_1 \overset{k_1}{\simeq} \tilde{P}_2\tilde{Q}_2,\ \cdots,\ \tilde{Q}_{n-1}\tilde{P}_{n-1} \overset{k_{n-1}}{\simeq} \tilde{P}_n\tilde{Q}_n,\ \tilde{Q}_n\tilde{P}_n \overset{k_n}{\simeq} \tilde{N}$$

gives a κ-ζ factorization of ϕ.

By Theorem 5.3, we know that if $\mathcal{M}$ and $\mathcal{N}$ are canonical, then every forward conjugacy ϕ of $(X_{\mathcal{M}}, \sigma_{\mathcal{M}})$ onto $(X_{\mathcal{N}}, \sigma_{\mathcal{N}})$ can be associated with some canonical strong shift equivalence of lag greater than 0 from $\mathcal{M}$ to $\mathcal{N}$ without changing the indexing of $\mathcal{M}$ or that of $\mathcal{N}$. (If ϕ is a symbolic conjugacy, then ϕ cannot always be associated with a canonical strong shift equivalence of lag 0 in this way.)

Two topological conjugacies ϕ and ϕ' of a sofic system (Y_1, σ_1) onto a sofic system (Y_2, σ_2) are said to be *similar* if there is a nonnegative integer m such that $\varphi\sigma_1^m$ and $\varphi'\sigma_1^m$ are forward conjugacies which are associated with the same canonical strong shift equivalence of representation matrices.

Remark 9.1. If ϕ and ϕ' are topological conjugacies of a topological Markov shift (X_1, σ_1) onto a topological Markov shift (X_2, σ_2), then ϕ and ϕ' are similar if and only if there is a nonnegative integer m such that $\phi\sigma_1^m$ and $\phi'\sigma_1^m$ are forward conjugacies which are associated with the same strong shift equivalence of nonnegative integral matrices.

Proof. If a forward conjugacy between topological Markov shifts is associated with a canonical strong shift equivalence of representation matrices, then it is also associated with its support. (cf. (5.4)~(5.6)). □

Clearly we have the following fact.

Fact 9.2. Let (Y_i, σ_i), $i = 1, 2, 3$, be sofic systems and let ϕ_1 and ϕ_1' be topological conjugacies of (Y_1, σ_1) onto (Y_2, σ_2) and ϕ_2 and ϕ_2' topological conjugacies of (Y_2, σ_2) onto (Y_3, σ_3). If ϕ_1 and ϕ_1' are similar and ϕ_2 and ϕ_2' are similar, then $\phi_2\phi_1$ and $\phi_2'\phi_1'$ are similar and so are ϕ_1^{-1} and ϕ_2^{-1}.

Let φ be an automorphism of a sofic system (Y, σ). We call φ a *sofic simple automorphism* if there are a canonical representation matrix $\mathcal{M}$, a topological conjugacy $\psi : (Y, \sigma) \to (X_{\mathcal{M}}, \sigma_{\mathcal{M}})$ and a specified equivalence

$$\mathcal{M} \overset{e}{\simeq} \mathcal{M}$$

such that

$$\varphi = \psi\epsilon\psi^{-1},$$

where ϵ is the symbolic automorphism of $(X_{\mathcal{M}}, \sigma_{\mathcal{M}})$ given by e. We call φ a *simple automorphism* if there are a square nonnegative integral matrix M, a topological conjugacy $\psi_0 : (Y, \sigma) \to (X_{\tilde{M}}, \sigma_{\tilde{M}})$ and a specified equivalence

$$\tilde{M} \overset{e_0}{\simeq} \tilde{M}$$

such that

$$\varphi = \psi_0\epsilon_0\psi_0^{-1},$$

where ϵ_0 is the symbolic automorphism of $(X_{\tilde{M}}, \sigma_{\tilde{M}})$ given by e_0 [N3]. Of course, a simple automorphism is a sofic simple automorphism. We note that for a canonical representation matrix $\mathcal{M}$, a specified equivalence $\mathcal{M} \overset{e}{\simeq} \mathcal{M}$ represents a λ-graph-automorphism which leaves the vertices fixed: if $\mathcal{G} = (G, \lambda)$ is the λ-graph with $\mathcal{M}_{\mathcal{G}} = \mathcal{M}$, then $\mathcal{M} \overset{e}{\simeq} \mathcal{M}$ represents the λ-graph-automorphism $(h, e) : \mathcal{G} \simeq \mathcal{G}$ such that $h_V = \mathrm{id}_{V_G}$ and

$\lambda(h_A(a)) = e(\lambda(a)), a \in A_G$. Similarly, for a square nonnegative integral matrix M, a specified equivalence $\tilde{M} \overset{e_0}{\simeq} \tilde{M}$ represents a graph-automorphism which does not move the vertices: if G is the graph with $M_G = M$, then it represents the graph-automorphism $h : G \to G$ with $h_A = e_0$ and $h_V = \mathrm{id}_{V_G}$.

Lemma 9.3. A sofic simple automorphism is similar to the identity automorphism.

Proof. Let φ be a simple automorphism of a sofic system $(X_{\mathcal{M}}, \sigma_{\mathcal{M}})$, where $\mathcal{M}$ is the canonical representation matrix. Then there are a canonical representation matrix $\mathcal{N}$, a topological conjugacy $\psi : (X_{\mathcal{M}}, \sigma_{\mathcal{M}}) \to (X_{\mathcal{N}}, \sigma_{\mathcal{N}})$ and a specified equivalence $\mathcal{N} \overset{e}{\simeq} \mathcal{N}$ such that $\varphi = \psi\epsilon\psi^{-1}$, where ϵ is the symbolic automorphism given by e. We may assume that ψ is a forward conjugacy. As stated above, there is a specified canonical strong shift equivalence

$$\mathcal{M} \overset{e_0}{\simeq} \mathcal{P}_1\mathcal{Q}_1,\ \mathcal{Q}_1\mathcal{P}_1 \overset{e_1}{\simeq} \mathcal{P}_2\mathcal{Q}_2,\ \cdots,\ \mathcal{Q}_{l-1}\mathcal{P}_{l-1} \overset{e_{l-1}}{\simeq} \mathcal{P}_l\mathcal{Q}_l,\ \mathcal{Q}_l\mathcal{P}_l \overset{e_l}{\simeq} \mathcal{N}$$

which gives a factorization of ψ with $l \geq 1$. Since $\mathrm{id}_{X_{\mathcal{M}}} = \psi\psi^{-1}$, a factorization of $\mathrm{id}_{X_{\mathcal{M}}}\sigma^l$ is given by

$$\mathcal{M} \overset{e_0}{\simeq} \mathcal{P}_1\mathcal{Q}_1,\ \mathcal{Q}_1\mathcal{P}_1 \overset{e_1}{\simeq} \mathcal{P}_2\mathcal{Q}_2,\ \cdots,\ \mathcal{Q}_{l-1}\mathcal{P}_{l-1} \overset{e_{l-1}}{\simeq} \mathcal{P}_l\mathcal{Q}_l,$$
$$\mathcal{Q}_l\mathcal{P}_l \overset{i}{\simeq} \mathcal{Q}_l\mathcal{P}_l,\ \mathcal{P}_l\mathcal{Q}_l \overset{e_{l-1}^{-1}}{\simeq} \mathcal{Q}_{l-1}\mathcal{P}_{l-1},\ \cdots,\ \mathcal{P}_2\mathcal{Q}_2 \overset{e_1^{-1}}{\simeq} \mathcal{Q}_1\mathcal{P}_1,\ \mathcal{P}_1\mathcal{Q}_1 \overset{e_0^{-1}}{\simeq} \mathcal{M},$$

where i is the identity specification. A factorization of $\varphi\sigma^l$ is given by the specified strong shift equivalence obtained from the above one by replacing $\mathcal{Q}_l\mathcal{P}_l \overset{i}{\simeq} \mathcal{Q}_l\mathcal{P}_l$ by

$$\mathcal{Q}_l\mathcal{P}_l \overset{f}{\simeq} \mathcal{Q}_l\mathcal{P}_l \quad \text{with} \quad f = e_lee_l^{-1}.$$

Thus φ and $\mathrm{id}_{X_{\mathcal{M}}}$ are similar. □

Theorem 9.4. Let ϕ and ϕ' be topological conjugacies between sofic systems. Then the following statements are equivalent.

(1) ϕ and ϕ' are similar.

(2) $\phi^{-1}\phi'$ is a composition of sofic simple automorphisms.

(3) $\phi'\phi^{-1}$ is a composition of sofic simple automorphisms.

In particular, if ϕ and ϕ' are topological conjugacies between topological Markov shifts, (1) is equivalent to each of the following (2′) and (3′).

(2′) $\phi^{-1}\phi'$ is a composition of simple automorphisms.

(3′) $\phi'\phi^{-1}$ is a composition of simple automorphisms.

Proof. Suppose that ϕ and ϕ' are topological conjugacies of a sofic system $(X_{\mathcal{M}}, \sigma_{\mathcal{M}})$ onto a sofic system $(X_{\mathcal{N}}, \sigma_{\mathcal{N}})$, where $\mathcal{M}$ and $\mathcal{N}$ are canonical.

Assume (1). Then there are integers $m \geq 0$ and $n \geq 1$, canonical strong shift equivalences

$$\mathcal{M} \overset{k_0}{\simeq} \mathcal{P}_1\mathcal{Q}_1,\ \mathcal{Q}_1\mathcal{P}_1 \overset{k_1}{\simeq} \mathcal{P}_2\mathcal{Q}_2,\ \cdots,\ \mathcal{Q}_{n-1}\mathcal{P}_{n-1} \overset{k_{n-1}}{\simeq} \mathcal{P}_n\mathcal{Q}_n,\ \mathcal{Q}_n\mathcal{P}_n \overset{k_n}{\simeq} \mathcal{N}$$

and

$$\mathcal{M} \overset{k'_0}{\simeq} \mathcal{P}_1\mathcal{Q}_1,\ \mathcal{Q}_1\mathcal{P}_1 \overset{k'_2}{\simeq} \mathcal{P}_2\mathcal{Q}_2,\ \cdots,\ \mathcal{Q}_{n-1}\mathcal{P}_{n-1} \overset{k'_{n-1}}{\simeq} \mathcal{P}_n\mathcal{Q}_n,\ \mathcal{Q}_n\mathcal{P}_n \overset{k'_n}{\simeq} \mathcal{N}$$

which give factorizations of $\phi\sigma_{\mathcal{M}}^m$ and $\phi'\sigma_{\mathcal{M}}^m$,respectively. Let these factorizations of $\phi\sigma_{\mathcal{M}}^m$ and $\phi'\sigma_{\mathcal{M}}^m$ be

$$\phi\sigma_{\mathcal{M}}^m = \kappa_n\zeta_n\kappa_{n-1}\cdots\kappa_1\zeta_1\kappa_0$$

and

$$\phi'\sigma_{\mathcal{M}}^m = \kappa'_n\zeta_n\kappa'_{n-1}\cdots\kappa'_1\zeta_1\kappa'_0,$$

respectively. Put

$$\hat{\phi} = \kappa_n\zeta_n\kappa_{n-1}\cdots\kappa_1\zeta_1\kappa_0$$

and put

$$\hat{\phi}_j = \kappa_n\zeta_n\cdots\kappa_{j+1}\zeta_{j+1}\kappa'_j\zeta_j\kappa_{j-1}\cdots\kappa_1\zeta_1\kappa_0.$$

for $j = 0, \cdots, n$. To see that (2) holds, it suffices to show that $\hat{\phi}^{-1}\hat{\phi}_j$ is a sofic simple automorphism of $(X_{\mathcal{M}}, \sigma_{\mathcal{M}})$ for any $0 \leq j \leq n$. We have

$$\hat{\phi}(\zeta_j\kappa_{j-1}\cdots\kappa_1\zeta_1\kappa_0)^{-1}\kappa_j^{-1}\kappa'_j(\zeta_j\kappa_{j-1}\cdots\kappa_1\zeta_1\kappa_0) = \hat{\phi}_j$$

for $j = 0, \cdots n$. Let $1 \leq j \leq n-1$. Since κ_j is given by the specified equivalence

$$\mathcal{Q}_j\mathcal{P}_j \overset{k_j}{\simeq} \mathcal{P}_{j+1}\mathcal{Q}_{j+1}$$

and κ'_j is given by the specified equivalence

$$\mathcal{Q}_j\mathcal{P}_j \overset{k'_j}{\simeq} \mathcal{P}_{j+1}\mathcal{Q}_{j+1},$$

$\kappa_j^{-1}\kappa'_j$ is given by the specified shift equivalence

$$\mathcal{Q}_j\mathcal{P}_j \overset{\hat{k}_j}{\simeq} \mathcal{Q}_j\mathcal{P}_j \quad \text{with } \hat{k}_j = k_j^{-1}k'_j.$$

Then

$$\hat{\phi}^{-1}\hat{\phi}_j = (\zeta_j\kappa_{j-1}\cdots\kappa_1\zeta_1\kappa_0)^{-1}\kappa_j^{-1}\kappa'_j(\zeta_j\kappa_{j-1}\cdots\kappa_1\zeta_1\kappa_0)$$

is a sofic simple automorphism. Similarly $\hat{\phi}^{-1}\hat{\phi}_0$ and $\hat{\phi}^{-1}\hat{\phi}_n$ are seen to be sofic simple automorphisms.

Therefore (2) holds.

In a similar way, we see that $\hat{\phi}_j\hat{\phi}^{-1}$ is a sofic simple automorphism of $(X_{\mathcal{N}}, \sigma_{\mathcal{N}})$ for any $0 \leq j \leq n$ so that (3) holds.

To prove that (2) implies (1), assume that

$$\phi^{-1}\phi' = \varphi_1 \cdots \varphi_l$$

where φ_i is a sofic simple automorphism of $(X_\mathcal{M}, \sigma_\mathcal{M})$ for $i = 1, \cdots, l$. Then $\phi = \phi\, \mathrm{id}_{X_\mathcal{M}} \cdots \mathrm{id}_{X_\mathcal{M}}$ and $\phi' = \phi\varphi_1 \cdots \varphi_l$. Hence ϕ and ϕ' are similar by Fact 9.2 and Lemma 9.3.

Similarly we see that (3) $\Rightarrow$ (1).

If ϕ and ϕ' are topological conjugacies between topological Markov shifts, then using Remark 9.1 it is similarly shown that (1),(2′) and (3′) are equivalent. □

Wagoner proved that simple automorphisms of a topological Markov shift generate a normal subgroup of the group of the automorphisms of the topological Markov shift. Statement (3) of the following corollary is closely related to Theorem 1.7 of [Wag 3].

Corollary 9.5. The following statements are valid.

(1) Similarity is an equivalence relation compatible with the multiplication on the group of automorphisms of a sofic system.

(2) The sofic simple automorphisms of a sofic system generate a normal subgroup of the group of the automorphisms of the sofic system and the normal subgroup is the kernel of the canonical surjection of similarity.

(3) The normal subgroup of the group of the automorphisms of a topological Markov shift generated by simple automorphisms is the kernel of the canonical surjection of similarity.

The quotient group of the group of automorphisms of a sofic system modulo similarity will be called the *strong shift equivalence group* of the sofic system.

Let $\mathcal{M}$ and $\mathcal{N}$ be representation matrices. A pair $(\mathcal{P}, \mathcal{Q})$ of representation matrices with the property that

$$\mathcal{M}\mathcal{P} \simeq \mathcal{P}\mathcal{N},\ \mathcal{N}\mathcal{Q} \simeq \mathcal{Q}\mathcal{M},\ \mathcal{M}^n \simeq \mathcal{P}\mathcal{Q},\ \mathcal{Q}\mathcal{P} \simeq \mathcal{N}^n$$

with $n \geq 0$, called a *shift equivalence* of *lag n from* $\mathcal{M}$ to $\mathcal{N}$ (see [BK2]). A shift equivalence of lag n from $\mathcal{M}$ to itself is called a *shift equivalence of lag n over* $\mathcal{M}$. A shift equivalence $(\mathcal{P}, \mathcal{Q})$ from $\mathcal{M}$ to $\mathcal{N}$ is said to be *canonical* if $\mathcal{M}$ and $\mathcal{N}$ are canonical.

Let $(\mathcal{P}, \mathcal{Q})$ be a canonical shift equivalence of lag n from $\mathcal{M}$ to $\mathcal{N}$ with $n \geq 1$. A forward conjugacy ϕ of the sofic system $(X_\mathcal{M}, \sigma_\mathcal{M})$ onto $(X_\mathcal{N}, \sigma_\mathcal{N})$ is said to be *associated with* $(\mathcal{P}, \mathcal{Q})$ if there is a canonical strong shift equivalence

$$\mathcal{M} \simeq \mathcal{P}_1\mathcal{Q}_1,\ \mathcal{Q}_1\mathcal{P}_1 \simeq \mathcal{P}_2\mathcal{Q}_2,\ \cdots,\ \mathcal{Q}_{n-1}\mathcal{P}_{n-1} \simeq \mathcal{P}_n\mathcal{Q}_n,\ \mathcal{Q}_n\mathcal{P}_n \simeq \mathcal{N}$$

such that ϕ is associated with it and

$$\mathcal{P} \simeq \mathcal{P}_1 \cdots \mathcal{P}_n \quad \text{and} \quad \mathcal{Q} \simeq \mathcal{Q}_n \cdots \mathcal{Q}_1.$$

Two topological conjugacies ϕ_1 and ϕ_2 of $(X_{\mathcal{M}}, \sigma_{\mathcal{M}})$ onto $(X_{\mathcal{N}}, \sigma_{\mathcal{N}})$ is said to be *weakly similar* if there is a nonnegative integer m such that $\phi_1\sigma_{\mathcal{M}}^m$ and $\phi_2\sigma_{\mathcal{M}}^m$ are forward conjugacies which are associated with the same canonical shift equivalence. We note that weak similarity as well as similarity is topological conjugacy invariant. (This easily follows from the definitions and Theorem 5.3.) Obviously we have

Fact 9.6. If two topological conjugacies between sofic systems are similar, then they are weakly similar.

For rectangular matrices $\mathcal{P}, \mathcal{Q}, \mathcal{P}'$ and $\mathcal{Q}'$, let $(\mathcal{P}, \mathcal{Q}) \simeq (\mathcal{P}', \mathcal{Q}')$ mean that $\mathcal{P} \simeq \mathcal{P}'$ and $\mathcal{Q} \simeq \mathcal{Q}'$.

Lemma 9.7 Let $\mathcal{M}, \mathcal{N}, \mathcal{M}'$ and $\mathcal{N}'$ be canonical representation matrices. Let $\phi : (X_{\mathcal{M}}, \sigma_{\mathcal{M}}) \to (X_{\mathcal{N}}, \sigma_{\mathcal{N}})$ and $\phi' : (X_{\mathcal{N}}, \sigma_{\mathcal{N}}) \to (X_{\mathcal{N}'}, \sigma_{\mathcal{N}'})$ be forward conjugacies. Assume that ϕ and ϕ' are associated with canonical shift equivalences $(\mathcal{P}, \mathcal{Q})$ and $(\mathcal{P}', \mathcal{Q}')$ of lag greater than 0, respectively. Then $\phi'\phi$ is a forward conjugacy associated with the canonical shift equivalence $(\mathcal{P}\mathcal{P}', \mathcal{Q}'\mathcal{Q})$ and $\sigma_{\mathcal{M}}^n\phi^{-1}$ is the forward conjugacy associated with the canonical shift equivalence $(\mathcal{Q}, \mathcal{P})$, where n is the lag of $(\mathcal{P}, \mathcal{Q})$.

Proof. There is a canonical specified strong shift equivalence

$$\mathcal{M} \overset{k_0}{\simeq} \mathcal{P}_1\mathcal{Q}_1,\ \mathcal{Q}_1\mathcal{P}_1 \overset{k_1}{\simeq} \mathcal{P}_2\mathcal{Q}_2,\ \cdots,\ \mathcal{Q}_{n-1}\mathcal{P}_{n-1} \overset{k_{n-1}}{\simeq} \mathcal{P}_n\mathcal{Q}_n,\ \mathcal{Q}_n\mathcal{P}_n \overset{k_n}{\simeq} \mathcal{N}$$

giving a factorization of ϕ with

$$\mathcal{P} \simeq \mathcal{P}_1 \cdots \mathcal{P}_n \quad \text{and} \quad \mathcal{Q} \simeq \mathcal{Q}_n \cdots \mathcal{Q}_1.$$

Also there is a canonical specified strong shift equivalence

$$\mathcal{M} \overset{k_0'}{\simeq} \mathcal{P}_1'\mathcal{Q}_1',\ \mathcal{Q}_1'\mathcal{P}_1' \overset{k_1'}{\simeq} \mathcal{P}_2'\mathcal{Q}_2',\ \cdots,\ \mathcal{Q}_{n'-1}'\mathcal{P}_{n'-1}' \overset{k_{n'-1}'}{\simeq} \mathcal{P}_{n'}\mathcal{Q}_{n'},\ \mathcal{Q}_{n'}\mathcal{P}_{n'} \overset{k_{n'}'}{\simeq} \mathcal{N}$$

giving a factorization of ϕ' with

$$\mathcal{P}' \simeq \mathcal{P}_1' \cdots \mathcal{P}_{n'}' \quad \text{and} \quad \mathcal{Q}' \simeq \mathcal{Q}_{n'}' \quad \mathcal{Q}_1'.$$

It is clear that $\phi'\phi$ is a forward conjugacy and the canonical specified strong shift equivalence

$$\mathcal{M} \overset{k_0}{\simeq} \mathcal{P}_1\mathcal{Q}_1,\ \mathcal{Q}_1\mathcal{P}_1 \overset{k_1}{\simeq} \mathcal{P}_2\mathcal{Q}_2,\ \cdots,\ \mathcal{Q}_{n-1}\mathcal{P}_{n-1} \overset{k_{n-1}}{\simeq} \mathcal{P}_n\mathcal{Q}_n,\ \mathcal{Q}_n\mathcal{P}_n \overset{k_0'k_n}{\simeq} \mathcal{P}_1'\mathcal{Q}_1',$$
$$\mathcal{Q}_1'\mathcal{P}_1' \overset{k_1'}{\simeq} \mathcal{P}_2'\mathcal{Q}_2',\ \cdots,\ \mathcal{Q}_{n'-1}'\mathcal{P}_{n'-1}' \overset{k_{n'-1}'}{\simeq} \mathcal{P}_{n'}\mathcal{Q}_{n'},\ \mathcal{Q}_{n'}\mathcal{P}_{n'} \overset{k_{n'}'}{\simeq} \mathcal{N}'$$

gives a factorization of $\phi'\phi$. Hence $\phi'\phi$ is associated with the shift equivalence

$$(\mathcal{P}_1 \cdots \mathcal{P}_n\mathcal{P}_1' \cdots \mathcal{P}_{n'}',\ \mathcal{Q}_{n'}' \cdots \mathcal{Q}_1'\mathcal{Q}_n \cdots \mathcal{Q}_1) \simeq (\mathcal{P}\mathcal{P}', \mathcal{Q}'\mathcal{Q}).$$

It is also clear that $\sigma_{\mathcal{M}}^n \phi^{-1}$ is a forward conjugacy and the canonical specified strong shift equivalence

$$\mathcal{N} \overset{k_n^{-1}}{\simeq} \mathcal{Q}_n \mathcal{P}_n,\ \mathcal{P}_n \mathcal{Q}_n \overset{k_{n-1}^{-1}}{\simeq} \mathcal{Q}_{n-1} \mathcal{P}_{n-1},\ \cdots,\ \mathcal{P}_2 \mathcal{Q}_2 \overset{k_1^{-1}}{\simeq} \mathcal{Q}_1 \mathcal{P}_1,\ \mathcal{P}_1 \mathcal{Q}_1 \overset{k_0^{-1}}{\simeq} \mathcal{M}$$

gives a factorization of $\sigma_{\mathcal{M}}^n \phi^{-1}$. Thus this is associated with the shift equivalence

$$(\mathcal{Q}_n \cdots \mathcal{Q}_1, \mathcal{P}_1 \cdots \mathcal{P}_n) \simeq (\mathcal{Q}, \mathcal{P}).$$

□

Proposition 9.8. Let (Y_i, σ_i), $i = 1, 2, 3$, be sofic systems and let ϕ_1 and ϕ_1' be topological conjugacies of (Y_1, σ_1) onto (Y_2, σ_2) and ϕ_2 and ϕ_2' topological conjugacies of (Y_2, σ_2) onto (Y_3, σ_3). If ϕ_i and ϕ_i' are weakly similar for $i = 1, 2$, then $\phi_2 \phi_1$ and $\phi_2' \phi_1'$ are weakly similar. If ϕ_1 and ϕ_1' are weakly similar, then so are ϕ_1^{-1} and $(\phi_1')^{-1}$.

Proof. This readily follows from Lemma 9.7. □

As is known (see [B4], [KMT], [Wag3]), two automorphisms of a topological Markov shift is weakly similar if and only if they induce the same automorphism of the dimension group of the defining matrix of the shift. An automorphism of a topological Markov shift is said to be inert if it induces the identity automorphism of the dimension group. Hence, let us say that an automorphism φ of a sofic system (Y, σ) is *inert* if φ and id_Y are weakly similar.

Corollary 9.9. Let ϕ and ϕ' be topological conjugacies between sofic systems. Then the following statements are equivalent:

(1) ϕ and ϕ' are weakly similar.

(2) $\phi^{-1}\phi'$ is inert

(3) $\phi'\phi^{-1}$ is inert.

Corollary 9.10. The following statements are valid.

(1) Weak similarity is an equivalence relation compatible with the multiplication on the group of automorphisms of a sofic system.

(2) The inert automorphisms of a sofic system generate a normal subgroup of the group of the automorphisms of a sofic system and the normal subgroup is the kernel of the canonical surjection of weak similarity.

The quotient group of the group of automorphisms of a sofic system modulo weak similarity will be called the *shift equivalence group* of the sofic system. Since similarity implies weak similarity, the shift equivalence group is a factor of the strong shift equivalence group.

It is easily seen that the following lemma holds.

Lemma 9.11. Let $\mathcal{M}$ be a canonical representation matrix. Then $\mathrm{id}_{X_{\mathcal{M}}}$ is associated with the canonical shift equivalence $(\mathcal{I}, \mathcal{M})$ and $\sigma_{\mathcal{M}}$ is associated with the canonical shift equivalence $(\mathcal{M}, \mathcal{I})$, where $\mathcal{I}$ is the representation matrix that represents the λ-graph (G_0, λ_0) such that $M_{G_0} = I$ (the identity matrix) and $\lambda_0(a) = \lambda_0(a')$ for any $a, a' \in A_{G_0}$.

Lemma 9.12. Let $\mathcal{M}$ be a canonical representation matrix. If $(\mathcal{P}_1, \mathcal{Q}_1)$ and $(\mathcal{P}_2, \mathcal{Q}_2)$ are canonical shift equivalences of lag greater than 0 over $\mathcal{M}$ such that there is a forward automorphism of $(X_{\mathcal{M}}, \sigma_{\mathcal{M}})$ associated with both $(\mathcal{P}_1, \mathcal{Q}_1)$ and $(\mathcal{P}_2, \mathcal{Q}_2)$, then there is a nonnegative integer m such that

$$\mathcal{P}_1\mathcal{M}^m \simeq \mathcal{P}_2\mathcal{M}^m.$$

Proof. Let $(\mathcal{P}, \mathcal{Q})$ be a canonical shift equivalence of lag n over $\mathcal{M}$ with $n \geq 1$. Let φ be an LR automorphism of $(X_{\mathcal{M}}, \sigma_{\mathcal{M}})$ associated with $(\mathcal{P}, \mathcal{Q})$. Then there is a canonical specified shift equivalence

$$\mathcal{M} \overset{k_0}{\simeq} \mathcal{P}_1\mathcal{Q}_1,\ \mathcal{Q}_1\mathcal{P}_1 \overset{k_1}{\simeq} \mathcal{P}_2\mathcal{Q}_2,\ \cdots,\ \mathcal{Q}_{n-1}\mathcal{P}_{n-1} \overset{k_{n-1}}{\simeq} \mathcal{P}_n\mathcal{Q}_n,\ \mathcal{Q}_n\mathcal{P}_n \overset{k_n}{\simeq} \mathcal{M} \tag{9.1}$$

such that it gives a κ-ζ factorization of φ and the following hold:

$$\mathcal{P} \simeq \mathcal{P}_1 \cdots \mathcal{P}_n \quad \text{and} \quad \mathcal{Q} \simeq \mathcal{Q}_n \cdots \mathcal{Q}_1.$$

Let

$$\mathcal{M}\mathcal{P} \overset{l}{\simeq} \mathcal{P}\mathcal{M} \tag{9.2}$$

be the specified equivalence naturally induced by (9.1). Let $\mathcal{T} = (T, \lambda, \mu)$ with $T = (p, q : \Gamma \to G)$, be the LR sofic textile system associated with (9.2). Then by Corollary 6.24, $\mathcal{T}$ is 1-1 and $\varphi_{\mathcal{T}} = \varphi$. Let

$$\tilde{M} \overset{k_0}{\simeq} \tilde{P}_1\tilde{Q}_1,\ \tilde{Q}_1\tilde{P}_1 \overset{k_1}{\simeq} \tilde{P}_2\tilde{Q}_2,\ \cdots,\ \tilde{Q}_{n-1}\tilde{P}_{n-1} \overset{\bar{k}_{n-1}}{\simeq} \tilde{P}_n\tilde{Q}_n,\ \tilde{Q}_n\tilde{P}_n \overset{k_n}{\simeq} \tilde{M} \tag{9.3}$$

be the support of (9.1). Let

$$\tilde{M}\tilde{P} \overset{\bar{l}}{\simeq} \tilde{P}\tilde{M} \tag{9.4}$$

be the specified equivalence naturally induced by (9.3) with $\tilde{P} = \tilde{P}_1 \cdots \tilde{P}_n$. Then since the derivation of (9.2) from (9.1) is accompanied by that of (9.4) from (9.3), we know that (9.4) is the support of (9.2) so that T is the LR textile system associated with (9.4). Therefore, by Corollary 6.11, T is 1-1 and (9.3) gives a κ-ζ factorization of φ_T. Since $\mathcal{M}$ is canonical, $\mathcal{M}$ is the representation matrix of the λ-graph $(\mathcal{K}_Y^-)_+ = (G, \lambda)$, where $Y = X_{\mathcal{M}}$. Put $\mathcal{K}_Y^- = (K, \lambda_0)$. Then by Krieger's theorem (Theorem 4.2), there is a unique automorphism φ_0 of (X_K, σ_K) such that $\pi_{\mathcal{K}_Y^-}\varphi_0 = \varphi\pi_{\mathcal{K}_Y^-}$. Let $(\varphi_0)_+$ be the unique automorphism of (X_G, σ_G) such that $\pi_{(\mathcal{K}_Y^-)_+}(\varphi_0)_+ = \varphi\pi_{(\mathcal{K}_Y^-)_+}$ and

$\gamma_+^{\kappa_Y^-}((\varphi_0)_+(x)) = \varphi_0(\gamma_+^{\kappa_Y^-}(x))$ for all $x \in X_G$ (cf. Theorem 4.3). Since (9.1) gives a κ-ζ factorization of φ and its support (9.3) gives a κ-ζ factorization of φ_T, we know, by the fact stated after the diagram (5.3), that $\varphi_T = (\varphi_0)_+$. By Corollary 7.25, a 1-1 LR textile system T with $\varphi_T = (\varphi_0)_+$ is unique so that T is unique for φ. Hence $\mathcal{T}$ is determined by φ except for μ.

Let $\alpha_1 \cdots \alpha_k \in L(\Gamma)$ be arbitrary with $k \geq 1$ and $\alpha_i \in A_\Gamma$. Since $\mathcal{T}$ is LR,

$$(\lambda(p(\alpha_1)) \cdots \lambda(p(\alpha_k)),\ \mu(t_\Gamma(\alpha_k))$$

uniquely determines

$$(\mu(i_\Gamma(\alpha_1)),\ \lambda(q(\alpha_1)) \cdots \lambda(q(\alpha_k)),$$

and the latter uniquely determines the former. Since $\mathcal{T}$ is 1-1, both ξ_T and η_T are 1-1. This implies that there is a nonnegative integer m such that if $k \geq m$, then $\lambda(p(\alpha_1)) \cdots \lambda(p(\alpha_k))$ uniquely determines $\mu(i_\Gamma(\alpha_1))$ and $\lambda(q(\alpha_1)) \cdots \lambda(q(\alpha_k))$ uniquely determines $\mu(t_\Gamma(\alpha_k))$. Therefore, if $k \geq m$, then there is a 1-1 correspondence between

$$(\mu(i_\Gamma(\alpha_1)),\ \lambda(q(\alpha_1)) \cdots \lambda(q(\alpha_k))$$

and

$$(\lambda(p(\alpha_1)) \cdots \lambda(p(\alpha_k)),\ \lambda(q(\alpha_1)) \cdots \lambda(q(\alpha_k))).$$

By the above, we conclude that if $(\mathcal{P}_1, \mathcal{Q}_1)$ and $(\mathcal{P}_2, \mathcal{Q}_2)$ are canonical shift equivalences with which φ is associated, then there is a nonnegative integer m with $\mathcal{P}_1\mathcal{M}^m \simeq \mathcal{P}_2\mathcal{M}^m$. □

Proposition 9.13. Let $\mathcal{M}$ be a canonical representation matrix. Let φ_1 and φ_2 be automorphisms of $(X_\mathcal{M}, \sigma_\mathcal{M})$. Let m be a nonnegative integer. Assume that $\varphi_i\sigma_\mathcal{M}^m$ is forward and associated with a canonical shift equivalence $(\mathcal{P}_i, \mathcal{Q}_i)$ of lag greater than 0 over $\mathcal{M}$ for $i = 1, 2$. Then φ_1 and φ_2 are weakly similar if and only if there is a nonnegative integer n such that $\mathcal{P}_1\mathcal{M}^n \simeq \mathcal{P}_2\mathcal{M}^n$.

Proof. Suppose that φ_1 and φ_2 are weakly similar. Then $\varphi_1\sigma_\mathcal{M}^m$ and $\varphi_2\sigma_\mathcal{M}^m$ are weakly similar. Hence there is $k \geq 0$ and a canonical shift equivalence $(\mathcal{P}, \mathcal{Q})$ of lag greater than 0 over $\mathcal{M}$ such that both $\varphi_1\sigma_\mathcal{M}^{m+k}$ and $\varphi_2\sigma_\mathcal{M}^{m+k}$ are associated with $(\mathcal{P}, \mathcal{Q})$. By Lemmas 9.7 and 9.11 $\varphi_i\sigma_\mathcal{M}^{m+k}$ is associated with $(\mathcal{P}_i\mathcal{M}^k, \mathcal{Q}_i)$ for $i = 1, 2$. By Lemma 9.12, there is $l \geq 0$ such that $\mathcal{P}_i\mathcal{M}^{k+l} \simeq \mathcal{P}\mathcal{M}^l$ for $i = 1, 2$. Thus $\mathcal{P}_1\mathcal{M}^{k+l} \simeq \mathcal{P}_2\mathcal{M}^{k+l}$.

Suppose that $\mathcal{P}_1\mathcal{M}^n \simeq \mathcal{P}_2\mathcal{M}^n$ with $n \geq 0$. Let l_i be the lag of $(\mathcal{P}_i, \mathcal{Q}_i)$ for $i = 1, 2$. Then

$$\begin{aligned} \mathcal{Q}_1\mathcal{M}^{n+l_2} &\simeq \mathcal{Q}_1\mathcal{M}^n\mathcal{P}_2\mathcal{Q}_2 \simeq \mathcal{Q}_1\mathcal{P}_2\mathcal{M}^n\mathcal{Q}_2 \\ &\simeq \mathcal{Q}_1\mathcal{P}_1\mathcal{M}^n\mathcal{Q}_2 \simeq \mathcal{M}^{l_1+n}\mathcal{Q}_2 \simeq \mathcal{Q}_2\mathcal{M}^{n+l_1}. \end{aligned}$$

By Lemma 9.7 and 9.11, $\varphi_1\sigma^{m+n}$ is associated with $(\mathcal{P}_1\mathcal{M}^n, \mathcal{Q}_1\mathcal{M}^{n+l_2})$ and $\varphi_2\sigma^{m+n}$ is associated with $(\mathcal{P}_2\mathcal{M}^n, \mathcal{Q}_2\mathcal{M}^{n+l_1})$. Since $(\mathcal{P}_1\mathcal{M}^n, \mathcal{Q}_1\mathcal{M}^{n+l_2}) \simeq (\mathcal{P}_2\mathcal{M}^n, \mathcal{Q}_2\mathcal{M}^{n+l_1})$, we conclude that φ_1 and φ_2 are weakly similar. $\square$

Let $\mathcal{M}$ be a canonical representation matrix. Let $\Sigma_0(\mathcal{M})$ be the set of all pairs $(\mathcal{P}, \mathcal{Q})$ of representation matrices such that $\mathcal{P}\mathcal{M} \simeq \mathcal{M}\mathcal{P}, \mathcal{Q}\mathcal{M} \simeq \mathcal{M}\mathcal{Q}$ and there are nonnegative integers m and n such that

$$\mathcal{M}^{m+2n} \simeq \mathcal{P}\mathcal{Q}\mathcal{M}^m \quad \text{and} \quad \mathcal{Q}\mathcal{P}\mathcal{M}^m \simeq \mathcal{M}^{m+2n}.$$

It is clear that if $(\mathcal{P}_1, \mathcal{Q}_1), (\mathcal{P}_2, \mathcal{Q}_2) \in \Sigma_0(\mathcal{M})$, then $(\mathcal{P}_1\mathcal{P}_2, \mathcal{Q}_2\mathcal{Q}_1) \in \Sigma_0(\mathcal{M})$. Therefore we can define the product of $(\mathcal{P}_1, \mathcal{Q}_1)$ and $(\mathcal{P}_2, \mathcal{Q}_2)$ in $\Sigma_0(\mathcal{M})$ to be $(\mathcal{P}_1\mathcal{P}_2, \mathcal{Q}_2\mathcal{Q}_1)$, and with this multiplication, $\Sigma_0(\mathcal{M})$ is a semigroup. We define an equivalence relation $\sim$ on $\Sigma_0(\mathcal{M})$ as follows: for $(\mathcal{P}_1, \mathcal{Q}_1), (\mathcal{P}_2, \mathcal{Q}_2)$ in $\Sigma_0(\mathcal{M})$,

$$(\mathcal{P}_1, \mathcal{Q}_1) \sim (\mathcal{P}_2, \mathcal{Q}_2)$$

means that there are nonnegative integers m_1 and m_2 such that

$$(\mathcal{P}_1\mathcal{M}^{m_1}, \mathcal{Q}_1\mathcal{M}^{m_1}) \simeq (\mathcal{P}_2\mathcal{M}^{m_2}, \mathcal{Q}_2\mathcal{M}^{m_2}).$$

Clearly $\sim$ is an equivalence relation compatible with the multiplication. For $(\mathcal{P}, \mathcal{Q}) \in \Sigma_0(\mathcal{M})$, let

$$[\mathcal{P}, \mathcal{Q}]$$

denote the equivalence class of $\sim$ containing $(\mathcal{P}, \mathcal{Q})$. Define

$$\Sigma(\mathcal{M}) \simeq \Sigma_0(\mathcal{M})/\sim.$$

Then $\Sigma(\mathcal{M})$ is a group, because $[\mathcal{I}, \mathcal{I}]$ is its identity element and the inverse of $[\mathcal{P}, \mathcal{Q}]$ in $\Sigma(\mathcal{M})$ is $[\mathcal{Q}, \mathcal{P}]$, where $\mathcal{I}$ is the same representation matrix as was defined in Lemma 9.11. We shall call $\Sigma(\mathcal{M})$ the *shift equivalence group* of $\mathcal{M}$.

Let $\mathcal{M}$ be a canonical representation matrix. Let $\mathrm{Aut}(\sigma_{\mathcal{M}})$ denote the automorphism group of the sofic system $(X_{\mathcal{M}}, \sigma_{\mathcal{M}})$. Let $g_{\mathcal{M}} : \mathrm{Aut}(\sigma_{\mathcal{M}}) \to \Sigma(\mathcal{M})$ be defined by

$$g_{\mathcal{M}}(\varphi) = [\mathcal{P}, \mathcal{P}'],$$

where $\mathcal{P}$ and $\mathcal{P}'$ are such that there are a nonnegative integer m and canonical shift equivalences $(\mathcal{P}, \mathcal{Q})$ and $(\mathcal{P}', \mathcal{Q}')$ of lag greater than 0 over $\mathcal{M}$ such that both $\varphi\sigma^m$ and $\varphi^{-1}\sigma^m$ are forward automorphisms, $\varphi\sigma^m$ is associated with $(\mathcal{P}, \mathcal{Q})$, and $\varphi^{-1}\sigma^m$ is associated with $(\mathcal{P}', \mathcal{Q}')$. By Lemma 9.7, σ^{2m} is associated with both $(\mathcal{P}\mathcal{P}', \mathcal{Q}'\mathcal{Q})$ and $(\mathcal{P}'\mathcal{P}, \mathcal{Q}\mathcal{Q}')$. By Lemmas 9.7 and 9.11, σ^{2m} is associated with $(\mathcal{M}^{2m}, \mathcal{I})$. Hence by Lemma 9.12, there is $n \geq 0$ such that $\mathcal{P}\mathcal{P}'\mathcal{M}^n \simeq \mathcal{M}^{2m+n}$ and $\mathcal{P}'\mathcal{P}\mathcal{M}^n \simeq \mathcal{M}^{2m+n}$. Hence $(\mathcal{P}, \mathcal{P}') \in \Sigma_0(\mathcal{M})$. By Lemma 9.12, we see that $g_{\mathcal{M}}(\varphi)$ is well-defined. If $\varphi_1, \varphi_2 \in \mathrm{Aut}(\sigma_{\mathcal{M}})$, all $\varphi_1\sigma^m_{\mathcal{M}}, \varphi_1^{-1}\sigma^m_{\mathcal{M}}, \varphi_2\sigma^n_{\mathcal{M}}$, and $\varphi_2^{-1}\sigma^n_{\mathcal{M}}$ are forward and they are associated with canonical shift equivalences $(\mathcal{P}_1, \mathcal{Q}_1), (\mathcal{P}'_1, \mathcal{Q}'_1), (\mathcal{P}_2, \mathcal{Q}_2)$, and $(\mathcal{P}'_2, \mathcal{Q}'_2)$, respectively, then $\varphi_2\varphi_1\sigma_{\mathcal{M}}^{m+n}$ and $\varphi_1^{-1}\varphi_2^{-1}\sigma_{\mathcal{M}}^{m+n}$ are forward and associated with $(\mathcal{P}_1\mathcal{P}_2, \mathcal{Q}_2\mathcal{Q}_1)$ and $(\mathcal{P}'_2\mathcal{P}'_1, \mathcal{Q}'_1\mathcal{Q}'_2)$, respectively. Therefore it follows that

$$g_{\mathcal{M}}(\varphi_2\varphi_1) = g_{\mathcal{M}}(\varphi_1)g_{\mathcal{M}}(\varphi_2).$$

Hence $g_{\mathcal{M}}$ is a group-homomorphism.

We shall know that the shift equivalence group of $(X_{\mathcal{M}}, \sigma_{\mathcal{M}})$ is embedded into $\Sigma_{\mathcal{M}}$.

Theorem 9.14. Let $\mathcal{M}$ be a canonical representation matrix with $h(\sigma_{\mathcal{M}}) > 0$. The shift equivalence group of $(X_{\mathcal{M}}, \sigma_{\mathcal{M}})$ is isomorphic to $g_{\mathcal{M}}(\mathrm{Aut}(\sigma_{\mathcal{M}}))$.

Proof. It suffices to show that for $\varphi \in \mathrm{Aut}(\sigma_{\mathcal{M}})$, $g_{\mathcal{M}}(\varphi) = [\mathcal{I}, \mathcal{I}]$ if and only if φ is inert.

Assume that $g_{\mathcal{M}}(\varphi) = [\mathcal{I}, \mathcal{I}]$. Then there are $k, l, m \geq 0$ and canonical shift equivalences $(\mathcal{P}, \mathcal{Q})$ and $(\mathcal{P}', \mathcal{Q}')$ of lag greater than 0 over $\mathcal{M}$ such that $\varphi\sigma^m$ and $\varphi^{-1}\sigma^m$ are forward and associated with $(\mathcal{P}, \mathcal{Q})$ and $(\mathcal{P}', \mathcal{Q}')$, respectively, and

$$(\mathcal{P}\mathcal{M}^k, \mathcal{P}'\mathcal{M}^k) \simeq (\mathcal{M}^l, \mathcal{M}^l).$$

It follows from Lemmas 9.7 and 9.11 that $\varphi\sigma_{\mathcal{M}}^{m+k}, \varphi^{-1}\sigma_{\mathcal{M}}^{m+k}$, and $\sigma_{\mathcal{M}}^l$ are associated with $(\mathcal{P}\mathcal{M}^k, \mathcal{Q}), (\mathcal{P}'\mathcal{M}^k, \mathcal{Q}')$ and $(\mathcal{M}^l, \mathcal{I})$, respectively. Hence, by Proposition 9.13, all $\varphi\sigma_{\mathcal{M}}^{m+k}, \varphi^{-1}\sigma_{\mathcal{M}}^{m+k}$ and $\sigma_{\mathcal{M}}^l$ are weakly similar. Thus by proposition 9.8, $\sigma_{\mathcal{M}}^{2m+2k}$ and $\sigma_{\mathcal{M}}^{2l}$ are weakly similar. By Lemmas 9.7 and 9.11, $\sigma_{\mathcal{M}}^{2m+2k}$ and $\sigma_{\mathcal{M}}^{2l}$ are associated with $(\mathcal{M}^{2m+2k}, \mathcal{I})$ and $(\mathcal{M}^{2l}, \mathcal{I})$, respectively. Therefore, by Proposition 9.13, there is $n \geq 0$ such that $\mathcal{M}^{2m+2k}\mathcal{M}^n \simeq \mathcal{M}^{2l}\mathcal{M}^n$. Hence $M^{2m+2k+n} = M^{2l+n}$, where M is the support of $\mathcal{M}$. Since $h(\sigma_{\mathcal{M}}) > 0$ and $\mathcal{M}$ represents a right-resolving λ-graph, we have $h(\sigma_M) > 0$. Thus we have $2m + 2k + n = 2l + n$, so that $m + k = l$. Since $\varphi\sigma_{\mathcal{M}}^{m+k}$ and $\sigma_{\mathcal{M}}^l$ are weakly similar, we conclude that φ is inert.

Conversely assume that φ is inert. Then φ^{-1} is inert. There is $m \geq 0$ such that $\varphi\sigma^m$ and $\varphi^{-1}\sigma^m$ are forward and all $\varphi\sigma^m, \varphi^{-1}\sigma^m$ and σ^m are weakly similar. Assume that $\varphi\sigma^m$ and $\varphi^{-1}\sigma^m$ are associated with canonical shift equivalences $(\mathcal{P}, \mathcal{Q})$ and $(\mathcal{P}', \mathcal{Q}')$ of lag greater than 0 over $\mathcal{M}$, respectively. By Lemmas 9.7 and 9.11, $\sigma_{\mathcal{M}}^m$ is associated with $(\mathcal{M}^m, \mathcal{I})$. By Proposition 9.13, there is $n \geq 0$ such that

$$(\mathcal{P}\mathcal{M}^n, \mathcal{P}'\mathcal{M}^n) \simeq (\mathcal{M}^{m+n}, \mathcal{M}^{m+n}).$$

Thus

$$g_{\mathcal{M}}(\varphi) = [\mathcal{P}, \mathcal{P}'] = [\mathcal{I}, \mathcal{I}].$$

□

If we restrict our consideration to the automorphisms of topological Markov shifts, the arguments above can be made more simply in the category of nonnegative integral matrices.

Let M be the adjacency matrix of a graph. A pair (P, Q) of (square) nonnegative matrices with the property that

$$MP = PM, \quad MQ = QM, \quad QP = M^n = PQ$$

with $n \geq 0$, called a *shift equivalence of lag n over M.* Let (P, Q) be a shift equivalence of lag greater than 0 over M. Recalling Theorem 6.29, an LR automorphism φ of the topological Markov shift (X_M, σ_M) is said to be *associated with* (P, Q) if φ is associated with a strong shift equivalence

$$M = P_1Q_1, \; Q_1P_1 = P_2Q_2, \; \cdots, \; Q_{n-1}P_{n-1} = P_nQ_n, \; Q_nP_n = M$$

such that

$$P = P_1 \cdots P_n \quad \text{and} \quad Q = Q_n \cdots Q_1.$$

It is clear that two automorphisms φ_1 and φ_2 of (X_M, σ_M) is weakly similar if and only if there is a nonnegative integer m such that $\varphi_1\sigma_M^m$ and $\varphi_2\sigma_M^m$ are LR automorphisms which are associated with the same shift equivalence.

Let M be the adjacency matrix of a graph. Let φ be an LR endomorphism of (X_M, σ_M). Then there is a one-sided 1-1 LR textile system T with $\varphi_T = \varphi$, which is uniquely determined by φ, by Corollary 7.25. Let $\tilde{M}\tilde{P} \overset{k}{\simeq} \tilde{P}\tilde{M}$ be the specified equivalence associated with T. We define

$$P(\varphi) = P.$$

For example, $P(\mathrm{id}_{X_M}) = I$, where I is the identity matrix, and $P(\sigma_M) = M$. By Observation 6.9, we know that if φ_1 and φ_2 are LR endomorphisms of a topological Markov shift, then $\varphi_2\varphi_1$ is also LR and $P(\varphi_2\varphi_1) = P(\varphi_1)P(\varphi_2)$. Corresponding to Proposition 9.13, we have the following proposition, which is a version of a result found in Section 3 of [KMT].

Proposition 9.15. Let M be the adjacency matrix of a graph. Two automorphisms φ_1 and φ_2 of (X_M, σ_M) are weakly similar if and only if there is $m \geq 0$ such that $\varphi_1\sigma^m$ and $\varphi_2\sigma^m$ are LR and $P(\varphi_1\sigma^m) = P(\varphi_2\sigma^m)$.

Let M be the adjacency matrix of a graph. Then $\tilde{M}$ is canonical. We can directly define $\Sigma(\tilde{M})$ as follows. Let $\Sigma_0(M)$ be the set of all shift equivalences with even lag over M. It is clear that if $(P_1, Q_1), (P_2, Q_2) \in \Sigma_0(M)$, then $(P_1P_2, Q_2Q_1) \in \Sigma_0(M)$. Therefore we can define the product of (P_1, Q_1) and (P_2, Q_2) in $\Sigma_0(M)$ to be (P_1P_2, Q_2Q_1). With this multiplication, $\Sigma_0(M)$ is a monoid with (I, I) as the identify element, where I is e identity matrix. We define an equivalence relation $\sim$ on $\Sigma_0(M)$ as follows: for $(P_1, Q_1), (P_2, Q_2)$,

$$(P_1, Q_1) \sim (P_2, Q_2)$$

means that there are nonnegative integers m_1 and m_2 such that

$$(P_1 M^{m_1}, Q_1 M^{m_1}) = (P_2 M^{m_2}, Q_2 M^{m_2}).$$

Clearly $\sim$ is a equivalence relation compatible with the multiplication. For $(P, Q) \in \Sigma_0(M)$, let $[P, Q]$ denote the equivalence class of $\sim$ containing (P, Q). Define

$$\Sigma(M) = \Sigma_0(M)/\sim.$$

Then $\Sigma(M)$ is a group, because the inverse of $[P, Q]$ in $\Sigma(M)$ is $[Q, P]$. Clearly, $\Sigma(\tilde{M})$ is isomorphic to $\Sigma(M)$. Corresponding to $g_{\tilde{M}}, g_M$ is given by

$$g_M(\varphi) = [P(\varphi\sigma^m), P(\varphi^{-1}\sigma^m)], \quad \varphi \in \mathrm{Aut}(\sigma_M),$$

where m is a nonnegative integer such that both $\varphi\sigma^m$ and $\varphi^{-1}\sigma^m$ are LR.

Here we review the notions of the dimension groups, the dimension triples, and the automorphisms of the dimension triples, associated with nonnegative integral matrices, according to an algebraic description given in [BMT]. For details and background, the reader is referred to [BMT] and [BLR].

Let M be an $r \times r$ nonnegative integral matrix with $r \in \mathbf{N}$. Let W_M denote the eventual range of the linear transformation $M : \boldsymbol{Q}^r \to \boldsymbol{Q}^r$, i.e., $W_M = M^r(\boldsymbol{Q}^r)$. We assume that M acts on row vectors. Define

$$D_M = \{v \in W_M \mid vM^k \in \mathbf{Z}^r \text{ for some } k \in \mathbf{N}\}$$

and define

$$D_M^+ = \{v \in W_M \mid vM^k \in \mathbf{Z}^r \text{ and } vM^k \geq 0 \text{ for some } k \in \mathbf{N}\}.$$

The *dimension group* associated with M is the ordered group (D_M, D_M^+). Let an automorphism $\widehat{M}$ of (D_M, D_M^+) be defined by

$$\widehat{M} = M|D_M.$$

Then $(D_M, D_M^+, \widehat{M})$ is called the *dimension triple* associated with M. An automorphism Φ of (D_M, D_M^+) such that $\Phi\widehat{M} = \widehat{M}\Phi$, is called an *automorphism* of the dimension triple $(D_M, D_M^+, \widehat{M})$. Let $\mathrm{Aut}(\widehat{M})$ denote the group of automorphisms of $(D_M, D_M^+, \widehat{M})$.

To justify doubly our terminology of an inert automorphism for sofic systems, we prove the following proposition, which is closely related to Proposition 4.17 of [Wag1].

Proposition 9.16. Let M be the adjacency matrix of a graph whose maximal eigenvalue is greater than 1. Then $\Sigma(M)$ is isomorphic to $\mathrm{Aut}(\widehat{M})$.

Proof. It readily follows that if (P, Q) is a shift equivalence of lag m over M with $m \geq 0$, then the mapping $\widehat{P} : D_M \to D_M$ defined by $\widehat{P} = P|D_M$ is an automorphism of $(D_M, D_M^+, \widehat{M})$ and

$$\widehat{P}\widehat{Q} = \widehat{Q}\widehat{P} = \widehat{M}^m.$$

Let $\bar{P}$ denote the nonsingular $\boldsymbol{Q}$-linear map of W_M defined by $\bar{P} = P|W_M$.

Let (P, Q) be a shift equivalence of even lag, say $2m$, over M. We define $h_M : \Sigma(M) \to \mathrm{Aut}(\widehat{M})$ by

$$h_M([P, Q]) = \widehat{P}\widehat{M}^{-m}.$$

To see that h is well defined, assume that (P_i, Q_i) is a shift equivalence of lag $2m_i$ for $i = 1, 2$ and that there are nonnegative integers k_1 and k_2 such that

$$P_1 M^{k_1} = P_2 M^{k_2} \quad \text{and} \quad Q_1 M^{k_1} = Q_2 M^{k_2}$$

Therefore

$$M^{2m_1+2k_1} = P_1 Q_1 M^{2k_1} = P_2 Q_2 M^{2k_2} = M^{2m_2+2k_2}$$

so that

$$m_1 + k_1 = m_2 + k_2,$$

because the maximal eigenvalue of M is greater than 1. Since $P_1 M^{k_1} = P_2 M^{k_2}$, we have $\widehat{P}_1 \widehat{M}^{k_1} = \widehat{P}_2 \widehat{M}^{k_2}$. Therefore

$$\widehat{P}_1 \widehat{M}^{-m_1} = \widehat{P}_1 \widehat{M}^{k_1} \widehat{M}^{-(m_1+k_1)} = \widehat{P}_2 \widehat{M}^{k_2} \widehat{M}^{-(m_2+k_2)} = \widehat{P}_2 \widehat{M}^{-m_2}.$$

Thus h is well defined.

It is easy to see that h_M is a group-homomorphism.

To see that h_M is 1-1, assume that (P_i, Q_i) is a shift equivalence of lag $2m_i$ over M for $i = 1, 2$ and that

$$\widehat{P}_1 \widehat{M}^{-m_1} = \widehat{P}_2 \widehat{M}^{-m_2}.$$

Since $\widehat{P}_i \widehat{Q}_i = \widehat{Q}_i \widehat{P}_i = \widehat{M}^{2m_i}$ for $i = 1, 2$, we have

$$\widehat{Q}_1 \widehat{M}^{-m_1} = \widehat{Q}_2 \widehat{M}^{-m_2}.$$

Since, as was proved in ([BMT], p.15), any group-automorphism of D_M extends uniquely to a nonsingular $\boldsymbol{Q}$-linear map of W_M, we have

$$\bar{P}_1 \bar{M}^{-m_1} = \bar{P}_2 \bar{M}^{-m_2} \quad \text{and} \quad \bar{Q}_1 \bar{M}^{-m_1} = \bar{Q}_2 \bar{M}^{-m_2}$$

so that

$$\bar{P}_1 \bar{M}^{m_2} = \bar{P}_2 \bar{M}^{m_1} \quad \text{and} \quad \bar{Q}_1 \bar{M}^{m_2} = \bar{Q}_2 \bar{M}^{m_1}.$$

Therefore

$$P_1 M^{m_2+r} = P_2 M^{m_1+r} \quad \text{and} \quad Q_1 M^{m_2+r} = Q_2 M^{m_1+r}$$

where M is $r \times r$. Thus $[P_1, Q_1] = [P_2, Q_2]$.

The proof that h_M is onto, is clear from the arguments given in the proof of Proposition 6.3 of [BLR]. We describe it for completeness.

Let $\Phi : D_M \to D_M$ be an automorphism of $(D_M, D_M^+, \widehat{M})$. Then as stated above, there is a nonsingular $\boldsymbol{Q}$-linear map $\bar{\Phi} : W_M \to W_M$ to which Φ extends. Let $K = \ker M^r$. Then $\boldsymbol{Q}^r = W_M \oplus K$. Let 0_K denote the zero map on K. Let P_0 and Q_0 be the matrices for $\bar{\Phi} \oplus 0_K$ and $\bar{\Phi}^{-1} \oplus 0_K$, respectively, with respect to the standard basis. Since

$$(\bar{\Phi} \oplus 0_K)(\bar{\Phi}^{-1} \oplus 0_K) = (\bar{\Phi}^{-1} \oplus 0_K)(\bar{\Phi} \oplus 0_K) = \mathrm{id}_{W_M} \oplus 0_K,$$

we have

$$P_0 Q_0 = Q_0 P_0 \quad \text{and} \quad \overline{P_0 Q_0} = \mathrm{id}_{W_M}$$

Since $\bar{\Phi}\bar{M} = \bar{M}\bar{\Phi}$, we have

$$MP_0 = P_0 M.$$

Therefore, since each row vector of M is in D_M^+ and P_0 maps each vector in D_M^+ to a vector in D_M^+, $P_0 M^l$ is a nonnegative integral matrix for all sufficiently large $l \in \mathbf{N}$. Similarly we have

$$MQ_0 = Q_0 M$$

and $Q_0 M^l$ is a nonnegative integral matrix for all sufficiently large $l \in \mathbf{N}$. Therefore, there is $n \in \mathbf{N}$ such that $P_0 M^n$ and $Q_0 M^n$ are nonnegative integral matrices. Let $P = P_0 M^n$ and $Q = Q_0 M^n$. Then

$$PM = MP, \quad QM = MQ, \quad PQ = QP = M^{2n}$$

and

$$h_M([P, Q]) = \widehat{P}\widehat{M}^{-n} = \Phi.$$

Therefore h_M is onto. □

Remark 9.17. Let φ be an automorphism of a topological Markov shift (X_M, σ_M) with $h(\sigma_M) > 0$. Then the induced action of φ on the dimension group (D_M, D_M^+) is given by

$$\widehat{P}\widehat{M}^{-n}$$

where $P = P(\varphi\sigma_M^n)$ and n is an integer such that $\varphi\sigma_M^n$ is an LR automorphism.

Similarity and weak similarity are different. In fact, Kim and Roush have given an example of an automorphism (of order 2) of a topological Markov shift which is not a composition of simple automorphisms but whose induced action on the dimension group is the identity [KR]. Recently, Kim, Roush, and Wagoner have shown that g_M is not necessarily onto for the adjacency matrix M of a graph with $h(\sigma_M) > 0$ [KRW].

Let φ_i be an automorphism of a sofic system (Y_i, σ_i) for $i = 1, 2$. We say that (Y_1, φ_1) and (Y_2, φ_2) are *shift equivalent* if (Y_1, φ_1) and (Y_2, φ_2) are topologically conjugate to sofic systems whose defining canonical representation matrices are shift equivalent to each other. If φ_i is an automorphism of a topological Markov shift (X_i, σ_i) for $i = 1, 2$, then (X_1, φ_1) and (X_2, φ_2) are shift equivalent if and only if they are topologically conjugate to topological Markov shifts whose defining matrices are shift equivalent to each other in the original sense of Williams [Wi].

Proposition 9.18. Let φ_1 and φ_2 be automorphisms of a sofic system (Y, σ). Then the following statements are valid.

(1) If φ_1 and φ_2 are forward and associated with the same shift equivalence, then $(Y, \varphi_1\sigma^l)$ and $(Y, \varphi_2\sigma^l)$ are topologically conjugate for all integers $l > 0$.

(2) If φ_1 and φ_2 are weakly similar, then for all sufficiently large integers l, $(Y, \varphi_1\sigma^l)$ and $(Y, \varphi_2\sigma^l)$ are topologically conjugate (and for all sufficiently large integers l, $(Y, \varphi_1\sigma^{-l})$ and $(Y, \varphi_2\sigma^{-l})$ are topologically conjugate).

(3) If φ_1 and φ_2 are forward and weakly similar to each other, then $(Y, \varphi_1\sigma)$ and $(Y, \varphi_2\sigma)$ are shift equivalent.

Proof. From Theorem 6.23, (1) follows. By (1) and the definition of weak similarity, (2) is proved.

To prove (3), assume that φ_1 and φ_2 are forward and weakly similar to each other. Let $\mathcal{M}$ be the canonical defining representation matrix of (Y, σ). Since φ_1 and φ_2 are forward, there are canonical shift equivalence $(\mathcal{P}_1, \mathcal{Q}_1)$ and $(\mathcal{P}_2, \mathcal{Q}_2)$ of lag greater than 0 over $\mathcal{M}$ such that φ_i is associated with $(\mathcal{P}_i, \mathcal{Q}_i)$ for $i = 1, 2$. Since φ_1 and φ_2 are weakly similar, it follows from Proposition 9.13 that there is an integer $m \geq 0$ such that $\mathcal{P}_1\mathcal{M}^m \simeq \mathcal{P}_2\mathcal{M}^m$. Hence, noting that $\mathcal{P}_i\mathcal{M} = \mathcal{M}\mathcal{P}_i$ for $i = 1, 2$, we have

$$(\mathcal{P}_1\mathcal{M})^n \simeq (\mathcal{P}_2\mathcal{M})^n$$

for all integers n with $n \geq m$. Therefore, by the theorem of Boyle and Krieger (Theorem 1.9 of [BK2]) stating that eventually conjugacy is equivalent to shift equivalence for representation matrices which define bounded-to-one covers, we conclude that $\mathcal{P}_1\mathcal{M}$ and $\mathcal{P}_2\mathcal{M}$ are shift equivalent. By Theorem 6.23, $(Y, \varphi_i\sigma)$ is topologically conjugate to the sofic system $(X_{\mathcal{P}_i\mathcal{M}}, \sigma_{\mathcal{P}_i\mathcal{M}})$ for $i = 1, 2$. Thus $(Y, \varphi_1\sigma)$ and $(Y, \varphi_2\sigma)$ are shift equivalent. □

If (Y, σ) is a topological Markov shift in the proposition above, then we can improve (1) and (3) in it.

Proposition 9.19. Let φ_1 and φ_2 be automorphisms of a topological Markov shift (X, σ). Then the following statements are valid.

(1) If φ_1 and φ_2 are forward, expansive, and associated with the same shift equivalence, then for all integers $l \geq 0$, $(X, \varphi_1\sigma^l)$ and $(X, \varphi_2\sigma^l)$ are topologically conjugate.

(2) If φ_1 and φ_2 are forward, expansive, and weakly similar to each other, then there are *eventually equal* nonnegative integral matrices P_1 and P_2 (i.e., $P_1^n = P_2^n$ for all sufficiently large n) such that (X, φ_i) is topologically conjugate to (X_{P_i}, σ_{P_i}) for $i = 1, 2$.

Proof. From Theorem 6.10, (1) follows.

To prove (2), assume that φ_1 and φ_2 are forward, expansive, and weakly similar to each other. Since φ_1 and φ_2 are LR and expansive, it follows from Corollary 6.5 and Lemmas 6.25 and 8.6 that there is $k \geq 1$ such that both $\varphi_1^k\sigma^{-1}$ and $\varphi_2^k\sigma^{-1}$ are essentially LR automorphisms of (X, σ). By Corollaries 3.18 and 6.5 we see that both $\varphi_1^{2k}\sigma^{-1} = (\varphi_1^k\sigma^{-1})^2\sigma$ and $\varphi_2^{2k}\sigma^{-1}$ are essentially LR expansive automorphisms of (X, σ). By Proposition 8.8, there is $l \geq 1$ such that both $\varphi_1^{2kl}\sigma^{-l}$ and $\varphi_2^{2kl}\sigma^{-l}$ are LR automorphisms of (X, σ). Put $2kl = n$. By Corollary 3.18, $\varphi_1^n\sigma^{-1}$ and $\varphi_2^n\sigma^{-1}$ are LR. We can let

$$N_i = P(\varphi_i^n\sigma^{-1}) \quad \text{and} \quad P_i = P(\varphi_i) \quad \text{for } i = 1, 2.$$

By Corollary 6.5, (X, φ_i) is topologically conjugate to (X_{P_i}, σ_{P_i}) for $i = 1, 2$. We have

$$P_i^n = N_i M \quad \text{for } i = 1, 2,$$

where M is the defining matrix of (X, σ). Since φ_1 and φ_2 are weakly similar, it follows form Proposition 9.15 that there is $m \geq 0$ such that

$$P_1 M^m = P_2 M^m.$$

Since $P_i M = M P_i$ and $N_i M = M N_i$ for $i = 1, 2$, it is derived from the above equations that

$$P_1^{j+(m+1)n} = P_2^{j+(m+1)n}$$

for all $j \geq 0$. □

Two eventually equal nonnegative integral matrices have the same dimension triple and hence they are shift equivalent by Krieger's theorem [Kr1]. The result (2) of Proposition 9.19 should be compared with the result of Boyle and Krieger (see [BK1], Theorem 2.17) that if φ_1 and φ_2 are two automorphisms of a topological Markov shift (X, σ) which have the same induced action on the dimension group and if l is an integer greater than a coding bound for $\varphi_1, \varphi_1^{-1}, \varphi_2$, and φ_2^{-1}, then $(X, \varphi_1\sigma^l)$ and $(X, \varphi_2\sigma^l)$ are shift equivalent.

By Proposition 9.19, we know that expansive forward automorphisms of topological Markov shifts are very degenerate in a dynamical sense: their dynamical behavior depends at most on the shift equivalences associated with them and moreover may only depend on their weak-similarity classes. We shall show in the following Proposition 9.21 and Corollary 9.22 that a stronger degeneration appears for expansive forward automorphisms of topological Markov shifts such that the inverses of their zeta functions are irreducible.

Lemma 9.20. Let M be an irreducible nonnegative integral matrix whose characteristic polynomial is of the form

$$x^m \hat{p}(x),$$

where $m \geq 0$ and $\hat{p}(x)$ is an irreducible polynomial with $\hat{p}(x) \neq x$. If P is a nonnegative integral matrix such that

$$MP = PM,$$

then the maximal eigenvalue of P is in $\boldsymbol{Q}[\lambda]$, where λ is the maximal eigenvalue of M. If $f(x) \in \boldsymbol{Q}[x]$ such that $f(\lambda)$ is equal to the maximal eigenvalue of P, then

$$PM^m = f(M)M^m.$$

Proof. Let l be the degree of $\hat{p}(x)$. Since $\hat{p}(x)$ is irreducible and $\hat{p}(x) \neq x$, its roots are all distinct and nonzero. There is a nonsingular matrix U such that UMU^{-1} is a Jordan form $\bar{M}$ of M of the form

$$\bar{M} = \begin{pmatrix} M_1 & 0 \\ 0 & M_2 \end{pmatrix},$$

where M_1 is a diagonal matrix of order l and M_2 is a nilpotent matrix of order m. Clearly

$$M_2^m = 0.$$

Let $\bar{P} = UPU^{-1}$. Then we have

$$\bar{M}\bar{P} = \bar{P}\bar{M}.$$

Since the diagonal elements of M_1 are all distinct and nonzero, it follows (see [G], Chapter VIII, Section 2) that $\bar{P}$ has the form

$$\bar{P} = \begin{pmatrix} P_1 & 0 \\ 0 & P_2 \end{pmatrix}$$

where P_1 is a diagonal matrix of order l and P_2 is a matrix of order m. Since $M_1 P_1 = P_1 M_1$ and all the elementary divisors of M_1 are co-prime in pairs, it follows from Corollary 1 to Theorem 2 in p.220 of [G] that there are complex numbers $c_0, \cdots, c_{l-1}$ such that

$$P_1 = c_{l-1} M_1^{l-1} + \cdots + c_1 M_1 + c_0 I.$$

Let

$$g(x) = c_{l-1} x^{l-1} + \cdots + c_1 x + c_0.$$

Since $M_2^m = 0$, we have

$$\begin{aligned} \bar{P}\bar{M}^m &= \begin{pmatrix} P_1 M_1^m & 0 \\ 0 & P_2 M_2^m \end{pmatrix} \\ &= \begin{pmatrix} g(M_1) M_1^m & 0 \\ 0 & g(M_2) M_2^m \end{pmatrix} = g(\bar{M})\bar{M}^m. \end{aligned}$$

Thus, we have

$$PM^m = g(M)M^m$$

so that we have

$$PM^m = c_{l-1}M^{m+l-1} + \cdots + c_0M^m.$$

This means that the system of $(l+m)^2$ integral-coefficient linear equations has a solution $(c_0, \cdots, c_{l-1}) \in \boldsymbol{C}^l$. Hence the system can have a solution in $\boldsymbol{Q}^l$. Thus we may assume that

$$g(x) \in \boldsymbol{Q}[x].$$

Let u be a maximal (row) eigenvector of M. Since M is irreducible, all the components of u is positive. Therefore, since $PM = MP$, it follows that u is a maximal eigenvector of P (cf. Corollary 1.12 of [BP]). Since $uM = \lambda u$,

$$\lambda^m uP = uM^m P = uM^m g(M) = \lambda^m g(\lambda)u$$

so that we have

$$uP = g(\lambda)u.$$

Hence $g(\lambda)$ is the maximal eigenvalue of P. Thus we conclude that the maximal eigenvalue of P is in $\boldsymbol{Q}[\lambda]$.

Assume that $f(x) \in \boldsymbol{Q}[x]$ such that $f(\lambda)$ is equal to the maximal eigenvalue of P. Then we have

$$f(\lambda) = g(\lambda).$$

Since $\hat{p}(x)$ is the minimal polynomial of λ, $f(x) - g(x)$ is divided by $\hat{p}(x)$, so that $x^m f(x) - x^m g(x)$ divided by the characteristic polynomial of M. Thus we have

$$f(M)M^m = g(M)M^m$$

so that we have

$$PM^m = f(M)M^m.$$

□

Proposition 9.21. Let M be an irreducible nonnegative integral matrix whose characteristic polynomial is of the form

$$x^m \hat{p}(x),$$

where $m \geq 0$ and $\hat{p}(x)$ is an irreducible polynomial with $\hat{p}(x) \neq x$. Let λ be the maximal eigenvalue of M. Let φ be an LR automorphism of (X_M, σ_M). Then

$$e^{h(\varphi)} \in \boldsymbol{Q}[\lambda].$$

If $e^{h(\varphi)} = f(\lambda)$ with $f(x) \in \boldsymbol{Q}[x]$, then the following statements are valid.

(1) $P(\varphi)M^m = f(M)M^m$.

(2) For $1 \le k \le m-1$, $P(\varphi)M^k$ is shift equivalent to $f(M)M^k$ if $f(M)M^k$ is a nonnegative integral matrix.

(3) $(X_M, \varphi\sigma_M^k)$ and $(X_{f(M)M^k}, \sigma_{f(M)M^k})$ are topologically conjugate for all k not less than $\max\{1, m\}$.

(4) If $m = 0$ and φ is expansive, then (X_M, φ) and $(X_{f(M)}, \sigma_{f(M)})$ are topologically conjugate.

(5) For $1 \le k \le m-1$, if $f(M)M^k$ is a nonnegative integral matrix, then $(X_M, \varphi\sigma_M^k)$ and $(X_{f(M)M^k}, \sigma_{f(M)M^k})$ are shift equivalent.

Proof. By Theorem 6.31, $e^{h(\varphi)}$ is the maximal eigenvalue of $P(\varphi)$. Hence $f(\lambda)$ is the maximal eigenvalue of $P(\varphi)$. By Proposition 6.1, $MP(\varphi) = P(\varphi)M$ so that (1) follows from Lemma 9.20.

It follows from (1) that

$$(P(\varphi)M^k)^n = (f(M)M^k)^n$$

for all $k \ge 1$ and $n \ge m$. Therefore, for $1 \le k \le m-1$, $P(\varphi)M^k$ and $f(M)M^k$ are eventually equal so that $P(\varphi)M^k$ is shift equivalent to $f(M)M^k$ if this is a nonnegative integral matrix. Thus (2) is proved.

By Corollary 6.5, $(X_M, \varphi\sigma_M^k)$ is topologically conjugate to $(X_{P(\varphi)M^k}, \sigma_{P(\varphi)M^k})$ for all $k \ge 1$, and if φ is expansive, then (X_M, φ) is topologically conjugate to $(X_{P(\varphi)}, \sigma_{P(\varphi)})$. Therefore (3) and (4) follow from (1), and (5) follows from (2).

□

The following remarks are in order.

(1) In the above proposition, we do not know whether $(X_M, \varphi\sigma_M^k)$ is topologically conjugate to $(X_{f(M)M^k}, \sigma_{f(M)M^k})$ or not (when $f(M)M^k$ is a nonnegative integral matrix), for $1 \le k \le m-1$. But, for all $k \ge 0$, we can subshift-identify $(X_M, \varphi\sigma_M^k)$ with $\varphi\sigma_M^k$ expansive, because $P(\varphi)$ can be explicitly obtained. The problem is that of "strong shift equivalence" versus "shift equivalence" for eventually equal nonnegative matrices.

(2) In [BK1], Boyle and Krieger gave the result that if (X, σ) is a topological Markov shift such that the inverse of the zeta function is an irreducible polynomial, and φ is an automorphism of finite order of (X, σ), then $(X, \varphi\sigma^n)$ and (X, σ^n) are shift equivalent for all $n \in \mathbf{Z}$, $n \ne 0$. They also showed that for the same topological Markov shift (X, σ), if an automorphism φ of (X, σ) has the property that $R(\varphi) = 1$, then $(X, \varphi\sigma^n)$ is shift equivalent to (X, σ^n) for all n greater than a coding bound for φ and φ^{-1}. These results are included in the proposition above, by virtue of Proposition 8.2 and Theorem 6.31.

Corollary 9.22. Let (X, σ) be a topological Markov shift such that the inverse of its zeta function is an irreducible polynomial. Let φ and φ' be any expansive forward automorphisms of (X,σ). If $h(\varphi) = h(\varphi')$, then φ and φ' are weakly similar and hence (X,φ) and (X,φ') are shift equivalent. If $n, n' \geq m$ and $h(\varphi\sigma^n) = h(\varphi'\sigma^{n'})$, then $(X,\varphi\sigma^n)$ and $(X,\varphi'\sigma^{n'})$ are topologically conjugate, where m is the order of the defining matrix of (X,σ) minus the degree of the inverse of the zeta function of (X,σ). In particular, if the characteristic polynomial of the defining matrix of (X,σ) is irreducible, then whenever $h(\varphi) = h(\varphi')$, (X,φ) and (X,φ') are topologically conjugate.

Proof. This corollary directly follows from the proposition above, Propositions 9.15 and 9.19, and Theorem 6.29. □

As stated, expansive forward automorphisms of topological Markov shifts are very degenerate in the dynamical sense. Their dynamical behavior is far from depending on the specifications of the specified strong shift equivalences giving their κ-ζ factorizations. But the situation is drastically different for automorphisms which are not expansive-forward. An example presented in the next section will show this.

Mike Boyle asked (in private communication) the question of whether similarity and weak similarity change as equivalence relations on the set of topological conjugacies of sofic systems if we change the choice of canonical λ-graph or if we consider those defined using non-canonical λ-graphs. Thanks to his question, we can add the knowledge that our notions of similarity and weak similarity are universal.

Let us say that two symbolic conjugacies $\kappa : (Y_1,\sigma_1) \to (Y_2,\sigma_2)$ and $\kappa' : (Y_1,\sigma_1) \to (Y_2,\sigma_2)$ between sofic systems are *noncanonically similar* if there are specified equivalences

$$\mathcal{M}_1 \overset{k}{\simeq} \mathcal{M}_2 \quad \text{and} \quad \mathcal{M}_1 \overset{k'}{\simeq} \mathcal{M}_2$$

between (not necessarily canonical) representation matrices such that $(X_{\mathcal{M}_i}, \sigma_{\mathcal{M}_i}) = (Y_i,\sigma_i)$ for $i = 1,2$, k gives κ, and k' gives κ', though it will turn out in a moment that noncanonical similarity is nothing more than similarity.

Lemma 9.23. Let

$$\mathcal{M}_1 \overset{k}{\simeq} \mathcal{M}_2 \quad \text{and} \quad \mathcal{M}_1 \overset{l}{\simeq} \mathcal{M}_2$$

be two specified equivalences between square representation matrices $\mathcal{M}_1$ and $\mathcal{M}_2$ which are not necessarily canonical. Let $(Y_i,\sigma_i) = (X_{\mathcal{M}_i}, \sigma_{\mathcal{M}_i})$ for $i = 1,2$. Let $\mathcal{N}_1$ and $\mathcal{N}_2$ be the representation matrices either of $\mathcal{K}^+_{Y_1}$ and $\mathcal{K}^+_{Y_2}$, of $\mathcal{K}^-_{Y_1}$ and $\mathcal{K}^-_{Y_2}$, of $(\mathcal{K}^+_{Y_1})_-$ and $(\mathcal{K}^+_{Y_2})_-$, or of $(\mathcal{K}^-_{Y_1})_+$ and $(\mathcal{K}^-_{Y_2})_+$. Then

$$\mathcal{N}_1 \overset{k}{\simeq} \mathcal{N}_2 \text{ and } \mathcal{N}_1 \overset{l}{\simeq} \mathcal{N}_2.$$

Proof. Let A_i be the set of symbols that appear in the entries of $\mathcal{M}_i$, for $i = 1,2$. Let $\{C_1,\cdots,C_r\}$ be the partition of A_1 such that each class C_i consists of the symbols belonging to the same cycle of the permutation $l^{-1}k : A_1 \to A_1$.

Let $1 \leq i \leq r$ and let $a \in C_i$. Then

$$C_i = \{a, l^{-1}k(a), (l^{-1}k)^2(a), \cdots, (l^{-1}k)^{s_i-1}(a)\},$$

where s_i is the length of the cycle containing a. We have $k(C_i) = l(C_i)$. Let D_i denote this set. Let $k_i : C_i \to D_i$ and $l_i : C_i \to D_i$ be the restrictions of k and l, respectively. If $na = \overbrace{a + \cdots + a}^{n}$ appears in an entry, say (u,v)-entry, of $\mathcal{M}_1$ with $n \in \mathbf{N}$, then $nk(a)$ must appear in the (u,v)-entry of $\mathcal{M}_2$ and hence $nl^{-1}k(a)$ must appear in the (u,v)-entry of $\mathcal{M}_1$. Therefore, for every index (u,v) of $\mathcal{M}_1$, if na appears in the (u,v)-entry of $\mathcal{M}_1$ with $n \in \mathbf{N}$, then for all $b \in C_i$, nb appears in the entry and $nk(b)$ appears in the (u,v)-entry of $\mathcal{M}_2$. We can suppress the multiplicity n and assume that every symbol in every entry of $\mathcal{M}_1$ and $\mathcal{M}_2$ appears once in the entry.

Let $\{c_1, \cdots, c_r\}$ and $\{d_1, \cdots, d_r\}$ be sets of symbols and let j be the bijection between them such that $j(c_i) = d_i$ for $i = 1, \cdots, r$. Then, by the above we can replace every formal sum of the form $\sum_{a \in C_i} a$ in the entries of $\mathcal{M}_1$ to c_i for all $i = 1, \cdots, r$ and obtain a representation matrix $\bar{\mathcal{M}}_1$. We can also replace each formal sum of the form $\sum_{b \in D_i} b$ in the entries of $\mathcal{M}_2$ to d_i for $i = 1, \cdots, r$ and obtain a representation matrix $\bar{\mathcal{M}}_2$. Clearly we have a specified equivalence

$$\bar{\mathcal{M}}_1 \overset{j}{\simeq} \bar{\mathcal{M}}_2.$$

Hence $\mathcal{M}_1$ is obtained from $\bar{\mathcal{M}}_1$ substituting $\sum_{a \in C_i} a$ for all $i = 1, \cdots, r$ and $\mathcal{M}_2$ is obtained from $\bar{\mathcal{M}}_2$ substituting $\sum_{b \in D_i} b$ for d_i for all $i = 1, \cdots, r$. We may say in the natural sense that $k : \sqcup_{i=1}^r C_i \to \sqcup_{i=1}^r D_i$ is obtained by substituting $k_i : C_i \to D_i$ for $j : c_i \mapsto d_i$ for all $i = 1, \cdots, r$ and that $l : \sqcup_{i=1}^r C_i \to \sqcup_{i=1}^r D_i$ is obtained by substituting $l_i : C_i \to D_i$ for $j : c_i \mapsto d_i$ for all $i = 1, \cdots, r$, where $\sqcup$ represents disjoint union. Thus $\mathcal{M}_1 \overset{k}{\simeq} \mathcal{M}_2$ is obtained from $\bar{\mathcal{M}}_1 \overset{j}{\simeq} \bar{\mathcal{M}}_2$ substituting $\sum_{a \in C_i} a$ for c_i, $\sum_{b \in D_i} b$ for d_i, and $k_i : C_i \to D_i$ for $j : c_i \mapsto d_i$, for all $i = 1, \cdots, r$; $\mathcal{M}_1 \overset{l}{\simeq} \mathcal{M}_2$ is obtained from $\bar{\mathcal{M}}_1 \overset{j}{\simeq} \bar{\mathcal{M}}_2$ substituting $\sum_{a \in C_i} a$ for c_i, $\sum_{b \in D_i} b$ for d_i, and $l_i : C_i \to D_i$ for $j : c_i \mapsto d_i$, for all $i = 1, \cdots, r$. Let $(\bar{Y}_i, \bar{\sigma}_i) = (X_{\bar{\mathcal{M}}_i}, \sigma_{\bar{\mathcal{M}}_i})$ for $i = 1, 2$. Let $\bar{\mathcal{N}}_1$ and $\bar{\mathcal{N}}_2$ be the representation matrices either of $\mathcal{K}^+_{\bar{Y}_1}$ and $\mathcal{K}^+_{\bar{Y}_2}$, of $\mathcal{K}^-_{\bar{Y}_1}$ and $\mathcal{K}^-_{\bar{Y}_2}$, of $(\mathcal{K}^+_{\bar{Y}_1})_-$ and $(\mathcal{K}^+_{\bar{Y}_2})_-$, or of $(\mathcal{K}^-_{\bar{Y}_1})_+$ and $(\mathcal{K}^-_{\bar{Y}_2})_+$ in accordance with the definition of $\mathcal{N}_1$ and $\mathcal{N}_2$. Then applying the construction of Krieger λ-graphs given in Section 5 of [N2] and the construction of induced resolving λ-graphs by definition, we can obtain $\bar{\mathcal{N}}_1 \overset{j}{\simeq} \bar{\mathcal{N}}_2$ from $\bar{\mathcal{M}}_1 \overset{j}{\simeq} \bar{\mathcal{M}}_2$. Those constructions also show that if we substitute $\sum_{a \in C_i} a$ for c_i, $\sum_{b \in D_i} b$ for d_i, and $k_i : C_i \to D_i$ for $j : c_i \mapsto d_i$ for all $i = 1, \cdots, r$ in $\bar{\mathcal{N}}_1 \overset{j}{\simeq} \bar{\mathcal{N}}_2$, then we obtain $\mathcal{N}_1 \overset{k}{\simeq} \mathcal{N}_2$ and that if we substitute $\sum_{a \in C_i} a$ for c_i, $\sum_{b \in D_i} b$ for d_i, and $l_i : C_i \to D_i$ for $j : c_i \mapsto d_i$ for all $i = 1, \cdots, r$ in $\bar{\mathcal{N}}_1 \overset{j}{\simeq} \bar{\mathcal{N}}_2$, then we obtain $\mathcal{N}_1 \overset{l}{\simeq} \mathcal{N}_2$. $\square$

Proposition 9.24. Let $\phi : (Y_1, \sigma_1) \to (Y_2, \sigma_2)$ and $\phi' : (Y_1, \sigma_1) \to (Y_2, \sigma_2)$ be two topological conjugacies between sofic systems. Then ϕ and ϕ' are similar if and only if they have factorizations of the form

$$\phi = \kappa_n \zeta_n \kappa_{n-1} \cdots \kappa_1 \zeta_1 \kappa_0$$
$$\phi' = \kappa'_n \zeta_n \kappa'_{n-1} \cdots \kappa'_1 \zeta_1 \kappa'_0,$$

where n is a nonnegative integer, ζ_i is a forward or backward bipartite conjugacy for $i = 1, \cdots, n$, and κ_i and κ'_i are noncanonically similar symbolic conjugacies for $i = 0, \cdots, n$. Moreover, similarity is unchanged if we adopt any one of $\mathcal{K}_Y^+, \mathcal{K}_Y^-$, and $(\mathcal{K}_Y^+)_-$ instead of $(\mathcal{K}_Y^-)_+$ as the canonical λ-graph of a sofic system (Y, σ).

Proof. By Lemma 9.23, we know that if two symbolic conjugacies are noncanonically similar, then they are similar. Therefore the first claim is proved by Fact 9.2.

Lemma 9.23 and Fact 9.2 also show that the first claim also holds for the case where we adopt any one of $\mathcal{K}_Y^-, \mathcal{K}_Y^+$, and $(\mathcal{K}_Y^+)^-$ as the canonical λ-graph of a sofic system (Y, σ). Therefore the second claim is proved. □.

Corollary 9.25. Two topological conjugacies $\phi : (Y_1, \sigma_1) \to (Y_2, \sigma_2)$ and $\phi' : (Y_1, \sigma_1) \to (Y_2, \sigma_2)$ between sofic systems are similar if and only if there is an integer m such that both $\phi_1 \sigma_1^m$ and $\phi_2 \sigma_1^m$ are forward and associated with the same strong shift equivalence (which is not necessarily canonical).

Proposition 9.26. Two topological conjugacies $\phi : (Y_1, \sigma_1) \to (Y_2, \sigma_2)$ and $\phi' : (Y_1, \sigma_1) \to (Y_2, \sigma_2)$ between sofic systems are weakly similar if and only if there are an integer m and (not necessarily canonical) strong shift equivalences

(9.5) $\mathcal{M} \simeq \mathcal{P}_1 \mathcal{Q}_1,\ \mathcal{Q}_1 \mathcal{P}_1 \simeq \mathcal{P}_2 \mathcal{Q}_2,\ \cdots,\ \mathcal{Q}_{n-1} \mathcal{P}_{n-1} \simeq \mathcal{P}_n \mathcal{Q}_n,\ \mathcal{Q}_n \mathcal{P}_n \simeq \mathcal{N}$

and

(9.6) $\mathcal{M} \simeq \mathcal{R}_1 \mathcal{S}_1,\ \mathcal{S}_1 \mathcal{R}_1 \simeq \mathcal{R}_2 \mathcal{S}_2,\ \cdots,\ \mathcal{S}_{n-1} \mathcal{R}_{n-1} \simeq \mathcal{R}_n \mathcal{S}_n,\ \mathcal{S}_n \mathcal{R}_n \simeq \mathcal{N}$

such that $n \geq 1$,

$$\mathcal{P}_1 \cdots \mathcal{P}_n \simeq \mathcal{R}_1 \cdots \mathcal{R}_n \quad \text{and} \quad \mathcal{Q}_n \cdots \mathcal{Q}_1 \simeq \mathcal{S}_n \cdots \mathcal{S}_1,$$

and $\phi \sigma_1^m$ and $\phi' \sigma_1^m$ are forward conjugacies which are associated with (9.5) and (9.6), respectively. Moreover weak-similarity is unchanged if we adopt any one of $\mathcal{K}_Y^+, \mathcal{K}_Y^-$, and $(\mathcal{K}_Y^+)_-$ instead of $(\mathcal{K}_Y^-)_+$ as the canonical λ-graph of a sofic system (Y, σ).

Proof. The only-if part of the first claim is clear by definition. To prove the if-part, put $\phi\sigma_1^m = \psi$ and $\phi'\sigma_1^m = \psi'$. Since ψ and ψ' are associated with (9.5) and (9.6), respectively, $\psi^{(n)}$ and $(\psi')^{(n)}$ are associated with the 1-step strong shift equivalences

$$(9.7) \qquad \mathcal{M}^n \simeq \mathcal{P}\mathcal{Q},\ \mathcal{Q}\mathcal{P} \simeq \mathcal{N}^n \qquad \text{with } \mathcal{P} = \mathcal{P}_1 \cdots \mathcal{P}_n \text{ and } \mathcal{Q} = \mathcal{Q}_n \cdots \mathcal{Q}_1$$

and

$$\mathcal{M}^n \simeq (\mathcal{R}_1 \cdots \mathcal{R}_n)(\mathcal{S}_n \cdots \mathcal{S}_1),\ \ (\mathcal{S}_n \cdots \mathcal{S}_1)(\mathcal{R}_1 \cdots \mathcal{R}_n) \simeq \mathcal{N}^n,$$

respectively (see Lemma 1 of [B4]). Since $\mathcal{P} \simeq \mathcal{R}_1 \cdots \mathcal{R}_n$ and $\mathcal{Q} \simeq \mathcal{S}_n \cdots \mathcal{S}_1$ by assumption, both $\psi^{(n)}$ and $(\psi')^{(n)}$ are associated with (9.7). Let $\bar{\mathcal{M}}$ and $\bar{\mathcal{N}}$ be the canonical representation matrices which define $(X_{\mathcal{M}}, \sigma_{\mathcal{M}})$ and $(X_{\mathcal{N}}, \sigma_{\mathcal{N}})$, respectively. We assume without loss of generality that the set of symbols that appear in the entries of $\mathcal{P}$ and the set of symbols that appear in the entries of $\mathcal{Q}$ are disjoint. Hence the canonical representation matrix which defines the same sofic system that $\begin{pmatrix} 0 & \mathcal{P} \\ \mathcal{Q} & 0 \end{pmatrix}$ defines, is written in the form $\begin{pmatrix} 0 & \bar{\mathcal{P}} \\ \bar{\mathcal{Q}} & 0 \end{pmatrix}$. All $\bar{\mathcal{M}}^n, \bar{\mathcal{N}}^n, \bar{\mathcal{P}}\bar{\mathcal{Q}}$, and $\bar{\mathcal{Q}}\bar{\mathcal{P}}$ are canonical representation matrices (see Section 5). Therefore since $\psi^{(n)}$ and $(\psi')^{(n)}$ are associated with (9.7), it follows from Lemma 9.23 that $\psi^{(n)}$ and $(\psi')^{(n)}$ are associated with the canonical 1-step strong shift equivalence

$$\bar{\mathcal{M}}^n \simeq \bar{\mathcal{P}}\bar{\mathcal{Q}},\ \bar{\mathcal{Q}}\bar{\mathcal{P}} \simeq \bar{\mathcal{N}}^n,$$

so that $\psi^{(n)}$ and $(\psi')^{(n)}$ are associated with canonical shift equivalence

$$(\bar{\mathcal{P}}, \bar{\mathcal{Q}})$$

of lag 1 from $\bar{\mathcal{M}}^n$ to $\bar{\mathcal{N}}^n$.

Since there are specifications of (9.5) and (9.6) which give κ-ζ factorizations of ψ and ψ', respectively, it follows from Theorem 5.2 (and the discussion preceding Theorem 5.3) that there are canonical strong shift equivalences

$$(9.8) \qquad \bar{\mathcal{M}} \simeq \bar{\mathcal{P}}_1\bar{\mathcal{Q}}_1,\ \bar{\mathcal{Q}}_1\bar{\mathcal{P}}_1 \simeq \bar{\mathcal{P}}_2\bar{\mathcal{Q}}_2,\ \cdots,\ \bar{\mathcal{Q}}_{n-1}\bar{\mathcal{P}}_{n-1} \simeq \bar{\mathcal{P}}_n\bar{\mathcal{Q}}_n,\ \bar{\mathcal{Q}}_n\bar{\mathcal{P}}_n \simeq \bar{\mathcal{N}}$$

and

$$(9.9) \qquad \bar{\mathcal{M}} \simeq \bar{\mathcal{R}}_1\bar{\mathcal{S}}_1,\ \bar{\mathcal{S}}_1\bar{\mathcal{R}}_1 \simeq \bar{\mathcal{R}}_2\bar{\mathcal{S}}_2,\ \cdots,\ \bar{\mathcal{S}}_{n-1}\bar{\mathcal{R}}_{n-1} \simeq \bar{\mathcal{R}}_n\bar{\mathcal{S}}_n,\ \bar{\mathcal{S}}_n\bar{\mathcal{R}}_n \simeq \bar{\mathcal{N}}$$

with which ψ and ψ' are associated, respectively. Hence $\psi^{(n)}$ and $(\psi')^{(n)}$ are associated with the 1-step strong shift equivalences

$$\bar{\mathcal{M}}^n \simeq (\bar{\mathcal{P}}_1 \cdots \bar{\mathcal{P}}_n)(\bar{\mathcal{Q}}_n \cdots \bar{\mathcal{Q}}_1),\ \ (\bar{\mathcal{Q}}_n \cdots \bar{\mathcal{Q}}_1)(\bar{\mathcal{P}}_1 \cdots \bar{\mathcal{P}}_n) \simeq \bar{\mathcal{N}}^n$$

and

$$\bar{\mathcal{M}}^n \simeq (\bar{\mathcal{R}}_1 \cdots \bar{\mathcal{R}}_n)(\bar{\mathcal{S}}_n \cdots \bar{\mathcal{S}}_1),\ \ (\bar{\mathcal{S}}_n \cdots \bar{\mathcal{S}}_1)(\bar{\mathcal{R}}_1 \cdots \bar{\mathcal{R}}_n) \simeq \bar{\mathcal{N}}^n$$

respectively, so that $\psi^{(n)}$ and $(\psi')^{(n)}$ are associated with the canonical shift equivalences

$$(\bar{\mathcal{P}}_1 \cdots \bar{\mathcal{P}}_n,\ \bar{\mathcal{Q}}_n \cdots \bar{\mathcal{Q}}_1) \quad \text{and} \quad (\bar{\mathcal{R}}_1 \cdots \bar{\mathcal{R}}_n,\ \bar{\mathcal{S}}_n \cdots \bar{\mathcal{S}}_1)$$

of lag 1 from $\bar{\mathcal{M}}^n$ to $\bar{\mathcal{N}}^n$, respectively. Using "LR sofic textile relation system" (use the approach written in the last paragraph of Section 6) and the generalization of Corollary 7.25 remarked in the last paragraph of Section 7, we can straightforwardly generalize Lemma 9.12 to canonical shift equivalences between different canonical representation matrices. Since $(\bar{\mathcal{P}}_1 \cdots \bar{\mathcal{P}}_n,\ \bar{\mathcal{Q}}_n \cdots \bar{\mathcal{Q}}_1)$ and $(\bar{\mathcal{P}}, \bar{\mathcal{Q}})$ are canonical shift equivalences from $\bar{\mathcal{M}}^n$ to $\bar{\mathcal{N}}^n$ and $\psi^{(n)}$ is associated with both of them, it follows from the generalization of Lemma 9.12 that there is an integer $r_1 \geq 0$ such that

$$(\bar{\mathcal{P}}_1 \cdots \bar{\mathcal{P}}_n)(\bar{\mathcal{N}}^n)^{r_1} \simeq \bar{\mathcal{P}}(\bar{\mathcal{N}}^n)^{r_1}.$$

Since $(\bar{\mathcal{Q}}_n \cdots \bar{\mathcal{Q}}_1, \bar{\mathcal{P}}_1 \cdots \bar{\mathcal{P}}_n)$ and $(\bar{\mathcal{Q}}, \bar{\mathcal{P}})$ are canonical shift equivalences from $\bar{\mathcal{N}}^n$ to $\bar{\mathcal{M}}^n$ and $(\psi^{(n)})^{-1}\sigma_2^{(n)}$ is associated with both of them, it also follows from the generalization of Lemma 9.12 that there is an integer $r_2 \geq 0$ such that

$$(\bar{\mathcal{Q}}_n \cdots \bar{\mathcal{Q}}_1)(\bar{\mathcal{M}}^n)^{r_2} \simeq \bar{\mathcal{Q}}(\bar{\mathcal{M}}^n)^{r_2}.$$

In the same way we know that there are integers $s_1, s_2 \geq 0$ such that

$$(\bar{\mathcal{R}}_1 \cdots \bar{\mathcal{R}}_n)(\bar{\mathcal{N}}^n)^{s_1} \simeq \bar{\mathcal{P}}(\bar{\mathcal{N}}^n)^{s_1}$$

and

$$(\bar{\mathcal{S}}_n \cdots \bar{\mathcal{S}}_1)(\bar{\mathcal{M}}^n)^{s_2} \simeq \bar{\mathcal{Q}}(\bar{\mathcal{M}}^n)^{s_2}.$$

Therefore, letting $l_1 = \max\{nr_1, ns_1\}$ and $l_2 = \max\{nr_2, ns_2\}$, we have

$$\bar{\mathcal{P}}_1 \cdots \bar{\mathcal{P}}_n \bar{\mathcal{N}}^{l_1} \simeq \bar{\mathcal{R}}_1 \cdots \bar{\mathcal{R}}_n \bar{\mathcal{N}}^{l_1} \tag{9.10}$$

and

$$\bar{\mathcal{Q}}_n \cdots \bar{\mathcal{Q}}_1 \bar{\mathcal{M}}^{l_2} \simeq \bar{\mathcal{S}}_n \cdots \bar{\mathcal{S}}_1 \bar{\mathcal{M}}^{l_2}. \tag{9.11}$$

Since ψ and ψ' are associated with the canonical strong shift equivalences (9.8) and (9.9), ψ and ψ' are associated with the canonical shift equivalences $(\bar{\mathcal{P}}_1 \cdots \bar{\mathcal{P}}_n, \bar{\mathcal{Q}}_n \cdots \bar{\mathcal{Q}}_1)$ and $(\bar{\mathcal{R}}_1 \cdots \bar{\mathcal{R}}_n, \bar{\mathcal{S}}_n \cdots \bar{\mathcal{S}}_1)$ from $\bar{\mathcal{M}}$ to $\bar{\mathcal{N}}$, respectively. Hence, by Lemmas 9.7 and 9.11 and by (9.10) and (9.11), we know that $\sigma_2^{l_1}\psi(\mathrm{id}_{Y_1})^{l_2}$ and $\sigma_2^{l_1}\psi'(\mathrm{id}_{Y_1})^{l_2}$ are associated with the same canonical shift equivalence from $\bar{\mathcal{M}}$ to $\bar{\mathcal{N}}$. Thus ϕ and ϕ' are weakly similar.

The proof of the first claim is completed.

The proof above remains valid if we change the definition of canonical λ-graph as in the second claim (cf. Section 5 and the proof of Lemma 9.12). Therefore the second claim is proved. □

Let $\mathcal{M}$ be the canonical representation matrix. Let $(Y, \sigma) = (X_{\mathcal{M}}, \sigma_{\mathcal{M}})$. Let $\mathcal{M}'$ be the representation matrix of $(\mathcal{K}_Y^+)_-$. We can define $\Sigma(\mathcal{M}')$ and $g_{\mathcal{M}'}$ in the same way as $\Sigma(\mathcal{M})$ and $g_{\mathcal{M}}$ were defined. By Theorem 9.14 and Proposition 9.26, we know that the subgroup $g_{\mathcal{M}}(\mathrm{Aut}(\sigma))$ of $\Sigma(\mathcal{M})$ is isomorphic to the subgroup $g_{\mathcal{M}'}(\mathrm{Aut}(\sigma))$ of $\Sigma(\mathcal{M}')$. But we have not known whether $\Sigma(\mathcal{M})$ and $\Sigma(\mathcal{M}')$ are isomorphic or not.

10. EXAMPLES

Example 1. Let

$$M = \begin{pmatrix} 2 & 1 \\ 1 & 1 \end{pmatrix}.$$

Let us consider the following strong shift equivalence from $\tilde{M}$ to itself.

$$\tilde{M} \simeq \tilde{P}\tilde{Q},\ \tilde{P}\tilde{Q} \simeq \tilde{M} \tag{10.1}$$

with

$$P = Q = \begin{pmatrix} 1 & 1 \\ 1 & 0 \end{pmatrix}.$$

The strong shift equivalence (10.1) is essentially the same as

$$\tilde{M} = \tilde{P}\tilde{Q},\ \tilde{Q}\tilde{P} \simeq \tilde{P}\tilde{Q}. \tag{10.2}$$

Putting

$$\tilde{P} = \begin{pmatrix} u & v \\ w & 0 \end{pmatrix} \quad \text{and} \quad \tilde{Q} = \begin{pmatrix} x & y \\ z & 0 \end{pmatrix}$$

we have

$$\tilde{P}\tilde{Q} = \begin{pmatrix} ux + vz & uy \\ wx & wy \end{pmatrix} \quad \text{and} \quad \tilde{Q}\tilde{P} = \begin{pmatrix} xu + yw & xv \\ zu & zv \end{pmatrix}.$$

Therefore there are the following two specified strong shift equivalences associated with (10.2):

$$\tilde{M} = \tilde{P}\tilde{Q},\ \tilde{Q}\tilde{P} \overset{k_0}{\simeq} \tilde{P}\tilde{Q} \tag{10.3}$$

with k_0 given by

$$xu \mapsto ux,\ yw \mapsto vz,\ xv \mapsto uy,\ zu \mapsto wx,\ zv \mapsto wy,$$

and

$$\tilde{M} = \tilde{P}\tilde{Q},\ \tilde{Q}\tilde{P} \overset{k}{\simeq} \tilde{P}\tilde{Q} \tag{10.4}$$

with k given by

$$xu \mapsto vz,\ yw \to ux,\ xv \mapsto uy,\ zu \mapsto wx,\ zv \mapsto wy.$$

Let φ_0 and φ be the forward automorphisms whose factorizations are given by (10.3) and (10.4), respectively. Let

$$\tilde{M}\tilde{P} = (\tilde{P}\tilde{Q})\tilde{P} \overset{i}{\simeq} \tilde{P}(\tilde{Q}\tilde{P}) \overset{k_0'}{\simeq} \tilde{P}(\tilde{P}\tilde{Q}) = \tilde{P}\tilde{M}$$

be the sequence of specified equivalences naturally derived by (10.3) and let $l_0 = k_0' i$. Then we get a specified equivalence

$$(10.5) \qquad \tilde{M}\tilde{P} \overset{l_0}{\simeq} \tilde{P}\tilde{M}.$$

Let

$$(10.6) \qquad \tilde{M}\tilde{P} \overset{l}{\simeq} \tilde{P}\tilde{M}$$

be the specified equivalence induced from (10.4) in the same way as above. Let T_0 and T be the LR textile systems associated with (10.5) and (10.6), respectively. Then T_0 and T are 1-1 and $\varphi_{T_0} = \varphi_0$ and $\varphi_T = \varphi$. If we put

$$ux = a, \ vz = b, \ uy = c, \ wx = d, \ wy = e,$$

then

$$\tilde{M} = \begin{pmatrix} a+b & c \\ d & e \end{pmatrix}$$

and T_0 and T are given by the following representation matrices:

$$\mathcal{M}_{\mathcal{G}_{T_0}} = \begin{array}{c} \\ u \\ v \\ w \end{array} \begin{array}{c} \begin{array}{ccc} u & v & w \end{array} \\ \begin{pmatrix} \frac{a}{a} & \frac{a}{c} & \frac{c}{b} \\ \frac{b}{d} & \frac{b}{e} & 0 \\ \frac{d}{a} & \frac{d}{c} & \frac{e}{b} \end{pmatrix} \end{array}$$

and

$$\mathcal{M}_{\mathcal{G}_T} = \begin{array}{c} \\ u \\ v \\ w \end{array} \begin{array}{c} \begin{array}{ccc} u & v & w \end{array} \\ \begin{pmatrix} \frac{a}{b} & \frac{a}{c} & \frac{c}{a} \\ \frac{b}{d} & \frac{b}{e} & 0 \\ \frac{d}{b} & \frac{d}{c} & \frac{e}{a} \end{pmatrix} \end{array},$$

respectively. Let $T_0 = (p_0, q_0 : \Gamma_0 \to G_0)$ with $T_0^* = (p_0^{T_0}, q_0^{T_0} : \Gamma_0^{T_0} \to G_0^{T_0})$. By definition, T_0^* is given by the following representation matrix

$$\mathcal{M}_{\mathcal{G}_{T_0^*}} = \begin{array}{c} \\ a \\ b \\ c \\ d \\ e \end{array} \begin{array}{c} \begin{array}{ccccc} a & b & c & d & e \end{array} \\ \begin{pmatrix} \frac{u}{u} & 0 & \frac{u}{v} & 0 & 0 \\ 0 & 0 & 0 & \frac{v}{u} & \frac{v}{v} \\ 0 & \frac{u}{w} & 0 & 0 & 0 \\ \frac{w}{u} & 0 & \frac{w}{v} & 0 & 0 \\ 0 & \frac{w}{w} & 0 & 0 & 0 \end{pmatrix} \end{array}.$$

It is easy to see that $p_0^{T_0}$ is a definite left-resolving graph-homomorphism and $q_0^{T_0}$ is a definite right-resolving graph-homomorphism so that T_0^* is 1-1. Therefore, by Corollary 6.5, (X_M, φ_0) is topologically conjugate to (X_P, σ_P) with $P = \begin{pmatrix} 1 & 1 \\ 1 & 0 \end{pmatrix}$. In fact, by computing T_0^2, we know that $\varphi_0^2 = \sigma_M$, that is, φ_0 is a root of σ_M.

Now let us consider φ. Let $T = (p, q : \Gamma \to G)$ with $T^* = (p^T, q^T : \Gamma^T \to G^T)$. By definition, T^* is given by the following representation matrix:

$$\mathcal{M}_{\mathcal{G}_{T^*}} = \begin{array}{c} \\ a \\ b \\ c \\ d \\ e \end{array} \begin{array}{c} \begin{array}{ccccc} a & b & c & d & e \end{array} \\ \begin{pmatrix} 0 & \frac{u}{u} & \frac{u}{v} & 0 & 0 \\ 0 & 0 & 0 & \frac{v}{u} & \frac{v}{v} \\ \frac{u}{w} & 0 & 0 & 0 & 0 \\ 0 & \frac{w}{u} & \frac{w}{v} & 0 & 0 \\ \frac{w}{w} & 0 & 0 & 0 & 0 \end{pmatrix} \end{array}.$$

Of course, T^* is LR because T is LR. But T^* is not 1-1. (In fact, p^T is not definite, and neither is q^T.) Therefore by Theorem 2.5, φ is not expansive.

From (10.4), we naturally derive

$$\tilde{M}\tilde{Q} = (\tilde{P}\tilde{Q})\tilde{Q} \overset{(k^{-1})'}{\simeq} (\tilde{Q}\tilde{P})\tilde{Q} \overset{i'}{\simeq} \tilde{Q}(\tilde{P}\tilde{Q}) = \tilde{Q}\tilde{M}.$$

Let $l' = i'(k^{-1})'$. Then we get a specified equivalence

$$\tilde{M}\tilde{Q} \overset{l'}{\simeq} \tilde{Q}\tilde{M}.$$

Let T' be the LR textile system associated with this. Then T' is 1-1, $\varphi_{T'} = \varphi^{-1}\sigma_M$ and T' is represented by

$$\mathcal{M}_{\mathcal{G}_{T'}} = \begin{array}{c} \\ x \\ y \\ z \end{array} \begin{array}{c} \begin{array}{ccc} x & y & z \end{array} \\ \begin{pmatrix} \frac{b}{a} & \frac{b}{c} & \frac{c}{b} \\ \frac{a}{d} & \frac{a}{e} & 0 \\ \frac{d}{a} & \frac{d}{c} & \frac{e}{b} \end{pmatrix} \end{array}.$$

It is also seen that $(T')^*$ is not 1-1. Therefore we know that $\varphi^{-1}\sigma_M$ is not expansive.

Thus we get an ELR cone $CK_\varphi(\sigma_M)$ for φ such that $\varphi^k\sigma_M^n$ is an essentially LR automorphism of (X_M, σ_M) if and only if (n, k) is in the cone

$$CK_\varphi(\sigma_M) = \{(x, y) \in \mathbf{R}^2 | x \geq 0, y \geq -x\}.$$

(Cf. Theorem 8.16.) Note that $CK_\varphi(\sigma_M)$ is closed.

Let us consider $\varphi^2\sigma_M^{-1}$. We note that $(-1, 2)$ is in the outside of $CK_\varphi(\sigma_M) \cup (-CK_\varphi(\sigma_M))$. Hence $\varphi^2\sigma_M^{-1}$ is not an essentially LR automorphism of (X_M, σ_M) and moreover it is not an essentially LR automorphism of $(X_M, \varphi^k\sigma_M^n)$ for all lattice points (n, k) in the interior of $CK_\varphi(\sigma_M) \cup (-CK_\varphi(\sigma_M))$. Let

$$T_1 = \overline{\text{trim}}\,(T \circ (T')^{-1}).$$

Then $\varphi_{T_1} = \varphi^2\sigma_M^{-1}$ and T_1 is nondegenerate and represented by

$$\mathcal{M}_{\mathcal{G}_{T_1}} = \begin{array}{c} \\ 1 \\ 2 \\ 3 \\ 4 \end{array} \begin{array}{c} \begin{array}{cccc} 1 & 2 & 3 & 4 \end{array} \\ \begin{pmatrix} \frac{e}{b} & \frac{d}{c} & \frac{d}{b} & 0 \\ \frac{c}{d} & \frac{a}{e} & \frac{a}{d} & 0 \\ 0 & 0 & \frac{b}{a} & \frac{b}{a} \\ \frac{c}{b} & \frac{a}{c} & \frac{a}{b} & 0 \end{pmatrix} \end{array}.$$

The dual T_1^* is represented by

$$\mathcal{M}_{\mathcal{G}_{T_1^*}} = \begin{array}{c} \\ a \\ b \\ c \\ d \\ e \end{array} \begin{array}{c} \begin{array}{ccccc} a & b & c & d & e \end{array} \\ \begin{pmatrix} 0 & \frac{4}{3} & \frac{4}{2} & \frac{2}{3} & \frac{2}{2} \\ \frac{3}{3} + \frac{3}{4} & 0 & 0 & 0 & 0 \\ 0 & \frac{4}{1} & 0 & \frac{2}{1} & 0 \\ 0 & \frac{1}{3} & \frac{1}{2} & 0 & 0 \\ 0 & \frac{1}{1} & 0 & 0 & 0 \end{pmatrix} \end{array},$$

One can see that T_1^* is degenerate. Let

$$T_2^* = \overline{\text{trim}}((T_1^*)^{[3]})$$

Let $T_2 = (T_2^*)^*$ with $T_2 = (p_2, q_2 : \Gamma_2 \to G_2)$ and let $T_2^* = (p_2^{T_2}, q_2^{T_2} : \Gamma_2^{T_2} \to G_2^{T_2})$ following our notation. Then one can see that the representation matrices of T_2^* and T_2 are given below.

	A	B	C	D	E	F	G	H	I	J	K	L	M	N	P	Q	R	S	T
A					$\frac{434}{343}$	$\frac{434}{342}$	$\frac{432}{343}$	$\frac{432}{342}$											
B													$\frac{421}{213}$	$\frac{421}{212}$					
C																$\frac{214}{321}$	$\frac{212}{321}$		
D																		$\frac{213}{214}$	$\frac{213}{213}$
E	$\frac{343}{434}$																		
F		$\frac{342}{421}$																	
G			$\frac{321}{432}$																
H				$\frac{321}{421}$															
I		$\frac{342}{321}$																	
J				$\frac{321}{321}$															
K					$\frac{434}{143}$		$\frac{432}{143}$												
L									$\frac{434}{132}$	$\frac{432}{132}$									
M															$\frac{213}{134}$				
N																$\frac{214}{121}$	$\frac{212}{121}$		
P					$\frac{134}{343}$	$\frac{134}{342}$	$\frac{132}{343}$	$\frac{132}{342}$											
Q											$\frac{143}{214}$	$\frac{143}{213}$							
R													$\frac{121}{213}$	$\frac{121}{212}$					
S					$\frac{134}{143}$		$\frac{132}{143}$												
T									$\frac{134}{132}$	$\frac{132}{132}$									

The representation matrix of T_2^*

	121	132	134	143	212	213	214	321	342	343	421	432	434
121					$\frac{R}{N}$	$\frac{R}{M}$							
132		$\frac{T}{J}$		$\frac{S}{G}$					$\frac{P}{H}$	$\frac{P}{G}$			
134		$\frac{T}{I}$		$\frac{S}{E}$					$\frac{P}{F}$	$\frac{P}{E}$			
143						$\frac{Q}{L}$	$\frac{Q}{K}$						
212	$\frac{N}{R}$							$\frac{C}{R}$					
213			$\frac{M}{P}$			$\frac{D}{T}$	$\frac{D}{S}$						
214	$\frac{N}{Q}$							$\frac{C}{Q}$					
321								$\frac{J}{D}$			$\frac{H}{D}$	$\frac{G}{C}$	
342								$\frac{I}{B}$			$\frac{F}{B}$		
343													$\frac{E}{A}$
421					$\frac{B}{N}$	$\frac{B}{M}$							
432		$\frac{L}{J}$		$\frac{K}{G}$					$\frac{A}{H}$	$\frac{A}{G}$			
434		$\frac{L}{I}$		$\frac{K}{E}$					$\frac{A}{F}$	$\frac{A}{E}$			

The representation matrix of T_2

Therefore, $G_2^{T_2}$ and G_2 are given by the following representation matrices:

$$\tilde{M}_{G_2^{T_2}} = \begin{array}{c} \\ 12 \\ 13 \\ 14 \\ 21 \\ 32 \\ 34 \\ 42 \\ 43 \end{array} \begin{array}{c} \begin{array}{cccccccc} 12 & 13 & 14 & 21 & 32 & 34 & 42 & 43 \end{array} \\ \begin{pmatrix} 0 & 0 & 0 & 121 & 0 & 0 & 0 & 0 \\ 0 & 0 & 0 & 0 & 132 & 134 & 0 & 0 \\ 0 & 0 & 0 & 0 & 0 & 0 & 0 & 143 \\ 212 & 213 & 214 & 0 & 0 & 0 & 0 & 0 \\ 0 & 0 & 0 & 321 & 0 & 0 & 0 & 0 \\ 0 & 0 & 0 & 0 & 0 & 0 & 342 & 343 \\ 0 & 0 & 0 & 421 & 0 & 0 & 0 & 0 \\ 0 & 0 & 0 & 0 & 432 & 434 & 0 & 0 \end{pmatrix} \end{array}$$

$$\tilde{M}_{G_2} = \begin{array}{c} \\ 12 \\ 13 \\ 14 \\ 21 \\ 32 \\ 34 \\ 42 \\ 43 \end{array} \begin{array}{c} \begin{array}{cccccccc} 12 & 13 & 14 & 21 & 32 & 34 & 42 & 43 \end{array} \\ \begin{pmatrix} 0 & 0 & 0 & R & 0 & 0 & 0 & 0 \\ 0 & T & S & 0 & 0 & P & 0 & 0 \\ 0 & 0 & 0 & Q & 0 & 0 & 0 & 0 \\ N & M & 0 & D & C & 0 & 0 & 0 \\ 0 & 0 & 0 & 0 & J & 0 & H & G \\ 0 & 0 & 0 & 0 & I & 0 & F & E \\ 0 & 0 & 0 & B & 0 & 0 & 0 & 0 \\ 0 & L & K & 0 & 0 & A & 0 & 0 \end{pmatrix} \end{array}.$$

One can see that T_2^* is 1-1 and nondegenerate by checking that the induced right and left resolving homomorphisms of $p_2^{T_2}$ and $q_2^{T_2}$ are all definite (cf. p.413 of [N1]). (Here, the *induced right resolving graph-homomorphism* of a graph-homomorphism $h : \Gamma \to G$ means the right-resolving graph-homomorphism $h_+ : \Gamma_+ \to G$ given as the induced right resolving λ-graph of the λ-graph (Γ, h_A). The meaning of the *induced left resolving graph-homomorphism* is similar.)

Since T_1 and T_2 are topologically conjugate by Lemma 2.8(2) and T_2 is nondegenerate by Lemma 2.8(5), there is a homeomorphism $\psi : X_M \to X_{G_2}$ such that

$$\psi(\varphi^2\sigma^{-1})\psi^{-1} = \varphi_{T_2} \quad \text{and} \quad \psi\sigma_M\psi^{-1} = \sigma_{G_2}.$$

Since T_2^* is 1-1 and nondegenerate, it follows from Theorem 2.5 that there is a homeomorphism $\chi_{T_2} : X_{G_2} \to X_{G_2}^{T_2}$ such that

$$\chi_{T_2}\varphi_{T_2}\chi_{T_2}^{-1} = \sigma_{G_2^{T_2}} \quad \text{and} \quad \chi_{T_2}\sigma_{G_2}\chi_{T_2}^{-1} = \varphi_{T_2^*}.$$

Therefore $(X_M, \varphi^2\sigma_M^{-1})$ is topologically conjugate to $(X_{G_2^{T_2}}, \sigma_{G_2^{T_2}})$. Thus we have subshift-identified $(X_M, \varphi^2\sigma_M^{-1})$. (We also know that $(X_{G_2^{T_2}}, \varphi_{T_2^*})$ is topologically conjugate to (X_M, σ_M)). As seen below, the two topological Markov shifts (X_M, σ_M) and $(X_{G_2^{T_2}}, \sigma_{G_2^{T_2}})$ have very different entropies and very different zeta-functions.

Let H be the graph whose representation matrix is given by

$$\begin{pmatrix} 0 & f & h & 0 & i \\ g & 0 & 0 & 0 & 0 \\ 0 & j & 0 & k & l \\ 0 & m & 0 & 0 & 0 \\ 0 & 0 & n & 0 & 0 \end{pmatrix}.$$

Let $\hat{h} : G_2^{T_2} \to H$ be the graph-homomorphism whose arc map is given by

$$\begin{array}{lllll} 121 \mapsto f, & 212 \mapsto g, & 213 \mapsto g, & 214 \mapsto g, & 132 \mapsto h, \\ 134 \mapsto h, & 143 \mapsto i, & 321 \mapsto j, & 342 \mapsto k, & 343 \mapsto l, \\ 421 \mapsto m, & 432 \mapsto n, & 434 \mapsto n. & & \end{array}$$

Then $\hat{h}$ gives a topological conjugacy $\psi_{\hat{h}} : (X_{G_2^{T_2}}, \sigma_{G_2^{T_2}}) \to (X_H, \sigma_H)$. Hence $(X_M, \varphi^2\sigma_M^{-1})$ is topologically conjugate to the topological Markov shift (X_H, σ_H). The characteristic polynomial of M_H is

$$(x+1)^2(x^3 - 2x^2 + x - 1).$$

If we construct the textile system

$$\hat{T} = (\hat{h}p_2^{T_2}, \hat{h}q_2^{T_2} : \Gamma_2^{T_2} \to H),$$

then we see that $\mathcal{G}_{\hat{T}}$ is left resolving. Let E be the state-equivalence of $\mathcal{G}_{\hat{T}}^{-1}$ whose equivalence classes are $\{A\}, \{B\}, \{C\}, \{D\}, \{E, G\}, \{F, H\}, \{I, J\}, \{K, L\}, \{M, N\}, \{P\}, \{Q, R\}$ and $\{S, T\}$. Let $\bar{T}$ be the textile system such that $\mathcal{G}_{\bar{T}}^{-1}$ is the quotient of $\mathcal{G}_{\hat{T}}^{-1}$ with respect to E, i.e.,

$$\mathcal{G}_{\bar{T}} = ((\mathcal{G}_{\hat{T}}^{-1})_E)^{-1}.$$

Then we get

$$\mathcal{M}_{\mathcal{G}_T} = \begin{array}{c} \\ A \\ B \\ C \\ D \\ E \\ F \\ I \\ K \\ M \\ P \\ Q \\ S \end{array} \begin{array}{c} \begin{array}{cccccccccccc} A & B & C & D & E & F & I & K & M & P & Q & S \end{array} \\ \left(\begin{array}{cccccccccccc} & & & & \frac{n}{l} & \frac{n}{k} & & & & & & \\ & & & & & & & & \frac{m}{g} & & & \\ & & & & & & & & & & \frac{g}{j} & \\ & & & & & & & & & & & \frac{g}{g} \\ \frac{l}{n} & & \frac{j}{n} & & & & & & & & & \\ & \frac{k}{m} & & \frac{j}{m} & & & & & & & & \\ & \frac{k}{j} & & \frac{j}{j} & & & & & & & & \\ & & & & \frac{n}{i} & & \frac{n}{h} & & & & & \\ & & & & & & & & & \frac{g}{h} & \frac{g}{f} & \\ & & & & \frac{h}{l} & \frac{h}{k} & & & & & & \\ & & & & & & & \frac{i}{g} & \frac{f}{g} & & & \\ & & & & \frac{h}{i} & & \frac{h}{h} & & & & & \end{array} \right) \end{array} .$$

One can find that $\varphi_{\bar{T}}\sigma_{\bar{T}}^3$ and $\varphi_{\bar{T}}^{-1}\sigma_{\bar{T}}^3$ are nonexpansive LR automorphisms of $(X_{\bar{T}}, \sigma_{\bar{T}})$ and that

$$P(\varphi_{\bar{T}}\sigma_{\bar{T}}^3) = \begin{pmatrix} 2 & 2 & 1 & 2 & 2 \\ 0 & 2 & 2 & 0 & 1 \\ 1 & 2 & 2 & 0 & 1 \\ 1 & 0 & 0 & 1 & 1 \\ 1 & 1 & 0 & 1 & 1 \end{pmatrix}$$

and

$$P(\varphi_{\bar{T}}^{-1}\sigma_{\bar{T}}^3) = \begin{pmatrix} 2 & 2 & 2 & 1 & 1 \\ 0 & 2 & 1 & 1 & 2 \\ 2 & 1 & 2 & 0 & 0 \\ 1 & 0 & 1 & 0 & 0 \\ 0 & 2 & 0 & 1 & 2 \end{pmatrix} .$$

(To check these, it is useful to note that for any textile system T, the textile system T' with $\varphi_{T'} = \varphi_T \sigma_T$ is easily constructed from $T^{[2]}$.) Put $P(\varphi_{\bar{T}}\sigma_{\bar{T}}^3) = \bar{P}$ and put $P(\varphi_{\bar{T}}^{-1}\sigma_{\bar{T}}^3) = \bar{Q}$. One can see that the maximal eigenvalues of $\bar{P}$ and $\bar{Q}$ are equal to λ^3, where λ is the maximal eigenvalue of M_H. Hence $h(\varphi_{\bar{T}}\sigma_{\bar{T}}^3) = h(\varphi_{\bar{T}}^{-1}\sigma_{\bar{T}}^3) = 3\log\lambda$ by Theorem 6.1. We know, by Corollary 6.5, that $(X_H, (\varphi_{\bar{T}}\sigma_{\bar{T}}^3)^l\sigma_{\bar{T}}^m)$ is topologically conjugate to the topological Markov shift whose defining matrix is $\bar{P}^l M_H^m$ and $(X_H, (\varphi_{\bar{T}}^{-1}\sigma_{\bar{T}}^3)^l\sigma_{\bar{T}}^m)$ is topologically conjugate to the topological Markov shift whose defining matrix is $\bar{Q}^l M_H^m$, for all integers $l \geq 0$ and $m > 0$.

Let $\chi = \psi_{\hat{h}}\chi_T\psi$. Then χ is a homeomorphism of X_M onto $X_{\bar{T}} = X_H$ such that

$$\chi(\varphi^2\sigma_M^{-1})\chi^{-1} = \sigma_{\bar{T}} \quad \text{and} \quad \chi\sigma_M\chi^{-1} = \varphi_{\bar{T}}.$$

Since $\chi(\varphi^6\sigma_M^{-2})\chi^{-1} = \varphi_{\bar{T}}\sigma_{\bar{T}}^3$ and $\chi(\varphi^6\sigma_M^{-4})\chi^{-1} = \varphi_{\bar{T}}^{-1}\sigma_{\bar{T}}^3$, $\varphi^3\sigma_M^{-1}$ and $\varphi^3\sigma_M^{-2}$ are nonexpansive, essentially LR automorphisms of $(X_M, \varphi^2\sigma_M^{-1})$. Therefore the ELR cone of $\varphi^2\sigma_M^{-1}$ for φ is given as

$$CK_\varphi(\varphi^2\sigma_M^{-1}) = \{(x,y) \in \mathbf{R}^2 \mid -\frac{3}{2}x \le y \le -3x\}.$$

By the above and Proposition 8.11, we find that if (n,k) is a lattice point in $CK_\varphi(\varphi^2\sigma_M^{-1})$, then

$$h(\varphi^k\sigma_M^n) = \frac{k}{2}\log\lambda.$$

If $(n, 2k)$ is a lattice point in the interior of $CK_\varphi(\varphi^2\sigma_M^{-1})$, then $(X_M, \varphi^{2k}\sigma_M^n)$ is topologically conjugate to the topological Markov shift whose defining matrix is $\bar{P}^{k+n}M_H^{-2k-3n}$ if $-2n \le 2k < -3n$, and $\bar{Q}^{-k-n}M_H^{4k+3n}$ if $-\frac{3}{2}n < 2k \le -2n$. Notice that $\lambda \notin \boldsymbol{Q}(\frac{1+\sqrt{5}}{2})$. Every lattice point in $CK_\varphi(\sigma_M)$ is written in the form

$$(ms,\ m - 2ms)$$

for some integer $m \ge 0$ and some rational number $0 \le s \le 1$. For this lattice point, we have

$$h(\varphi^{m-2ms}\sigma_M^{ms}) = m\log\frac{1+\sqrt{5}}{2},$$

by Proposition 8.11, and if it is the interior point of the cone, then $(X_M, \varphi^{m-2ms}\sigma_M^{ms})$ is topologically conjugate to the topological Markov shift whose defining matrix is

$$\begin{pmatrix} 1 & 1 \\ 1 & 0 \end{pmatrix}^m,$$

by Corollary 6.7 or Proposition 9.21(4).

We note that $\varphi^2\sigma_M^{-1}$ is a composition of simple automorphisms. (The possibility of this was suggested by M. Boyle.) For φ and φ_0 are similar and $\varphi_0^2\sigma_M^{-1} = \mathrm{id}_{X_M}$. Therefore $\varphi^2\sigma_M^{-1}$ is similar to id_{X_M}, by Fact 9.2. Thus by Theorem 9.4, $\varphi^2\sigma_M^{-1}$ is a composition of simple automorphisms.

Thus we find that similar automorphisms φ_1 and φ_2 of a topological Markov shift (X,σ) can be very different in their dynamical behavior, although for all sufficiently large n, $\varphi_1\sigma^n$ and $\varphi_2\sigma^n$ have the same dynamical behavior and so do $\varphi_1\sigma^{-n}$ and $\varphi_2\sigma^{-n}$ (by Proposition 9.19). If $(X,\sigma) = (X_M, \sigma_M)$, $\varphi_1 = \mathrm{id}_{X_M}$ and $\varphi_2 = \varphi^2\sigma_M^{-1}$, then φ_1 and φ_2 are quite different in their dynamical behavior, but $(X, \varphi_1^k\sigma^n)$ and $(X, \varphi_2^k\sigma^n)$ are topologically conjugate for all $(n,k) \in \mathbf{Z}^2$ with $|n| > |k|$.

Example 2. Let (X,σ) be the full 2-shift over the alphabet $\{0,1\}$. Let $\varphi : X \to X$ be the block map of $(1,2)$ type given by $f : \{0,1\}^4 \to \{0,1\}$ such that $f(1001) =$

$0, f(1101) = 1$, and if $abc \neq 101$ with $a, b, c \in \{0, 1\}$, then $f(a0bc) = 1$ and $f(a1bc) = 0$. We note that φ is the well-known automorphism of infinite order of the 2-shift given in Section 20 of [H]. Passing through the higher block system of order 2 and the coding of the symbols such that $00 \mapsto 0, 01 \mapsto 1, 10 \mapsto 2$ and $11 \mapsto 3$, the automorphism becomes equal to φ_{T_0} for the following textile system T_0 over G with $\tilde{M_G} = \begin{pmatrix} 0 & 1 \\ 2 & 3 \end{pmatrix}$ (note that 0 as well as $1, 2$, and 3 is a symbol which represents an arc, here):

$$\mathcal{M}_{\mathcal{G}_{T_0}} = \begin{array}{c} \\ A \\ B \\ C \\ D \\ E \\ F \\ G \end{array} \begin{array}{c} \begin{array}{ccccccc} A & B & C & D & E & F & G \end{array} \\ \begin{pmatrix} \frac{0}{3} & \frac{0}{3} & & & & & \\ & & \frac{0}{3} & & & & \\ & & & \frac{1}{2} & \frac{1}{2} & & \\ & & \frac{2}{1} & & & & \\ \frac{2}{1} & & & & \frac{3}{0} & \frac{2}{0} & \frac{3}{1} \\ & & \frac{0}{1} & & & & \\ & & \frac{2}{3} & & & & \end{pmatrix} \end{array}.$$

One can easily see that T_0^* is not 1-1 so that φ is not expansive.

Let $T_2 = (p_2, q_2 : \Gamma_2 \to G)$ be the textile system given by

$$\mathcal{M}_{\mathcal{G}_{T_2}} = \begin{array}{c} \\ a \\ b \\ c \\ d \\ e \\ f \\ g \\ h \end{array} \begin{array}{c} \begin{array}{cccccccc} a & b & c & d & e & f & g & h \end{array} \\ \begin{pmatrix} \frac{0}{3} & \frac{0}{2} & & & & & & \\ & & \frac{0}{1} & \frac{0}{0} & & & & \\ & & & & \frac{1}{2} & \frac{1}{3} & & \\ & & & & & & \frac{1}{1} & \frac{1}{0} \\ & & \frac{2}{1} & \frac{2}{0} & & & & \\ \frac{2}{3} & & & & \frac{3}{2} & & & \\ & \frac{2}{2} & & & & \frac{3}{3} & & \\ & & & & & & \frac{3}{1} & \frac{3}{0} \end{pmatrix} \end{array}.$$

One can see that $\varphi_{T_2} = \varphi_{T_0}\sigma_G^2$. Noting that T_2 is LR and hence T_2^* is nondegenerate, we find that T_2 can weave the two textiles $(\alpha_{ij})_{i,j\in\mathbf{Z}}$ and $(\beta_{ij})_{i,j\in\mathbf{Z}}$ as follows:

$$\begin{aligned}
\alpha_{2i,0} &= e\overset{2}{\underset{1}{}}c \quad\text{and}\quad \alpha_{2i-1,0} = c\overset{1}{\underset{2}{}}e \quad\text{for } i \le 0,\\
\alpha_{10} &= c\overset{1}{\underset{2}{}}e,\\
\alpha_{2i,0} &= f\overset{2}{\underset{3}{}}a \quad\text{and}\quad \alpha_{2i+1,0} = f\overset{3}{\underset{2}{}}e \quad\text{for } i \ge 1;\\
\beta_{2i,0} &= e\overset{2}{\underset{1}{}}c \quad\text{and}\quad \beta_{2i-1,0} = c\overset{1}{\underset{2}{}}e \quad\text{for } i \le 0,\\
\beta_{10} &= c\overset{1}{\underset{3}{}}f,\\
\beta_{2i,0} &= f\overset{3}{\underset{2}{}}e \quad\text{and}\quad \beta_{2i+1,0} = f\overset{2}{\underset{3}{}}a \quad\text{for } i \ge 1;
\end{aligned}$$

$$(\alpha_{ij})_{i\in\mathbf{Z},j\le -1} = (\beta_{ij})_{i\in\mathbf{Z},j\le -1} \tag{10.7}$$

(this is possible because $i_{\Gamma_2}(\alpha_{i0}) = i_{\Gamma_2}(\beta_{i0})$ for all $i \in \mathbf{Z}$), where, for example, $e\overset{2}{\underset{1}{}}c$ denotes the square whose upper and lower sides are 2 and 1, respectively, and whose left and right sides are e and c, respectively. We note that

$$\alpha_{10} \ne \beta_{10} \tag{10.8}$$

but that $\alpha_{i0} = \beta_{i0}$ for all $i \le 0$. It is easily checked that $(t_{\Gamma_2}(\alpha_{i0}))_{i\le 0}$ uniquely determines

$$(\alpha_{-i,j})_{i\ge 1, 1\le j\le i}$$

so that

$$(\alpha_{-i,j})_{i\ge 0, 0\le j\le i} = (\beta_{-i,j})_{i\ge 0, 0\le j\le i}. \tag{10.9}$$

By (10.7),(10.8), and (10.9), we conclude that $\check{\theta}_{T_2}^{(k,-n)} : U_{T_2} \to \check{X}_{T_2}^{(k,-n)}$ is not 1-1 for $k, n \in \mathbf{Z}$ with $0 \le n \le k$ and $(n,k) \ne (0,0)$. Hence, by Proposition 2.3, $\boldsymbol{\sigma}_{T_2}^{(k,-n)} : U_{T_2} \to U_{T_2}$ is not expansive for $k, n \ge 0$ with $n \le k$. Thus, by Proposition 2.4, $\varphi_{T_2}^k \sigma_{T_2}^{-n}$ is not expansive for $k, n \ge 0$ with $n \le k$.

Since T_2 is LR and $(X_{T_2^*}, \sigma_{T_2^*}) = (X_{G^2}, \sigma_{G^2})$, it follows from Corollary 6.7 that $(X_G, \varphi_{T_2}^k \sigma_G^n)$ is topologically conjugate to (X_G, σ_G^{2k+n}) for all $k \ge 0$ and $n > 0$.

Let $\epsilon : (X_G, \sigma_G) \to (X_G, \sigma_G)$ be the symbolic automorphism of order 2 given by the graph-automorphism defined by $0 \mapsto 3, 1 \mapsto 2, 2 \mapsto 1$ and $3 \mapsto 0$. Then we have

$$\varphi_{T_0}^{-1} = \epsilon\varphi_{T_0}\epsilon.$$

Using this and the above, we conclude that the ELR cone $CK_\varphi(\sigma)$ is given by

$$CK_\varphi(\sigma) = \{(x,y) \in \mathbf{R}^2 \mid x \ge 0,\ |y| \le \frac{1}{2}x\}$$

and that $\varphi^k\sigma^n$ is not expansive for (n,k) with $\frac{1}{2}|n| \le |k| \le |n|$. For a lattice point (n,k) in the interior of $CK_\varphi(\sigma)$ (i.e., $n > 0$, $|k| < \frac{1}{2}n$), $(X, \varphi^k\sigma^n)$ is topologically conjugate to (X, σ^n). Of course, $CK_\varphi(\sigma^{-1}) = -CK_\varphi(\sigma)$.

Curtis, Hedlund and Lyndon [H] proved that φ is of infinite order though φ is the composition of two automorphisms of order 2. In fact, these two are simple automorphisms, one is the symbolic automorphism ϵ_0 given by the map defined by $0 \mapsto 1$ and $1 \mapsto 0$, and the other is $\epsilon_0\varphi$, which was known to be simple by the observation of Boyle and Lind [B3].

As stated, φ is not expansive. But we have not yet known any more about the dynamics of $\varphi^k\sigma^n$ for $k, n \in \mathbf{Z}$ with $|k| > |n|$.

To add the following example, a discussion with Teturo Kamae was helpful.

Example 3. Let A be an alphabet. Let $r, s \in \mathbf{N}$. Let $B = A^{r+s}$. Let $\varphi : X_A \to X_A$ be a bipermutive block map of (r,s) type. Let $f : A^{r+s+1} \to A$ be the mapping which gives φ. Let $T = T_\varphi$, where T_φ is the textile system as in Fact 2.2. Then $(X_T, \sigma_T) = (X_A, \sigma_A)$ and, as seen in the proof of Proposition 3.14, T^* is 1-1 and LL with $(X_{T^*}, \sigma_{T^*}) = (X_B, \sigma_B)$. We shall show that the ELR cones of the LL automorphism $\psi = \varphi_{T^*}$ of (X_B, σ_B) are $\Delta_1, \bar{\Delta}_1, \Delta_2, \bar{\Delta}_2, \Delta_3$, and $\bar{\Delta}_3$, where

$$\begin{aligned} \Delta_1 &= \{(x,y) \in \mathbf{R}^2 \mid x \le 0,\ y \ge -sx\}, \\ \Delta_2 &= \{(x,y) \in \mathbf{R}^2 \mid rx \le y \le -sx\}, \\ \Delta_3 &= \{(x,y) \in \mathbf{R}^2 \mid x \le 0,\ y \le rx\}, \end{aligned}$$

and $\bar{\Delta}_i = \{(-x,-y) \mid (x,y) \in \Delta_i\}$ for $i = 1,2,3$. Moreover, we can subshift-identify $(X_B, \psi^k\sigma_B^n)$ for all $(n,k) \in \mathbf{Z}^2$ such that $\psi^k\sigma_B^n$ is expansive.

Let $e : B \to A^r$ and $e' : B \to A^s$ be defined by $e(b) = a_1 \cdots a_r$ and $e'(b) = a_{r+1} \cdots a_{r+s}$, where $b = a_1 \cdots a_{r+s}$ with $a_i \in A$. Let

$$C = \{be(b) \mid b \in B\} \quad \text{and} \quad D = \{be'(b) \mid b \in B\}.$$

Then it is not difficult to observe that we may write

$$(\check{X}_T^{(-1,-r)}, \check{\sigma}_T^{(-1,-r)}) = (X_C, \sigma_C) \quad \text{and} \quad (\check{X}_T^{(-1,s)}, \check{\sigma}_T^{(-1,s)}) = (X_D, \sigma_D)$$

and $\check{\theta}_T^{(-1,-r)}$ and $\check{\theta}_T^{(-1,s)}$ are not 1-1. Clearly θ_T is not 1-1. Let $\tilde{M}_A$ denote the 1×1 representation matrix with $\sum_{a \in A} a$ as its (only one) component. Let $\tilde{M}_C$ and $\tilde{M}_D$ be defined similarly.

We define $k_1 : DA \to AD$ as follows: for $d \in D$ and $a \in A$ if $d = be'(b)$ with $b = a_1 \cdots a_{r+s}, a_i \in A$, then $k_1(da) = a'(b'e'(b'))$, where $a' = f(a_1 \cdots a_{r+s}a)$ and $b' = a_2 \cdots a_{r+s}a$. Using the property of f that it gives a left-most permutive block map, we easily see that k_1 is 1-1. Hence we have a specified equivalence $\tilde{M}_D\tilde{M}_A \overset{k_1}{\simeq} \tilde{M}_A\tilde{M}_D$. Let T_1 be the LR textile system associated with this specified equivalence. Let Ψ :

$U_T \to U_{T_1}$ be the natural "superposition" homeomorphism with $\check{\theta}_T^{(-1,s)}(t) = \theta_{T_1}(\Psi(t))$ and $\theta_T(t) = \theta^*_{T_1}(\Psi(t))$ for all $t \in U_T$. Then it is clear that $\boldsymbol{\sigma}_{T_1}^{(k,l)}\Psi = \Psi\boldsymbol{\sigma}_T^{(-l,k+sl)}$. Let $k_0, l_0 > 0$. Then for all integers $k, l \geq 0$, $\boldsymbol{\sigma}_{T_1}^{(k,l)}$ is an essentially LR automorphism of $(U_{T_1}, \boldsymbol{\sigma}_{T_1}^{(k_0,l_0)})$ (Proposition 8.17). Therefore for all integers $k, l \geq 0$, $\boldsymbol{\sigma}_T^{(-l,k+sl)}$ is an essentially LR automorphism of $(U_T, \boldsymbol{\sigma}_T^{(-l_0,k_0+sl_0)})$. Since θ_T and $\check{\theta}_T^{(-1,s)}$ are not 1-1, $\boldsymbol{\sigma}_T^{(0,1)}$ and $\boldsymbol{\sigma}_T^{(-1,s)}$ are nonexpansive (by Proposition 2.3). Thus, by Proposition 2.4, for all $k, l \geq 0$ $\psi^{k+sl}\sigma_B^{-l}$ is an essentially LR automorphism of $(X_B, \psi^{k_0+sl_0}\sigma_B^{-l_0})$, and ψ and $\psi^s\sigma_B^{-1}$ are nonexpansive, so that Δ_1 and $\bar{\Delta}_1$ are ELR cones for ψ.

Let $\tilde{M}_C\tilde{M}_D \overset{k_2}{\simeq} \tilde{M}_D\tilde{M}_C$ be the specified equivalence with $k_2 : CD \to DC$ defined as follows : for $c \in C$ and $d \in D$, if $c = be(b)$ and $d = b'e'(b')$ with $b = a_1 \cdots a_{r+s}$ and $b' = a'_1 \cdots a'_{r+s}, a_i, a'_i \in A$, then $k_2(cd) = (be'(b))(b''e(b''))$, where $b'' = a''_1 \cdots a''_{r+s}$ with $a''_1, \cdots, a''_{r+s} \in A$ such that

$$a_1a_2 \cdots a_{r+s} = f(a'_1 \cdots a'_{r+s}a''_1)f(a'_2 \cdots a'_{r+s}a''_1a''_2) \cdots f(a'_{r+s}a''_1 \cdots a''_{r+s}).$$

By using the property of f that it gives a bipermutive block map, it is easily seen that k_2 is a mapping and 1-1. Using that specified equivalence, a similar discussion to the above shows that Δ_2 and $\bar{\Delta}_2$ are ELR cones for ψ.

The proof that Δ_3 and $\bar{\Delta}_3$ are ELR cones for ψ is similar to the proof that Δ_1 and $\bar{\Delta}_1$ are ELR cones for ψ.

Let $(n, k) \in \mathbf{Z}^2$. As is easily seen, the following hold: if $(n, k) \in \Delta_1$, then there is a symbolic conjugacy of $(\check{X}_T^{(n,k)}, \check{\sigma}_T^{(n,k)})$ onto (X_{E_1}, σ_{E_1}), where $E_1 = B^{|n|}A^{k-s|n|}$; if $(n, k) \in \Delta_2$, then there is a symbolic conjugacy of $(\check{X}_T^{(n,k)}, \check{\sigma}_T^{(n,k)})$ onto (X_{E_2}, σ_{E_2}), where $E_2 = B^{|n|}$; if $(n, k) \in \Delta_3$, then there is a symbolic conjugacy of $(\check{X}_T^{(n,k)}, \check{\sigma}_T^{(n,k)})$ onto (X_{E_3}, σ_{E_3}), where $E_3 = B^{|n|}A^{|k|-r|n|}$. Therefore, using Proposition 2.4, we know the following: if (n, k) is an interior point of $\Delta_1 \cup \bar{\Delta}_1$, then $(X_B, \psi^k\sigma_B^n)$ is topologically conjugate to $(X_{A^{|k|+r|n|}}, \sigma_{A^{|k|+r|n|}})$; if (n, k) is an interior point of $\Delta_2 \cup \bar{\Delta}_2$, then $(X_B, \psi^k\sigma_B^n)$ is topologically conjugate to $(X_{A^{(r+s)|n|}}, \sigma_{A^{(r+s)|n|}})$; if (n, k) is an interior point of $\Delta_3 \cup \bar{\Delta}_3$, then $(X_B, \psi^k\sigma_B^n)$ is topologically conjugate to $(X_{A^{|k|+s|n|}}, \sigma_{A^{|k|+s|n|}})$; if (n, k) is on either of the lines $y = -sx$ and $y = rx$, then $h(\psi^k\sigma^n) = (r + s)|n|\log(\#A)$; if (n, k) is on the line $x = 0$, then $h(\psi^k\sigma_B^n) = |k|\log(\#A)$.

REFERENCES

[AM] R. L. Adler and B. Marcus, Topological entropy and equivalence of dynamical systems, Memoirs Amer. Math. Soc. 219 (1979).

[B1] M. Boyle, Topological orbit equivalence and factor maps in symbolic dynamics, Ph. D. Thesis, University of Washington, Seattle (1983).

[B2] M. Boyle, Constraints on the degree of a sofic homomorphism and the induced multiplication of measures on unstable sets, Israel J. Math. 53 (1986), 52-68.

[B3] M. Boyle, Nasu's simple automorphisms, *Dynamical Systems: Proceedings, University of Maryland 1986-87*, ed. J. C. Alexander, Springer Lecture Notes in Math. 1342, Springer-Verlag, 1988.

[B4] M. Boyle, Eventual extensions of finite codes, Proc. Amer. Math. Soc. 104 (1988), 965-972.

[BK1] M. Boyle and W. Krieger, Periodic points and automorphisms of the shift, Trans. Amer. Math. Soc. 302 (1987), 125-149.

[BK2] M. Boyle and W. Krieger, Almost Markov and shift equivalent sofic systems, *Dynamical Systems: Proceedings, University of Maryland 1986-87*, ed. J. C. Alexander, Springer Lecture Notes in Math. 1342, Springer-Verlag, 1988.

[BLR] M. Boyle, D. Lind, and D. Rudolph, The automorphism group of a subshift of finite type, Trans. Amer. Math. Soc. 306 (1988),71-114.

[BMT] M. Boyle, B. Marcus, and P. Trow, Resolving maps and the dimension group for shifts of finite type, Memoirs Amer. Math. Soc. 377 (1987).

[BP] A. Berman and R. J. Plemmons, *Nonnegative Matrices in The Mathematical Sciences*, Academic Press, New York, 1979.

[C] E. M. Coven, Topological entropy of block maps, Proc. Amer. Math. Soc. 78 (1980), 590-594.

[CP] E. M. Coven and M. E. Paul, Endomorphisms of irreducible subshifts of finite type, Math. Systems Theory 8 (1974), 167-175.

[DGS] M. Denker, C. Grillenberger, and K. Sigmund, *Ergodic Theory on Compact Spaces*, Springer Lecture Notes in Math. 527, Springer-Verlag, Berlin, 1976.

[F1] R. Fischer, Sofic systems and graphs, Monats. für Math. 80 (1975), 179-186.

[F2] R. Fischer, Graphs and symbolic dynamics, In *Colloq. Math. Soc. János Bolyai* 16, *Topics in Information Theory*, Keszthely, Hungary, 1975.

[G] F. R. Gantmacher, *The Theory of Matrices*, Vol. I, Chelsea, New York, 1959.

[GS] B. Grünbaum and G. Shephard, *Tiling and Patterns*, Freeman, New York, 1987.

[H] G. A. Hedlund, Endomorphisms and automorphisms of the shift dynamical system, Math. Systems Theory 3 (1969), 320-375.

[HN] T. Hamachi and M. Nasu, Topological conjugacy for 1-block factor maps of subshifts and sofic covers, *Dynamical Systems: Proceedings, University of Maryland 1986-87*, ed. J. C. Alexander, Springer Lecture Notes in Math. 1342, Springer-Verlag, 1988.

[Ki] B. Kitchens, Continuity properties of factor maps in ergodic theory, Ph. D. Thesis, University of North Carolina, Chapel Hill (1981).

[Kr1] W. Krieger, On dimension functions and topological Markov chains, Inventiones Math. 56 (1980), 239-250.

[Kr2] W. Krieger, On sofic systems I, Israel J. Math. 48 (1984), 305-330.

[Kr3] W. Krieger, On sofic systems II, Israel J. Math. 60 (1987), 167-176.

[KMT] W. Krieger, B. Marcus and S. Tuncel, On automorphisms of Markov chains, Trans. Amer. Math. Soc. 333 (1992), 531-565.

[KR] K. H. Kim and F. W. Roush, Solution of two conjectures in symbolic dynamics, Proc. Amer. Math. Soc. 112 (1991), 1163-1168.

[KRW] K. H. Kim, F. W. Roush, and J. B. Wagoner, Automorphisms of the dimension group and gyration numbers, J. Amer. Math. Soc. 5 (1992), 191-212.

[L] D. A. Lind, Entropies of automorphisms of a topological Markov shift, Proc. Amer. Math. Soc. 99 (1987), 589-595.

[M] J. Milnor, Directional entropies of cellular automaton-maps, NATO ASI Series, F 20, *Disordered Systems and Biological Organization*, ed. E. Bienenstoc et al. Springer-Verlag, Berlin, 1986.

[MT] B. Marcus and S. Tuncel, The weight-per-symbol polytope and scaffolds of invariants associated with Markov chains, Ergod. Th. & Dynam. Sys. 11 (1991), 129-180.

[N1] M. Nasu, Constant-to-one and onto global maps of homomorphisms between strongly connected graphs, Ergod. Th. & Dynam. Sys., 3 (1983), 387-413.

[N2] M. Nasu, Topological conjugacy for sofic systems, Ergod. Th. & Dynam. Sys. 6 (1986), 265-280.

[N3] M. Nasu, Topological conjugacy for sofic systems and extensions of automorphisms of finite subsystems of topological Markov shifts, *Dynamical Systems: Proceedings, University of Maryland 1986-87*, ed. J. C. Alexander, Springer Lecture Notes in Math. 1342, Springer-Verlag, 1988.

[P] W. Parry, A finitary classification of topological Markov chains and sofic systems, Bull. London Math. Soc., 9 (1977), 86-92.

[PRS] M. Perles, M. O. Rabin, and E. Shamir, The theory of definite automata, IEEE Trans. Electr. Comp. EC-12 (1963), 233-243.

[S] J. Smillie, Properties of the directional entropy function for cellular automata, *Dynamical Systems: Proceedings, University of Maryland 1986-87*, ed. J. C. Alexander, Springer Lecture Notes in Math. 1342, Springer-Verlag, 1988.

[SA] M. A. Shereshevsky and V.S. Afraimovich, Bipermutative cellular automata are topologically conjugate to the one-sided Bernoulli shift, Random & Computational Dynamics 1 (1992), 91-98.

[T1] P. Trow, Degrees of constant-to-one factor maps, Proc. Amer. Math. Soc. 103 (1988), 184-188.

[T2] P. Trow, Degrees of finite-to-one factor maps, Israel J. Math. 71 (1990), 229-238.

[Wag1] J. B. Wagoner, Markov partitions and K_2, Inst. Hautes Études Sci. Publ. Math. 65 (1987), 91-129.

[Wag2] J. B. Wagoner, Triangle identities and symmetries of a subshift of finite type, Pacific J. Math. 144 (1990), 181-205.

[Wag3] J. B. Wagoner, Eventual finite order generation for the kernel of the dimension group representation, Tans. Amer. Math. Soc. 317 (1990), 331-350.

[Wal] P. Walters, *An Introduction to Ergodic Theory*, Springer-Verlag, New York, 1982.

[Wan] H. Wang, Notes on a class of tiling problems, Fundam. Math., 82 (1975), 295-305.

[We] B. Weiss, Subshifts of finite type and sofic systems, Monats. für Math. 77 (1973), 462-474.

[Wi] R. F. Williams, Classification of subshifts of finite type, Ann. of Math. 98 (1973), 120-153; Errata: Ann. of Math. 99 (1974), 380-381.

LIST OF NOTATION

(X_A, σ_A)	full shift over A, 10
(X, σ)	subshift, 10
A_G	arc-set of G, 10
V_G	vertex-set of G, 10
i_G	mapping specifying the initial vertices of the arcs of G, 10
t_G	mapping specifying the terminal vertices of the arcs of G, 10
(X_G, σ_G)	topological Markov shift defined by G, 10
M_G	adjacency matrix of G, 10
(X_M, σ_M)	topological Markov shift defined by M, 11
$L_n(X)$	set of the words of length n appearing on the bisequences of subshift (X, σ), 11
h_A	arc-map of h, 11
h_V	vertex-map of h, 11
$\mathcal{G} = (G, \lambda)$	λ-graph, 12
$(X_{\mathcal{G}}, \sigma_{\mathcal{G}})$	sofic system defined by $\mathcal{G}$, 12
$(X^{[n]}, \sigma^{[n]})$	higher block system of order n of (X, σ), 12
$(X^{(n)}, \sigma^{(n)})$	nth power system of (X, σ), 12
$\varphi^{(n)}$	endomorphism of $(X^{(n)}, \sigma^{(n)})$ which is naturally induced by endomorphism φ of (X, σ), 12
$L_n(G)$	set of all paths of length n of G, 12
$L(G)$	set of all paths of G, 12
$G^{[n]}$	higher block system of order n of G, 12
G^n	nth power of G, 13
G^{-1}	transpose of G, 13

INDEX

Faculty of Engineering
Mie University
Tsu 514
Japan

Editorial Information

To be published in the *Memoirs*, a paper must be correct, new, nontrivial, and significant. Further, it must be well written and of interest to a substantial number of mathematicians. Piecemeal results, such as an inconclusive step toward an unproved major theorem or a minor variation on a known result, are in general not acceptable for publication. *Transactions* Editors shall solicit and encourage publication of worthy papers. Papers appearing in *Memoirs* are generally longer than those appearing in *Transactions* with which it shares an editorial committee.

As of December 7, 1994, the backlog for this journal was approximately 3 volumes. This estimate is the result of dividing the number of manuscripts for this journal in the Providence office that have not yet gone to the printer on the above date by the average number of monographs per volume over the previous twelve months, reduced by the number of issues published in four months (the time necessary for preparing an issue for the printer). (There are 6 volumes per year, each containing at least 4 numbers.)

A Copyright Transfer Agreement is required before a paper will be published in this journal. By submitting a paper to this journal, authors certify that the manuscript has not been submitted to nor is it under consideration for publication by another journal, conference proceedings, or similar publication.

Information for Authors and Editors

Memoirs are printed by photo-offset from camera copy fully prepared by the author. This means that the finished book will look exactly like the copy submitted.

The paper must contain a *descriptive title* and an *abstract* that summarizes the article in language suitable for workers in the general field (algebra, analysis, etc.). The *descriptive title* should be short, but informative; useless or vague phrases such as "some remarks about" or "concerning" should be avoided. The *abstract* should be at least one complete sentence, and at most 300 words. Included with the footnotes to the paper, there should be the 1991 *Mathematics Subject Classification* representing the primary and secondary subjects of the article. This may be followed by a list of *key words and phrases* describing the subject matter of the article and taken from it. A list of the numbers may be found in the annual index of *Mathematical Reviews*, published with the December issue starting in 1990, as well as from the electronic service e-MATH [**telnet e-MATH.ams.org** (or **telnet 130.44.1.100**). Login and password are **e-math**]. For journal abbreviations used in bibliographies, see the list of serials in the latest *Mathematical Reviews* annual index. When the manuscript is submitted, authors should supply the editor with electronic addresses if available. These will be printed after the postal address at the end of each article.

Electronically prepared manuscripts. The AMS encourages submission of electronically prepared manuscripts in AMS-TEX or AMS-LATEX because properly prepared electronic manuscripts save the author proofreading time and move more quickly through the production process. To this end, the Society has prepared "preprint" style files, specifically the amsppt style of AMS-TEX and the amsart style of AMS-LATEX, which will simplify the work of authors and of the production staff. Those authors who make use of these style files from the beginning of the writing process will further reduce their own effort. Electronically submitted manuscripts prepared in plain TEX or LATEX do not mesh properly with the AMS production systems and cannot, therefore, realize the same kind of expedited processing. Users of plain TEX should have little difficulty learning AMS-TEX, and LATEX users will find that AMS-LATEX is the same as LATEX with additional commands to simplify the typesetting of mathematics.

Guidelines for Preparing Electronic Manuscripts provides additional assistance and is available for use with either AMS-TEX or AMS-LATEX. Authors with FTP access may obtain *Guidelines* from the Society's Internet node e-MATH.ams.org (130.44.1.100). For those without FTP access *Guidelines* can be obtained free of charge from the e-mail address guide-elec@math.ams.org (Internet) or from the Customer Services Department, American Mathematical Society, P.O. Box 6248, Providence, RI 02940-6248. When requesting *Guidelines*, please specify which version you want.

At the time of submission, authors should indicate if the paper has been prepared using AMS-TEX or AMS-LATEX. The *Manual for Authors of Mathematical Papers* should be consulted for symbols and style conventions. The *Manual* may be obtained free of charge from the e-mail address cust-serv@math.ams.org or from the Customer Services Department, American Mathematical Society, P.O. Box 6248, Providence, RI 02940-6248. The Providence office should be supplied with a manuscript that corresponds to the electronic file being submitted.

Electronic manuscripts should be sent to the Providence office immediately after the paper has been accepted for publication. They can be sent via e-mail to pub-submit@math.ams.org (Internet) or on diskettes to the Publications Department, American Mathematical Society, P. O. Box 6248, Providence, RI 02940-6248. When submitting electronic manuscripts please be sure to include a message indicating in which publication the paper has been accepted.

Two copies of the paper should be sent directly to the appropriate Editor and the author should keep one copy. The *Guide for Authors of Memoirs* gives detailed information on preparing papers for *Memoirs* and may be obtained free of charge from the Editorial Department, American Mathematical Society, P. O. Box 6248, Providence, RI 02940-6248. For papers not prepared electronically, model paper may also be obtained free of charge from the Editorial Department.

Any inquiries concerning a paper that has been accepted for publication should be sent directly to the Editorial Department, American Mathematical Society, P. O. Box 6248, Providence, RI 02940-6248.

Editors

This journal is designed particularly for long research papers (and groups of cognate papers) in pure and applied mathematics. Papers intended for publication in the *Memoirs* should be addressed to one of the following editors:

Recent Titles in This Series

(Continued from the front of this publication)

518 **René A. Carmona and S. A. Molchanov,** Parabolic Anderson problem and intermittency, 1994
517 **Takashi Shioya,** Behavior of distant maximal geodesics in finitely connected complete 2-dimensional Riemannian manifolds, 1994
516 **Kevin W. J. Kadell,** A proof of the q-Macdonald-Morris conjecture for BC_n, 1994
515 **Krzysztof Ciesielski, Lee Larson, and Krzysztof Ostaszewski,** $\mathcal{I}$-density continuous functions, 1994
514 **Anthony A. Iarrobino,** Associated graded algebra of a Gorenstein Artin algebra, 1994
513 **Jaume Llibre and Ana Nunes,** Separatrix surfaces and invariant manifolds of a class of integrable Hamiltonian systems and their perturbations, 1994
512 **Maria R. Gonzalez-Dorrego,** $(16, 6)$ configurations and geometry of Kummer surfaces in $\mathbb{P}^3$, 1994
511 **Monique Sablé-Tougeron,** Ondes de gradients multidimensionnelles, 1993
510 **Gennady Bachman,** On the coefficients of cyclotomic polynomials, 1993
509 **Ralph Howard,** The kinematic formula in Riemannian homogeneous spaces, 1993
508 **Kunio Murasugi and Jozef H. Przytycki,** An index of a graph with applications to knot theory, 1993
507 **Cristiano Husu,** Extensions of the Jacobi identity for vertex operators, and standard $A_1^{(1)}$-modules, 1993
506 **Marc A. Rieffel,** Deformation quantization for actions of R^d, 1993
505 **Stephen S.-T. Yau and Yung Yu,** Gorenstein quotient singularities in dimension three, 1993
504 **Anthony V. Phillips and David A. Stone,** A topological Chern-Weil theory, 1993
503 **Michael Makkai,** Duality and definability in first order logic, 1993
502 **Eriko Hironaka,** Abelian coverings of the complex projective plane branched along configurations of real lines, 1993
501 **E. N. Dancer,** Weakly nonlinear Dirichlet problems on long or thin domains, 1993
500 **David Soudry,** Rankin-Selberg convolutions for $\mathrm{SO}_{2\ell+1} \times \mathrm{GL}_n$: Local theory, 1993
499 **Karl-Hermann Neeb,** Invariant subsemigroups of Lie groups, 1993
498 **J. Nikiel, H. M. Tuncali, and E. D. Tymchatyn,** Continuous images of arcs and inverse limit methods, 1993
497 **John Roe,** Coarse cohomology and index theory on complete Riemannian manifolds, 1993
496 **Stanley O. Kochman,** Symplectic cobordism and the computation of stable stems, 1993
495 **Min Ji and Guang Yin Wang,** Minimal surfaces in Riemannian manifolds, 1993
494 **Igor B. Frenkel, Yi-Zhi Huang, and James Lepowsky,** On axiomatic approaches to vertex operator algebras and modules, 1993
493 **Nigel J. Kalton,** Lattice structures on Banach spaces, 1993
492 **Theodore G. Faticoni,** Categories of modules over endomorphism rings, 1993
491 **Tom Farrell and Lowell Jones,** Markov cell structures near a hyperbolic set, 1993
490 **Melvin Hochster and Craig Huneke,** Phantom homology, 1993
489 **Jean-Pierre Gabardo,** Extension of positive-definite distributions and maximum entropy, 1993
488 **Chris Jantzen,** Degenerate principal series for symplectic groups, 1993
487 **Sagun Chanillo and Benjamin Muckenhoupt,** Weak type estimates for Cesaro sums of Jacobi polynomial series, 1993

(See the AMS catalog for earlier titles)